Ergebnisse der Mathematik und ihrer Grenzgebiete

Band 40

Herausgegeben von
P. R. Halmos · P. J. Hilton · R. Remmert · B. Szőkefalvi-Nagy
Unter Mitwirkung von
L. V. Ahlfors · R. Baer · F. L. Bauer · R. Courant · A. Dold
J. L. Doob · S. Eilenberg · M. Kneser · M. M. Postnikov
H. Rademacher · B. Segre · E. Sperner
Geschäftsführender Herausgeber: P. J. Hilton

G. B. Seligman

Modular Lie Algebras

Springer-Verlag Berlin Heidelberg New York 1967

Prof. Dr. G. B. Seligman
Yale University
Department of Mathematics
New Haven, Conn. 06520 / USA

ISBN-13: 978-3-642-94987-6 e-ISBN-13: 978-3-642-94985-2
DOI: 10.1007/978-3-642-94985-2

Softcover reprint of the hardcover 1st edition 1967
Library of Congress Catalog Card Number 67-28452

Titel-Nr. 4584

Foreword

The study of the structure of Lie algebras over arbitrary fields is now a little more than thirty years old. The first papers, to my knowledge, which undertook this study as an end in itself were those of JACOBSON ("Rational methods in the theory of Lie algebras") in the *Annals*, and of LANDHERR ("Über einfache Liesche Ringe") in the Hamburg *Abhandlungen*, both in 1935. Over fields of characteristic zero, these thirty years have seen the ideas and results inherited from LIE, KILLING, E. CARTAN and WEYL developed and given new depth, meaning and elegance by many contributors. Much of this work is presented in [47, 64, 128 and 234] of the bibliography. For those who find the rationalization for the study of Lie algebras in their connections with Lie groups, satisfying counterparts to these connections have been found over general non-modular fields, with the substitution of the formal groups of BOCHNER [40] (see also DIEUDONNÉ [108]), or that of the algebraic linear groups of CHEVALLEY [71], for the usual Lie group. In particular, the relation with algebraic linear groups has stimulated the study of Lie algebras of linear transformations.

When one admits to consideration Lie algebras over a base field of positive characteristic (such are the algebras to which the title of this monograph refers), he encounters a new and initially confusing scene. It is not simply the case that new methods must be found to establish analogues of the theorems for characteristic zero, but rather that almost the only analogues which remain true (with the same degree of generality) are those whose traditional proofs turn out to have been independent of the characteristic anyway. Chapter V of this report deals with a number of analogues of fundamental classical theorems, and attempts in particular (Chap. V, § 4) to organize the rather awkward array of simple modular Lie algebras which would be totally unexpected to one acquainted only with the non-modular case.

Chapter VI is an indication of some ways in which Lie algebras, especially those of prime characteristic, have arisen in other areas of mathematics; indeed, §§ 1 and 4 would be meaningless except for modular Lie algebras. Present indications seem to be that the Lie algebra is assured a lasting and prominent place in the theories of formal groups and algebraic groups of arbitrary characteristic, even though it does not serve as handily for general fields as for non-modular ones. An attempt has been made in §§ 2 and 3 to sketch its status in these theories at present. It is quite likely that progress, especially concerning group schemes, has already made my comments obsolete at this writing; such obsolescence is probably inevitable by the time this monograph reaches public view. I beg the patience of my better-informed readers with my efforts to give a hint of these theories to readers who may be totally uninitiated.

One setting in which a rather full modular analogue of the classical theory exists, without constituting a word-for-word translation of that theory, is that of Lie algebras with non-singular Killing forms. I have been influenced by the fact of my own participation in the development of this analogue to give it a considerable amount of space (Chap. II, III, IV and part of Chap. I). I have tried to make the exposition of this material nearly complete and self-contained, whereas the rest of the book consists more often than not of summary comments with references to the appropriate literature.

It has been my intention to include in the bibliography all papers known to me to be relevant, with the occasional exception of short research announcements (as in *Comptes Rendus* or *Doklady*) whose results have since been published in more complete form. I have leaned heavily on *Mathematical Reviews* for guidance to these papers, and may therefore be less than complete as to quite recent work. My unfamiliarity with Chinese and Japanese, and my inadequacy in Russian, may have caused me to give insufficient notice to work published in these languages. Joint papers are listed under the name of the (lexicographically) first author only; I hope that my colleagues who share with me the tail of the alphabet will not feel neglected thereby.

No attempt has been made to set or to follow fixed procedures as to notation. Rather, I have chosen notation and terminology on the basis of my own previous conditioning to the matter at hand and on the basis of the usage in original sources. Whenever a conflict between these guiding principles has arisen, I have not hesitated to follow my own preferences. As the reader will soon notice (perhaps to his annoyance) these include a choice of what I regard as local clarity over global consistency and a quite conservative attitude toward terminology.

I regret that Professor NAKAYAMA, who solicited this work for the *Ergebnisse* series, has been taken from us while it was in progress. He will be remembered with gratitude and deepest respect.

It seems certain that I should not be authoring this volume, and quite likely that much of the material presented here would not yet have been developed, if it were not for the continuing contributions, both to mathematics and to my education, of Professor NATHAN JACOBSON. His methods and ideas are for me models of elegance and imagination. As teacher, colleague and friend, he has every right to such honor as I may do him.

A multitude of colleagues and students have given valuable advice and assistance in the preparation of this work. Among the former, I cite P. CARTIER, WALTER FEIT, J. PETER MAY, T. TAMAGAWA, MAGUERITE FRANK, RICHARD BLOCK and JOHN WALTER. The last three have generously supplied me with reports of their work prior to publication. My students, RICHARD POLLACK and JAMES HUMPHREYS, have read portions of the manuscript and have taken me to task for some obscurities and errors. HARRY ALLEN and JOSEPH FERRAR have kept me abreast of developments concerning Lie algebras related to the exceptional Jordan algebras. As student and later as colleague, DAVID J. WINTER has been a frequent and very helpful critic and contributor.

The secretarial staff of the Yale mathematics department and the editorial staff of the Springer-Verlag have been most generous and gracious in the typing and further preparation of the manuscript for publication. Financial support during several summers from grants AFOSR-402-63, from the Air Force Office of Scientific Research, and NSF-GP-4017, from the National Science Foundation, as well as a Senior Faculty Fellowship from Yale University, assisted by National Science Foundation grant NSF-GP-6558, during the academic year 1966—67, has contributed to the completion of the work while the author is still relatively young.

New Haven (Conn.), August 1967

G. B. SELIGMAN

Contents

Chapter I

Fundamentals

§ 1. Definitions

Let $\mathfrak{F}$ be a commutative ring with unit; by a *Lie algebra* over $\mathfrak{F}$ we understand a unitary $\mathfrak{F}$-module $\mathfrak{L}$, together with a mapping $(x, y) \to [x\,y]$ from $\mathfrak{L} \times \mathfrak{L}$ into $\mathfrak{L}$ which is a homomorphism in each of its variables when the other is fixed, and which satisfies in addition the following conditions:

Anticommutativity: $[x\,x] = 0$;

Jacobi identity: $[[x\,y]\,z] + [[y\,z]\,x] + [[z\,x]\,y] = 0$;

for all $x, y, z \in \mathfrak{L}$. If $\mathfrak{F} = \boldsymbol{Z}$, the integers, $\mathfrak{L}$ is called a Lie ring; clearly every Lie algebra may be regarded as a Lie ring.

Consideration of the quantity $[(x+y)\,(x+y)] - [x\,x] - [y\,y]$, which is zero in any Lie ring $\mathfrak{L}$, shows that

$$[x\,y] + [y\,x] = 0 \tag{1}$$

holds for all $x, y \in \mathfrak{L}$, and conversely if the additive group of $\mathfrak{L}$ is without 2-torsion, the condition (1) applied to $y = x$ implies anticommutativity.

If $\mathfrak{L}$ and $\mathfrak{M}$ are Lie algebras over the commutative ring $\mathfrak{F}$, we understand by a *homomorphism* of $\mathfrak{L}$ into $\mathfrak{M}$ a mapping $\eta\colon \mathfrak{L} \to \mathfrak{M}$ which is a homomorphism of $\mathfrak{F}$-modules and which satisfies $[x\,y]\,\eta = [x\,\eta, y\,\eta]$ for all $x, y \in \mathfrak{L}$. (We follow here the convention of writing mappings on the right of elements of their domains, as well as the concomitant convention that the product written $\varphi\,\eta$ of two mappings φ, η represents the result of applying first φ, then η. Thus $x(\varphi\,\eta) = (x\,\varphi)\,\eta$, if x is an element of the domain of φ such that $x\,\varphi$ is in the domain of η.) By an *ideal* in $\mathfrak{L}$ is meant a submodule $\mathfrak{K}$ such that $[x\,y] \in \mathfrak{K}$ for all $x \in \mathfrak{L}$, $y \in \mathfrak{K}$; by (1), all ideals are two-sided. In this case, the quotient module $\mathfrak{L}/\mathfrak{K}$ carries the structure of a Lie algebra over $\mathfrak{F}$, the product being specified by requiring that the canonical mapping of $\mathfrak{L}$ onto $\mathfrak{L}/\mathfrak{K}$ be a homomorphism of Lie algebras.

The fundamental homomorphism theorems of group and ring theory have their counterparts for Lie algebras. We cite:

(A) If $\eta\colon \mathfrak{L} \to \mathfrak{M}$ is a homomorphism of Lie algebras, then the kernel $\mathfrak{K}$ of η is an ideal in $\mathfrak{L}$, the image $\mathfrak{L}\eta$ is a subalgebra of $\mathfrak{M}$, and there is a unique isomorphism η' of $\mathfrak{L}/\mathfrak{K}$ onto $\mathfrak{L}\eta$ such that the diagram

$$\begin{array}{ccc} \mathfrak{L} & \xrightarrow{\eta} & \mathfrak{L}\eta \\ & {}_{\text{canon.}}\searrow \quad \nearrow_{\eta'} & \\ & \mathfrak{L}/\mathfrak{K} & \end{array}$$

is commutative. If φ is any homomorphism of $\mathfrak{L}$ into a Lie algebra $\mathfrak{N}$ (over the same ring) such that the kernel of φ contains $\mathfrak{K}$, there is a unique homomorphism ψ of $\mathfrak{L}\eta$ onto $\mathfrak{L}\varphi$ such that the diagram

$$\begin{array}{ccc} \mathfrak{L} & \xrightarrow{\eta} & \mathfrak{L}\eta \\ & {}_{\varphi}\searrow \quad \swarrow_{\psi} & \\ & \mathfrak{L}\varphi & \end{array}$$

is commutative. The mapping $\mathfrak{I} \to \mathfrak{I}\eta$ is a bijection of the set of subalgebras of $\mathfrak{L}$ containing $\mathfrak{K}$ onto the set of subalgebras of $\mathfrak{L}\eta$, under which ideals in $\mathfrak{L}$ and ideals in $\mathfrak{L}\eta$ correspond.

(B) If $\mathfrak{M}$ is a subalgebra of $\mathfrak{L}$ and if $\mathfrak{K}$ is an ideal in $\mathfrak{L}$, then $\mathfrak{M} + \mathfrak{K}$ is a subalgebra of $\mathfrak{L}$, $\mathfrak{M} \cap \mathfrak{K}$ is an ideal in $\mathfrak{M}$, and there is a unique isomorphism φ of $\mathfrak{M}/(\mathfrak{M} \cap \mathfrak{K})$ onto $(\mathfrak{M} + \mathfrak{K})/\mathfrak{K}$ making the diagram

$$\begin{array}{ccc} \mathfrak{M} & \longrightarrow & \mathfrak{M} + \mathfrak{K} \\ {\scriptstyle\text{canon.}}\downarrow & & \downarrow{\scriptstyle\text{canon.}} \\ \mathfrak{M}/(\mathfrak{M} \cap \mathfrak{K}) & \xrightarrow{\varphi} & (\mathfrak{M} + \mathfrak{K})/\mathfrak{K} \end{array}$$

commutative. Here the mapping of $\mathfrak{M}$ into $\mathfrak{M} + \mathfrak{K}$ is the inclusion mapping.

If $\mathfrak{R}$ is an associative algebra over $\mathfrak{F}$, one verifies at once that the definition $[x\,y] = x\,y - y\,x$ gives $\mathfrak{R}$ the structure of a Lie algebra. If $\mathfrak{L}$ is a Lie algebra over $\mathfrak{F}$, and if $\mathfrak{V}$ is an $\mathfrak{F}$-module, a *representation* of $\mathfrak{L}$ in $\mathfrak{V}$ is a homomorphism of $\mathfrak{L}$ into the set of endomorphisms $\mathfrak{E}(\mathfrak{V})$ of $\mathfrak{V}$, where $\mathfrak{E}(\mathfrak{V})$ has the Lie algebra structure resulting as above from its structure as associative algebra (the associative product of endomorphisms being their composite). An *associative embedding* of $\mathfrak{L}$ is an isomorphism of $\mathfrak{L}$ onto a Lie subalgebra of an associative algebra $\mathfrak{R}$ over $\mathfrak{F}$. Since every associative $\mathfrak{F}$-algebra $\mathfrak{R}$ ($1 \in \mathfrak{R}$ is assumed) is mapped isomorphically onto a subalgebra of $\mathfrak{E}(\mathfrak{R})$ by the map $x \to R_x$, where $y\,R_x = y\,x$ for all $y \in \mathfrak{R}$, we see that if $\mathfrak{L}$ has an associative embedding, then $\mathfrak{L}$ is isomorphic to a Lie subalgebra of some $\mathfrak{E}(\mathfrak{V})$, $\mathfrak{V}$ an $\mathfrak{F}$-module; i.e., $\mathfrak{L}$ has a *faithful* representation.

Certain Lie algebras and their subalgebras will be especially important in this exposition; they should also serve as examples to illustrate the preceding concepts. For each x in a given Lie algebra $\mathfrak{L}$, the mapping $y \to [y\,x]$ of $\mathfrak{L}$ into $\mathfrak{L}$ has been traditionally denoted by *ad* x. The defining identities in $\mathfrak{L}$ yield:

$$(2) \qquad [z[x\,y]] = [[z\,x]\,y] - [[z\,y]\,x];$$

$$(3) \qquad [[y\,z]\,x] = [[y\,x]\,z] + [y[z\,x]].$$

From (2), ad $[x\,y] = (\text{ad } x)\,(\text{ad } y) - (\text{ad } y)\,(\text{ad } x)$, the multiplication on the right being the associative multiplication in $\mathfrak{E}(\mathfrak{L})$. Thus $x \to \text{ad } x$ is a representation of $\mathfrak{L}$ in $\mathfrak{L}$, called the *adjoint representation*. The kernel of this representation is the set of all $x \in \mathfrak{L}$ such that $[y\,x] = 0$ for all $y \in \mathfrak{L}$; this ideal is called the *center* of $\mathfrak{L}$. The identity (3) says that ad x is a *derivation* of $\mathfrak{L}$, where the notion of derivation may be defined more generally as follows: If $\mathfrak{B}$ is an arbitrary, not-necessarily-associative $\mathfrak{F}$-algebra (no identity element in $\mathfrak{B}$ being assumed), a derivation of $\mathfrak{B}$ is an endomorphism D of $\mathfrak{B}$ as $\mathfrak{F}$-module, which satisfies in addition the condition $(b\,c)\,D = (b\,D)\,c + b(c\,D)$ for all $b, c \in \mathfrak{B}$.

Now if $\mathfrak{B}$ is as above, the totality of derivations of $\mathfrak{B}$ is a Lie subalgebra $\mathfrak{D}(\mathfrak{B})$ of $\mathfrak{E}(\mathfrak{B})$; for $\mathfrak{D}(\mathfrak{B})$ is clearly a submodule of $\mathfrak{E}(\mathfrak{B})$, and if $D, E \in \mathfrak{D}(\mathfrak{B})$, we have

$$(b\,c)\,(DE - ED) = (b(DE - ED))\,c + b(c(DE - ED)),$$

i.e., $(b\,c)\,[DE] = (b[DE])\,c + b(c[DE])$, so that $[DE] \in \mathfrak{D}(\mathfrak{B})$ as required. In particular, $\mathfrak{D}(\mathfrak{L})$ is a Lie subalgebra of $\mathfrak{E}(\mathfrak{L})$ containing the image $\text{ad}(\mathfrak{L})$ of $\mathfrak{L}$ under the adjoint representation. If $D \in \mathfrak{D}(\mathfrak{L})$, and if $x, y \in \mathfrak{L}$, then

$$y[(\text{ad } x), D] = [y\,x]\,D - [(y\,D)\,x] = [y(x\,D)] = y \ \text{ad}(x\,D);$$

thus $[(\text{ad } x), D] = \text{ad}(x\,D) \in \text{ad}(\mathfrak{L})$, so that $\text{ad}(\mathfrak{L})$ is an *ideal* in $\mathfrak{D}(\mathfrak{L})$. The derivations of $\mathfrak{L}$ belonging to $\text{ad}(\mathfrak{L})$ are called *inner* derivations.

If $\mathfrak{L}$ is a Lie algebra over $\mathfrak{F}$, and if $\mathfrak{M}$ is a subalgebra, then let $\mathfrak{N}(\mathfrak{M})$ be the set of $x \in \mathfrak{L}$ such that $[x\,\mathfrak{M}] \subseteqq \mathfrak{M}$; $\mathfrak{N}(\mathfrak{M})$ is readily seen to be a subalgebra containing $\mathfrak{M}$, and $\mathfrak{M}$ is an ideal in $\mathfrak{N}(\mathfrak{M})$; $\mathfrak{N}(\mathfrak{M})$ may also be characterized as the largest subalgebra of $\mathfrak{L}$ containing $\mathfrak{M}$ as an ideal, and is called the *normalizer* of $\mathfrak{M}$ in $\mathfrak{L}$. If $\mathfrak{K}$ and $\mathfrak{M}$ are ideals in $\mathfrak{L}$, so are $\mathfrak{K} + \mathfrak{M}$ and $[\mathfrak{K}\,\mathfrak{M}]$, the latter being defined as the smallest submodule of $\mathfrak{L}$ containing all products $[k\,m]$, where $k \in \mathfrak{K}$, $m \in \mathfrak{M}$. More concretely, we may realize $[\mathfrak{K}\,\mathfrak{M}]$ as the set of finite sums $\sum [k_i\,m_i]$, $k_i \in \mathfrak{K}$, $m_i \in \mathfrak{M}$. In particular, $[\mathfrak{L}\,\mathfrak{L}]$ is an ideal in $\mathfrak{L}$, called the *derived algebra* of $\mathfrak{L}$. If $[\mathfrak{L}\,\mathfrak{L}] = 0$ (in other words, if $\mathfrak{L}$ is its own center), we say that $\mathfrak{L}$ is *abelian*.

A *universal associative algebra* for $\mathfrak{L}$ consists of an associative $\mathfrak{F}$-algebra $\mathfrak{U}$ and a Lie homomorphism φ of $\mathfrak{L}$ into $\mathfrak{U}$ such that: if $\mathfrak{B}$ is any associative $\mathfrak{F}$-algebra and ψ a Lie homomorphism of $\mathfrak{L}$ into $\mathfrak{B}$, there is a unique homomorphism $\eta\colon \mathfrak{U} \to \mathfrak{B}$ of (associative) $\mathfrak{F}$-algebras such that the diagram

$$\begin{array}{ccc} \mathfrak{L} & \xrightarrow{\varphi} & \mathfrak{U} \\ & {\scriptstyle\psi}\searrow \quad \swarrow{\scriptstyle\eta} & \\ & \mathfrak{B} & \end{array}$$

is commutative. (For our purposes, associative algebras have an identity, and homomorphisms are unitary.) Being defined as a solution to a "universal problem", $\mathfrak{U}$ and φ are unique in the sense that if $\mathfrak{B}, \psi$ is a second such pair, then the unique mapping η as above is an isomorphism of $\mathfrak{U}$ onto $\mathfrak{B}$ (cf. [47, § 2; 234, Chap. 5]). The existence of a universal associative algebra for $\mathfrak{L}$ may be seen as follows: Ignoring for the time being the product in $\mathfrak{L}$, we form the tensor algebra $\mathfrak{T}(\mathfrak{L})$ [46, 74, 75, 222, 234]. This may be constructed as the direct sum of the $\mathfrak{F}$-modules $\mathfrak{F}, \mathfrak{L}, \mathfrak{L} \otimes \mathfrak{L}, \mathfrak{L} \otimes \mathfrak{L} \otimes \mathfrak{L}, \ldots$, with a product $u \cdot v$ defined by bilinearity and $\alpha \cdot x = x \cdot \alpha = \alpha x$, $\alpha \in \mathfrak{F}$, $x \in \mathfrak{T}(\mathfrak{L})$, $(x_1 \otimes \cdots \otimes x_m) \cdot (y_1 \otimes \cdots \otimes y_n) = x_1 \otimes \cdots \otimes x_m \otimes y_1 \otimes \cdots \otimes y_n$, where $x_i, y_j \in \mathfrak{L}$. What is important here is that $\mathfrak{T}(\mathfrak{L})$ is an associative $\mathfrak{F}$-algebra with an $\mathfrak{F}$-module homomorphism $\varphi'\colon \mathfrak{L} \to \mathfrak{T}(\mathfrak{L})$ such that if $\mathfrak{B}$ is a second associative $\mathfrak{F}$-algebra and $\psi\colon \mathfrak{L} \to \mathfrak{B}$ a homomorphism of $\mathfrak{F}$-modules, then there is a unique homomorphism of $\mathfrak{F}$-algebras $\eta'\colon \mathfrak{T}(\mathfrak{L}) \to \mathfrak{B}$ making the diagram

$$\begin{array}{ccc} \mathfrak{L} & \xrightarrow{\varphi'} & \mathfrak{T}(\mathfrak{L}) \\ & {\scriptstyle\psi}\searrow \quad \swarrow{\scriptstyle\eta'} & \\ & \mathfrak{B} & \end{array}$$

commutative. Now let $\mathfrak{J}$ be the two-sided ideal in $\mathfrak{T}(\mathfrak{L})$ generated by the elements $(x\varphi')(y\varphi') - (y\varphi')(x\varphi') - [x\,y]\varphi'$, $x, y \in \mathfrak{L}$. Let $\mathfrak{U}$ be the quotient $\mathfrak{F}$-algebra $\mathfrak{T}(\mathfrak{L})/\mathfrak{J}$, and let φ be the composite of φ' and the canonical homomorphism of $\mathfrak{T}(\mathfrak{L})$ onto $\mathfrak{U}$. Then if $\mathfrak{B}$ is a second associative algebra over $\mathfrak{F}$ and ψ a Lie homomorphism of $\mathfrak{L}$ into $\mathfrak{B}$, let $\eta'\colon \mathfrak{T}(\mathfrak{L}) \to \mathfrak{B}$ be as above; applying η' to $(x\varphi')(y\varphi') - (y\varphi')(x\varphi') - [x\,y]\varphi'$ gives $(x\psi)(y\psi) - (y\psi)(x\psi) - [x\,y]\psi$, which is zero by the fact that ψ is a Lie homomorphism. Hence the kernel of η' contains $\mathfrak{J}$, so that there is a unique homomorphism $\eta\colon \mathfrak{U} \to \mathfrak{B}$ such that

$$\begin{array}{ccc} \mathfrak{T}(\mathfrak{L}) & \xrightarrow{\text{canon.}} & \mathfrak{U} \\ & {\scriptstyle\eta'}\searrow \quad \swarrow{\scriptstyle\eta} & \\ & \mathfrak{B} & \end{array}$$

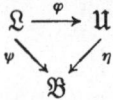

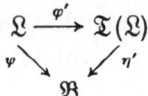

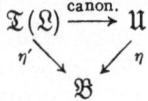

is commutative. Thus we have $\varphi\eta = \varphi'$ (canon.) $\eta = \varphi'\eta' = \psi$, and if $\zeta\colon \mathfrak{U} \to \mathfrak{B}$ is a second homomorphism with $\varphi\zeta = \psi$, then φ' (canon.) $\zeta = \psi$, or (canon.) $\zeta = \eta'$ by uniqueness of η'; then $\zeta = \eta$ by uniqueness of η in the last diagram above. The existence of a universal associative algebra is thus settled; this does not however show that $\mathfrak{L}$ has an associative *embedding*, since the mapping φ may have non-trivial kernel [59, 90, 366]. We consider this question in the next section.

§ 2. The Poincaré–Birkhoff–Witt theorem

Let $\mathfrak{L}$ be a Lie algebra over $\mathfrak{F}$ as before, and suppose further that $\mathfrak{L}$ is free as $\mathfrak{F}$-module, with basis $B \subseteqq \mathfrak{L}$. We assume B to be linearly ordered (e.g., by the well-ordering theorem). Let $\mathfrak{U}$ be the universal associative algebra of $\mathfrak{L}$, φ the mapping of $\mathfrak{L}$ into $\mathfrak{U}$, both constructed as in § 1. Then the Poincaré [321]–Birkhoff [31]–Witt [415] (or "P–B–W") theorem asserts the following:

Theorem I.2.1. $\mathfrak{U}$ is a free $\mathfrak{F}$-module, with basis consisting of 1 and the standard monomials $(b_1\varphi)\ldots(b_n\varphi)$, $b_i \in B$, $b_1 \leqq \cdots \leqq b_n$, $n = 1, 2, \ldots$; that is, the restriction to the (free) submodule of $\mathfrak{T}(\mathfrak{L})$ with basis 1 and the $b_1 \otimes \cdots \otimes b_n$, $b_1 \leqq \cdots \leqq b_n$, of the canonical homomorphism onto $\mathfrak{U}$ is an isomorphism of this $\mathfrak{F}$-module onto the $\mathfrak{F}$-module $\mathfrak{U}$. In particular, $\mathfrak{L}$ is mapped isomorphically into $\mathfrak{U}$.

Proofs of this theorem are to be found in [234, 64, 56, 47, 269] as well as in a number of research papers. The proofs in these references usually establish as well that $\mathfrak{U}$ admits a filtration such that the associated graded algebra is the symmetric algebra $\mathfrak{S}(\mathfrak{L})$ (i.e., the commutative polynomials in a basis) of the $\mathfrak{F}$-module $\mathfrak{L}$. The fact that the latter algebra has no zero-divisors if $\mathfrak{F}$ has none and is noetherian if B is finite and $\mathfrak{F}$ noetherian implies, under these conditions, that $\mathfrak{U}$ has no zero-divisors and is left (or right) noetherian. It follows by theorems of Goldie [152] and Ore [313] (see also [234, Chap. 5, and 387]) that $\mathfrak{U}$ may be embedded in a division ring of left or right quotients. For proofs of the embedding property of φ under other hypotheses as to the structure of $\mathfrak{L}$ as $\mathfrak{F}$-module, cf. [59] ($\mathfrak{F}$ is a Dedekind ring), [90] (the additive group of $\mathfrak{L}$ is torsion-free), [277] ($\mathfrak{L}$ is a direct limit of cyclic $\mathfrak{F}$-modules), [417] ($\mathfrak{F}$ is the integers).

It is by means of the associative algebra $\mathfrak{U}$ that the homology and cohomology of the Lie algebra $\mathfrak{L}$ can be defined so as to fit into a general theory of homological algebra [56, Chap. 13]. The question as to which filtered algebras generated by the module $\mathfrak{L}$ have $\mathfrak{S}(\mathfrak{L})$ as associated graded algebra has been studied by Sridharan [377]. He showed that these algebras all arise from a Lie algebra structure on $\mathfrak{L}$ and a 2-cocycle f on the Lie algebra $\mathfrak{L}$ with values in the trivial $\mathfrak{L}$-module $\mathfrak{F}$, by construction of an algebra $\mathfrak{U}_f$ which is analogous to $\mathfrak{U}$ in being universal for $\mathfrak{F}$-homomorphisms ϱ of $\mathfrak{L}$ into $\mathfrak{E}(\mathfrak{M})$, $\mathfrak{M}$ being an $\mathfrak{F}$-module, and satisfying

$$[x\varrho, y\varrho] = [x\,y]\,\varrho + f(x, y)\,I,$$

I being the identity map of $\mathfrak{M}$ (cf. also [89]). The algebra $\mathfrak{U}_f$ has many properties in common with $\mathfrak{U}$ $(= \mathfrak{U}_0)$. Relations between its left global dimension [56, Chap. 6], the rank of the free $\mathfrak{F}$-module $\mathfrak{L}$, and the global dimension of $\mathfrak{F}$ have been studied in [155, 319, 324]. BERNAT [30] has studied the center of the division ring of quotients of $\mathfrak{U}$ in some cases where such a division ring exists.

§ 3. Free Lie algebras. Restricted Lie algebras

Given a set M, one may define a *free* Lie algebra (over $\mathfrak{F}$) on M to be a Lie algebra $\mathfrak{L}$ over $\mathfrak{F}$ and a mapping $\varphi\colon M \to \mathfrak{L}$ such that whenever $\mathfrak{N}$ is a Lie algebra over $\mathfrak{F}$ and ψ a mapping of M into $\mathfrak{N}$, there is a unique homomorphism of Lie algebras $\eta\colon \mathfrak{L} \to \mathfrak{N}$, such that the following diagram is commutative:

$$\begin{array}{ccc} M & \xrightarrow{\varphi} & \mathfrak{L} \\ & {\scriptstyle\psi}\searrow \quad \swarrow{\scriptstyle\eta} & \\ & \mathfrak{N} & \end{array}$$

A free Lie algebra on M may be constructed by forming the free non-associative algebra $\mathfrak{B}$ (without 1) on M, then factoring out the two-sided ideal generated by all elements $x\,x$, $(x\,y)\,z + (y\,z)\,x + (z\,x)\,y$, $x, y, z \in \mathfrak{B}$. By appeal to the P–B–W theorem, it is possible to give a simpler description of a free Lie algebra in the cases where that theorem applies, as well as to show that the mapping φ is one-one. Namely, let $\mathfrak{T}$ be the free associative algebra with 1 on M, i.e., the tensor algebra of the free $\mathfrak{F}$-module $\mathfrak{M}$ with basis M. Then $\mathfrak{T}$ is a Lie algebra with $[x\,y] = x\,y - y\,x$, and since M is embedded in $\mathfrak{M}$, which in turn is embedded in $\mathfrak{T}$, we have a one-one mapping φ of M into the elements of degree one of $\mathfrak{T}$, such that the image $M\varphi$, together with 1, generates the $\mathfrak{F}$-algebra $\mathfrak{T}$. Let $\mathfrak{L}$ be the Lie subalgebra of $\mathfrak{T}$ generated by $M\varphi$. Then we have the following theorem, due to WITT [415]:

Theorem I.3.1. If $\mathfrak{F}$ is a field, then $\mathfrak{L}$ is the free Lie algebra on M and $\mathfrak{T}$ is its universal associative algebra.

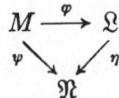

For if $\mathfrak{N}$ is a second Lie algebra over $\mathfrak{F}$ and ψ a mapping of M into $\mathfrak{N}$, let $\mathfrak{U}(\mathfrak{N})$ be the universal associative algebra of $\mathfrak{N}$. The mapping γ of $\mathfrak{N}$ into $\mathfrak{U}(\mathfrak{N})$ affords a mapping $\psi' = \psi\gamma$ of M into $\mathfrak{U}(\mathfrak{N})$; hence there is a unique homomorphism η' of $\mathfrak{F}$-algebras: $\mathfrak{T} \to \mathfrak{U}(\mathfrak{N})$, such that $\varphi\eta' = \psi'$. Since $\mathfrak{F}$ is a field, $\mathfrak{N}$ is a free $\mathfrak{F}$-module, so that γ is one-one by the P–B–W theorem; moreover, η' maps $\mathfrak{L}$ onto the Lie subalgebra of $\mathfrak{U}(\mathfrak{N})$ generated by $M\varphi\eta' = M\psi' = M\psi\gamma$. Since γ is a one-one Lie homomorphism and since $M\psi$ generates a Lie subalgebra of $\mathfrak{N}$, the Lie subalgebra of $\mathfrak{U}(\mathfrak{N})$ generated by $M\varphi\eta'$ may be identified via γ^{-1} with a Lie subalgebra of $\mathfrak{N}$. Let η be the mapping $\eta'\gamma^{-1}$, a Lie homomorphism of $\mathfrak{L}$ into $\mathfrak{N}$ such that $\varphi\eta = \psi$; by the fact that $M\varphi$ generates $\mathfrak{L}$, η is the only such homomorphism. Thus $\mathfrak{L}$ is the free Lie algebra on M. If $\mathfrak{B}$ is an arbitrary associative algebra (with 1), and if σ is a Lie homomorphism of $\mathfrak{L}$ into $\mathfrak{B}$, then $\varphi\sigma$ maps M into $\mathfrak{B}$, so that there is a unique homomorphism τ: $\mathfrak{T} \to \mathfrak{B}$, with $\varphi\tau = \varphi\sigma$. That is, σ and τ coincide on $M\varphi$, and since $M\varphi$ generates $\mathfrak{L}$, they coincide on $\mathfrak{L}$. Now any algebra-homomorphism of $\mathfrak{T}$ into $\mathfrak{B}$ is determined by its values on $\mathfrak{L}$, since $M\varphi \subseteqq \mathfrak{L}$ generates $\mathfrak{T}$. Thus τ is the unique extension of σ to $\mathfrak{T}$, so that $\mathfrak{T}$ is the universal associative algebra of $\mathfrak{L}$.

It will be noted that $\mathfrak{T}$ is always the universal associative algebra of $\mathfrak{L}$, without restriction on the commutative ring $\mathfrak{F}$, while the existence of an extension of ψ: $M \to \mathfrak{N}$ to a homomorphism of $\mathfrak{L}$ into $\mathfrak{N}$ depends only on the P–B–W property for $\mathfrak{N}$.

Free Lie algebras and Lie rings (i.e., algebras over $\boldsymbol{Z}$) have been studied by M. Hall [159], Shirshov [367—369] and Witt [418] with regard to the analogue of the Schreier–Nielsen theorem for groups, as well as to the determination of explicit bases. In particular, every subalgebra of a free Lie algebra over a field is free [367, 418].

If the ground ring $\mathfrak{F}$ is a field of prime characteristic p, and if the set M consists of the two elements x, y (which we identify with elements of $\mathfrak{T}$), then it is known [58, Exposé 3; 234, Chap. 5] that the element $\Lambda_p(x, y) = (x + y)^p - x^p - y^p$ of $\mathfrak{T}$ is in fact in $\mathfrak{L}$, and that $\Lambda_p(x, y)$ is a linear combination of monomials $x_1(\mathrm{ad}\, x_2) \ldots (\mathrm{ad}\, x_p)$, where each x_i is either x or y, and where one may always take $x_1 = x$. With this latter convention, the term in $x(\mathrm{ad}\, y)^{p-1}$ occurs with coefficient one. Since $\Lambda_p(x, y)$ is a uniquely determined element of $\mathfrak{L}$, it makes sense to define $\Lambda_p(u, v)$ whenever u and v are elements of a Lie algebra $\mathfrak{M}$ over $\mathfrak{F}$, as the image of $\Lambda_p(x, y)$ under the homomorphism of $\mathfrak{L}$ into $\mathfrak{M}$ sending x into u, y into v.

Now in any associative ring $\mathfrak{R}$, one has

$$x(\mathrm{ad}\, y)^k = \sum_{i=0}^{k} (-1)^i \binom{k}{i} y^i x y^{k-i}, \tag{4}$$

so that if $\mathfrak{R}$ is an $\mathfrak{F}$-algebra, $(\mathrm{ad}\, y)^p = \mathrm{ad}(y^p)$ for all $y \in \mathfrak{R}$. With the aid of the definition of $\Lambda_p(x, y)$, one can define a *restricted Lie algebra* (Jacobson) as a Lie algebra $\mathfrak{L}$ over a field $\mathfrak{F}$ of prime characteristic p, together with a mapping $z \to z^{[p]}$ of $\mathfrak{L}$ into $\mathfrak{L}$ satisfying the identities:

$$\begin{aligned} &\text{a)} \quad \mathrm{ad}(z^{[p]}) = (\mathrm{ad} z)^p; \\ &\text{b)} \quad (\alpha z)^{[p]} = \alpha^p z^{[p]}; \\ &\text{c)} \quad (y + z)^{[p]} = y^{[p]} + z^{[p]} + \Lambda_p(y, z); \end{aligned} \tag{5}$$

for all $y, z \in \mathfrak{L}$, all $\alpha \in \mathfrak{F}$.

In particular, any Lie subalgebra of an associative algebra $\mathfrak{N}$ over $\mathfrak{F}$ is a restricted Lie algebra if it is closed under p-th powers in $\mathfrak{N}$. As a special case, one sees from the Leibniz formula that the $\mathfrak{F}$-derivations of any (not necessarily associative) algebra over $\mathfrak{F}$ form a restricted Lie algebra [211, 215]. Furthermore, if $\mathfrak{L}$ is an abelian Lie algebra over $\mathfrak{F}$, so that $\Lambda_p(x, y)$ is always zero, then $\mathfrak{L}$ is made into a restricted Lie algebra by any p-semi-linear mapping, i.e., an additive mapping $P\colon \mathfrak{L} \to \mathfrak{L}$ such that $(\alpha x) P = \alpha^p (x P)$ for all $x \in \mathfrak{L}$, $\alpha \in \mathfrak{F}$. Such algebras have been studied in [225].

It will be noted that if $\mathfrak{F}$ is a commutative ring with $p\,\mathfrak{F} = 0$, every Lie algebra $\mathfrak{L}$ over $\mathfrak{F}$ may be regarded as a Lie algebra over the prime field $\boldsymbol{Z}_p$; thus the element $\Lambda_p(x, y) \in \mathfrak{L}$ is defined for $x, y \in \mathfrak{L}$, and we may define the notion of restricted Lie algebra over $\mathfrak{F}$ just as when $\mathfrak{F}$ is a field.

By a *restricted representation* of $\mathfrak{L}$ in the $\mathfrak{F}$-module $\mathfrak{B}$ we mean a representation φ of $\mathfrak{L}$ in $\mathfrak{B}$ with $(x^{[p]})\,\varphi = (x\,\varphi)^p$, the right-hand side being the p-th iterate of the endomorphism $x\,\varphi$ of $\mathfrak{B}$. The notions of restricted homomorphism, restricted subalgebra, restricted ideal are now easy to express, as are the counterparts of § 1 above; in particular, it will be noted from (5) and § 1 that the adjoint representation is a restricted representation of $\mathfrak{L}$ by derivations of $\mathfrak{L}$.

A *restricted universal associative algebra* for $\mathfrak{L}$ (restricted) is an associative $\mathfrak{F}$-algebra $\mathfrak{U}$ and a homomorphism $\varphi\colon \mathfrak{L} \to \mathfrak{U}$ of restricted Lie algebras, such that if $\mathfrak{B}$ is a second associative algebra and ψ a restricted Lie homomorphism of $\mathfrak{L}$ into $\mathfrak{B}$, there is a unique homomorphism $\eta\colon \mathfrak{U} \to \mathfrak{B}$ of $\mathfrak{F}$-algebras such that

$$\begin{array}{ccc} \mathfrak{L} & \xrightarrow{\varphi} & \mathfrak{U} \\ & {\scriptstyle\psi}\searrow \quad \swarrow{\scriptstyle\eta} & \\ & \mathfrak{B} & \end{array}$$

is commutative. The uniqueness of $\mathfrak{U}$ and φ is clear; following Jacobson [215], we shall refer to such an algebra $\mathfrak{U}$ as *the u-algebra* of $\mathfrak{L}$. A construction for $\mathfrak{U}$ and φ is now reasonably apparent; let $\mathfrak{U}'$, φ' be a universal associative algebra for $\mathfrak{L}$, and let $\mathfrak{K}$ be the two-sided ideal in $\mathfrak{U}'$ generated by the elements $(x^{[p]})\,\varphi' - (x\,\varphi')^p$, $x \in \mathfrak{L}$; let $\mathfrak{U} = \mathfrak{U}'/\mathfrak{K}$, and let φ be the composite of φ' and the canonical homomorphism of $\mathfrak{U}'$ onto $\mathfrak{U}$. One verifies easily that $\mathfrak{U}$, φ has the required properties. We also have the following analogue of the P–B–W theorem:

Theorem I.3.2 (Jacobson [215]). Let $\mathfrak{L}$ be a restricted Lie algebra over the commutative ring $\mathfrak{F}$ with $p\,\mathfrak{F} = 0$, and suppose $\mathfrak{L}$ is a free $\mathfrak{F}$-module with the linearly ordered basis B. Then the mapping φ of $\mathfrak{L}$ into its u-algebra $\mathfrak{U}$ is one-one and, identifying B with $B\,\varphi$, $\mathfrak{U}$ is a free $\mathfrak{F}$-module with basis the monomials $b_1^{s_1} \ldots b_k^{s_k}$, $b_1 < \cdots < b_k$,

$1 \leqq s_i < p$. In particular, if $B = \{b_1, \ldots, b_n\}$ is finite, $\mathfrak{U}$ has as basis the monomials $b_1^{s_1} \ldots b_n^{s_n}$, $0 \leqq s_i < p$, hence is a free $\mathfrak{F}$-module on p^n generators.

The idea of the proof of this and several other results to be given below lies in the fact that one can choose a basis for the universal associative algebra $\mathfrak{U}'$ which is compatible with relations of the type introduced in the formation of $\mathfrak{U}$. More precisely, the algebra $\mathfrak{U}'$ is *filtered*; $\mathfrak{U}'$ is the increasing union of $\{\mathfrak{U}'_n\}$, where $\mathfrak{U}'_n$ is the set of linear combinations of products of at most n factors from $B\varphi'$; one has $\mathfrak{U}'_0 = \mathfrak{F}$, $\mathfrak{U}'_i \mathfrak{U}'_j \subseteqq \mathfrak{U}'_{i+j}$. We identify B and $B\varphi'$. Now suppose that for each $b \in B$ there is an element $z(b)$ in the center of $\mathfrak{U}'$ and an integer $n(b)$ such that $b^{n(b)} \equiv z(b) \bmod \mathfrak{U}'_{n(b)-1}$. (For example, $b^p \equiv b^p - b^{[p]} \bmod \mathfrak{U}'_{p-1}$, and $b^p - b^{[p]}$ centralizes $\mathfrak{L}$, hence $\mathfrak{U}'$, in the Theorem; it will be noted, though, that no assumption of restrictedness is made on $\mathfrak{L}$ in formulating the general principle.) Then one has the following

Lemma I.3.1 (Jacobson). A basis for $\mathfrak{U}'$ consists of the elements

$$z(b_1)^{r_1} \ldots z(b_m)^{r_m} b_1^{s_1} \ldots b_m^{s_m}, \tag{6}$$

where $b_1 < \cdots < b_m$, $0 \leqq r_j$, $0 \leqq s_j < n(b_j)$, $r_j + s_j > 0$, $m = 0, 1, 2, \ldots$

For by the condition $b^{n(b)} \equiv z(b) \bmod \mathfrak{U}'_{n(b)-1}$ and the P–B–W theorem, the above elements are independent; moreover, each standard monomial $b_1^{t_1} \ldots b_m^{t_m}$ is congruent modulo $\mathfrak{U}'_{t-1}$, where $t = \sum t_i$, to $z(b_1)^{r_1} \ldots z(b_m)^{r_m} b_1^{s_1} \ldots b_m^{s_m}$, where $t_i = r_i n(b_i) + s_i$, $0 \leqq s_i < n(b_i)$, from which the lemma follows by induction on t and the P–B–W theorem. (For more detail cf. [234, p. 189].)

With the notations preceding the statement of Th. 2, we see that the ideal $\mathfrak{K}$ contains all the basis elements for $\mathfrak{U}'$ involving at least one factor $z(b) = b^p - b^{[p]}$; furthermore, it follows from the properties of the p-power in $\mathfrak{U}'$ and the map $x \to x^{[p]}$ in $\mathfrak{L}$ that $\mathfrak{K}$ is generated as an ideal by the $z(b)$, $b \in B$. Hence the basis elements (6) involving a factor $z(b)$ must span $\mathfrak{K}$. That is, the remaining basis elements have cosets which form a basis for $\mathfrak{U}'/\mathfrak{K} = \mathfrak{U}$. This is the assertion of Th. 2.

Theorem I.3.3 (Jacobson). Let $\mathfrak{L}$ be a Lie algebra over the commutative ring $\mathfrak{F}$ with $p\mathfrak{F} = 0$, and let $\mathfrak{L}$ be a free $\mathfrak{F}$-module with basis B. Suppose given a mapping $b \to b^{[p]}$ of B into $\mathfrak{L}$ such that $\operatorname{ad}(b^{[p]}) = (\operatorname{ad} b)^p$ for all $b \in B$. Then there is a unique p-power operation in $\mathfrak{L}$ extending the map on B and relative to which $\mathfrak{L}$ is a restricted Lie algebra.

Uniqueness is immediate from the identities (5), b) and c). To see the existence, we may assume B linearly ordered, and set $z(b) = b^p - b^{[p]}$ in $\mathfrak{U}'$, as in the proof of Th. 2. Letting $\mathfrak{K}'$ be the ideal in $\mathfrak{U}'$ generated by these $z(b)$, it follows as above that $\mathfrak{L}$ is embedded in the associative algebra $\mathfrak{U}'/\mathfrak{R}' = \mathfrak{U}''$, and that if $y = \sum \alpha_i b_i \in \mathfrak{L} \subseteqq \mathfrak{U}''$ ($\alpha_i \in \mathfrak{F}$, $b_i \in B$), then $y^p - \sum \alpha_i^p b_i^p \in \mathfrak{L}$, i.e., $y^p - \sum \alpha_i^p b_i^{[p]} \in \mathfrak{L}$, or $y^p \in \mathfrak{L}$. The map $x \to x^p$ of $\mathfrak{U}''$ thus maps $\mathfrak{L}$ into $\mathfrak{L}$ and extends $b \to b^{[p]}$.

It is clear that any Lie algebra $\mathfrak{L}$ over $\mathfrak{F}$ which has an associative embedding can be mapped isomorphically into a restricted Lie algebra. This is of course the case if $\mathfrak{F}$ is a field. More generally, Cohn has shown that for the existence of an embedding of $\mathfrak{L}$ in a restricted Lie algebra it is sufficient that the mapping $\alpha \to \alpha^p$ of $\mathfrak{F}$ be an automorphism; also sufficient are the joint conditions that $\mathfrak{F}$ have no nilpotent elements and that $\mathfrak{F}^p$ be pure in $\mathfrak{F}$. He has also given examples of Lie algebras, over rings satisfying each (but not both) of the latter conditions, which are not embeddable in restricted Lie algebras, and of restricted Lie algebras without associative embeddings [91]; cf. also [215].

§ 4. Iwasawa's theorem

The P–B–W theorem has the consequence that a Lie algebra $\mathfrak{L}$ over a commutative ring $\mathfrak{F}$, which is free as $\mathfrak{F}$-module, has a faithful representation in a free $\mathfrak{F}$-module, namely the universal associative algebra. However, this module has no finite basis, even though $\mathfrak{L}$ may have a finite basis. The theorem of Ado asserts that if $\mathfrak{L}$ is finite-dimensional, where $\mathfrak{F}$ is the real field, then $\mathfrak{L}$ has a faithful finite-dimensional representation. Combined with the correspondence between closed connected subgroups of a Lie group and subalgebras of its Lie algebra, this theorem yields the conclusion that every real Lie algebra is the Lie algebra of a real Lie group of matrices. The theorem of Ado has been extended to Lie algebras over arbitrary fields, the proof being split into the case of characteristic zero, where Harish-Chandra [165] has given the proof which seems to enjoy (and deserve) the greatest popularity, and the modular case, originally treated by Iwasawa [208] and where a brief and elegant proof has been given by Jacobson [221]. In keeping with the central theme of this report, we give here only Jacobson's proof in the modular case, referring to the extensive literature for the case of characteristic zero; for once, the modular proof is more elementary than the non-modular one. For the latter, one may consult [4, 31, 47, 55, 64, 72, 234], in addition to the references above.

Theorem I.4.1 (Iwasawa). Let $\mathfrak{L}$ be a finite-dimensional Lie algebra over the field $\mathfrak{F}$ of prime characteristic p. Then $\mathfrak{L}$ has a faithful finite-dimensional representation.

Namely, let $b_1, \ldots, b_n$ be a basis for $\mathfrak{L}$. For each b_i, the space of linear transformations of $\mathfrak{L}$ spanned by ad b_i, $(\text{ad } b_i)^p$, $(\text{ad } b_i)^{p^2}, \ldots$ is finite-dimensional, from which it follows that there is a non-zero monic polynomial $m_i(X) \in \mathfrak{F}[X]$ of the form $m_i(X) = \sum_{j=0}^{n} \alpha_j X^{p^j}$ (a "p-

polynomial") with $m_i(\mathrm{ad}\ b_i) = 0$. By replacing $m_i(X)$ by $m_i(X)^p$ if necessary, we may assume its degree $d_i > 1$. Translating this condition to the universal associative algebra $\mathfrak{U}$ of $\mathfrak{L}$ we have that in $\mathfrak{U}$, $x m_i(\mathrm{ad}\ b_i) = 0 = [x, m_i(b_i)]$ for all $x \in \mathfrak{L}$. Thus each $m_i(b_i)$ is in the center of $\mathfrak{U}$.

Now let $z(b_i) = m_i(b_i)$ for each i, and let $\mathfrak{J}$ be the ideal in $\mathfrak{U}$ generated by these $z(b_i)$. As in § 3, it follows from Lemma 3.1 that the cosets of the monomials $b_1^{s_1} \ldots b_n^{s_n}$, $0 \leqq s_k < d_k$, form a basis for $\mathfrak{U}/\mathfrak{J}$, and (since each $d_k > 1$) that the canonical mapping is one-one on $\mathfrak{L}$. That is, the canonical mapping defines an embedding of $\mathfrak{L}$ in $\mathfrak{U}/\mathfrak{J}$. This completes the proof.

It may be remarked that Jacobson's analogue, for the u-algebra of a restricted Lie algebra, of the P–B–W theorem (Th. 3.2) proves the result analogous to the above for finite-dimensional restricted Lie algebras, namely:

Theorem I.4.2 (Jacobson). Let $\mathfrak{F}$ be a commutative ring with $p\,\mathfrak{F} = 0$, $\mathfrak{L}$ a restricted Lie algebra over $\mathfrak{F}$ which is a free $\mathfrak{F}$-module on a finite basis. Then $\mathfrak{L}$ has a faithful restricted representation in a free $\mathfrak{F}$-module with a finite basis, viz., in its u-algebra.

Hochschild [200] has shown that the conclusion of Th. I.4.1 can be sharpened to assert that there is a faithful finite-dimensional representation ϱ such that x^ϱ is nilpotent for each $x \in \mathfrak{L}$ with ad x nilpotent.

§ 5. Nilpotent Lie algebras. Engel's theorem

If $\mathfrak{L}$ is a Lie algebra over $\mathfrak{F}$, we set $\mathfrak{L}^2 = [\mathfrak{L}\,\mathfrak{L}]$, and in general, for $k \geqq 2$, $\mathfrak{L}^k = [\mathfrak{L}^{k-1}\,\mathfrak{L}]$, with $\mathfrak{L}^1 = \mathfrak{L}$. Then all $\mathfrak{L}^k$ are ideals in $\mathfrak{L}$, and $\mathfrak{L}^1 \supseteqq \mathfrak{L}^2 \supseteqq \cdots$. The ideals $\mathfrak{L}^k$ make up the *lower central series* of $\mathfrak{L}$. As with groups, $\mathfrak{L}$ is called *nilpotent* if $\mathfrak{L}^k = 0$ for some k.

Now let $\mathfrak{L}$ be a finite-dimensional Lie algebra over a field $\mathfrak{F}$. The classical criterion of Engel for the nilpotency of $\mathfrak{L}$ is valid over arbitrary fields; that is, it is clear from the definition of $\mathfrak{L}^k$ that for any $x \in \mathfrak{L}$, $\mathfrak{L}(\mathrm{ad}\ x)^k \subseteqq \mathfrak{L}^{k+1}$, hence that if $\mathfrak{L}$ is nilpotent then so is ad x (as linear transformation of $\mathfrak{L}$) for every $x \in \mathfrak{L}$. Engel's theorem asserts the converse, which follows immediately from the following version:

Theorem I.5.1 (Engel's Theorem). Let $\mathfrak{V}$ be a non-zero finite-dimensional vector space over a field $\mathfrak{F}$, and let $\mathfrak{L}$ be a Lie subalgebra of $\mathfrak{E}(\mathfrak{V})$ consisting entirely of nilpotent transformations. Then there is a non-zero $v \in \mathfrak{V}$ such that $vT = 0$ for all $T \in \mathfrak{L}$.

We prove the assertion by induction on the dimension of $\mathfrak{L}$, say d, there being no difficulty if $d = 0$ or 1. Thus let $d > 1$, and assume that the result holds for all Lie algebras of nilpotent linear transformations (of *any* finite-dimensional space) of dimension less than d. First we show that $\mathfrak{L}$ contains an ideal of codimension one. Namely, let $\mathfrak{B}$

be a proper subalgebra of $\mathfrak{L}$ (e.g., a one-dimensional subspace); then for each $b \in \mathfrak{B}$, it follows, by (4) and the hypotheses, that ad b is nilpotent, acting in $\mathfrak{L}$. Thus $\mathfrak{B}$ is an invariant subspace of $\mathfrak{L}$ relative to the adjoint representation of $\mathfrak{B}$ in $\mathfrak{L}$, so that a representation of $\mathfrak{B}$ by nilpotent linear transformations is induced in the quotient space $\mathfrak{L}/\mathfrak{B}$. Since the representing algebra for $\mathfrak{B}$ has dimension $<d$, this algebra annihilates a non-zero element of $\mathfrak{L}/\mathfrak{B}$, i.e., there is $c \in \mathfrak{L}$, $c \notin \mathfrak{B}$, such that $[c\,\mathfrak{B}] \subseteqq \mathfrak{B}$. Thus $\mathfrak{B}$ *is properly contained in its normalizer* in $\mathfrak{L}$. Taking $\mathfrak{N}$ to be a maximal proper subalgebra of $\mathfrak{L}$, one now sees easily that $\mathfrak{N}$ is an ideal of codimension one.

Now we may assume by induction that there is a vector $x \neq 0$ in $\mathfrak{V}$ such that $x\,\mathfrak{N} = 0$. Let $\mathfrak{W}$ be the set of all $x \in \mathfrak{V}$ such that $x\,\mathfrak{N} = 0$, a non-zero subspace of $\mathfrak{V}$. Let $T \in \mathfrak{L}$, $T \notin \mathfrak{N}$, so that $\mathfrak{L} = \mathfrak{F}T + \mathfrak{N}$; then $x \in \mathfrak{W}$, $N \in \mathfrak{N}$, implies $xTN = xNT + x[TN] = 0$, since $[TN]$ is in the ideal $\mathfrak{N}$. Thus $\mathfrak{W}T \subseteqq \mathfrak{W}$, so that by nilpotency of T there is $v \neq 0$ in $\mathfrak{W}$ with $vT = 0$, therefore with $v\,\mathfrak{L} = 0$. This completes the proof.

Corollary 1. Let $\mathfrak{V}$ and $\mathfrak{L}$ be as above. Then there is a chain of subspaces $\mathfrak{V} = \mathfrak{V}_0 \supset \mathfrak{V}_1 \supset \cdots \supset \mathfrak{V}_n = 0$ for $\mathfrak{V}$, such that $\mathfrak{V}_{i+1}$ is of codimension one in $\mathfrak{V}_i$ and such that $\mathfrak{V}_i \mathfrak{L} \subseteqq \mathfrak{V}_{i+1}$, $0 \leqq i < n$. That is, there is a basis for $\mathfrak{V}$ relative to which the matrices of all $T \in \mathfrak{L}$ are properly (upper) triangular.

Corollary 2. Let $\mathfrak{L}$ be a finite-dimensional Lie algebra over $\mathfrak{F}$. Then $\mathfrak{L}$ is nilpotent if and only if ad x is nilpotent for all $x \in \mathfrak{L}$, and in this case there is a chain of ideals $\mathfrak{L} = \mathfrak{L}_0 \supset \mathfrak{L}_1 \supset \cdots \supset \mathfrak{L}_n = 0$ in $\mathfrak{L}$, $\mathfrak{L}_{i+1}$ of codimension one in $\mathfrak{L}_i$ and $[\mathfrak{L}_i\,\mathfrak{L}] \subseteqq \mathfrak{L}_{i+1}$, $0 \leqq i < n$.

Corollary 1 follows by the fact that every composition factor of $\mathfrak{V}$ as $\mathfrak{L}$-module is one-dimensional and annihilated by $\mathfrak{L}$; the $\mathfrak{V}_i$ may be taken as an arbitrary composition series. The desired basis is obtained by choosing a basis for $\mathfrak{V}$ which contains a basis for each $\mathfrak{V}_i$. Corollary 2 follows from the theorem and from Cor. 1 by noting that these together imply the existence of such a chain of ideals when every ad x is nilpotent, and then that $\mathfrak{L}^k \subseteqq \mathfrak{L}_{k+1}$ by induction on k; thus $\mathfrak{L}^{n-1} = 0$.

It is clear from the above that if $\mathfrak{L}$ is a Lie subalgebra of $\mathfrak{E}(\mathfrak{V})$, the nilpotency of all $T \in \mathfrak{L}$ is a sufficient condition for the nilpotency of the Lie algebra $\mathfrak{L}$. This condition is by no means necessary; for example, any one-dimensional subspace of $\mathfrak{E}(\mathfrak{V})$ is a nilpotent Lie algebra. In case $\mathfrak{F}$ is algebraically closed, the following may be regarded as giving a characterization of nilpotent Lie subalgebras of $\mathfrak{E}(\mathfrak{V})$:

Theorem I.5.2 (ZASSENHAUS [419]). Let $\mathfrak{V}$ be a finite-dimensional vector space over a field $\mathfrak{F}$, and let $\mathfrak{L}$ be a nilpotent Lie subalgebra of $\mathfrak{E}(\mathfrak{V})$ such that, for each $T \in \mathfrak{L}$, all characteristic roots of T are in $\mathfrak{F}$. For each $\mathfrak{F}$-valued function φ on $\mathfrak{L}$, let $\mathfrak{V}_\varphi$ be the set of $v \in \mathfrak{V}$

such that for each $T \in \mathfrak{L}$, there is a positive integer n with $v(T - \varphi(T)\, I)^n = 0$. Then $\mathfrak{B}_\varphi\, \mathfrak{L} \subseteqq \mathfrak{B}_\varphi$ for all φ, and $\mathfrak{B}$ is the direct sum of the non-zero spaces $\mathfrak{B}_\varphi$. Conversely, if $\mathfrak{L}$ is a Lie subalgebra of $\mathfrak{E}(\mathfrak{B})$, and if $\mathfrak{B}$ is the sum of $\mathfrak{L}$-invariant subspaces $\mathfrak{B}_i$, where for each i there is a scalar-valued function φ_i on $\mathfrak{L}$ such that $T - \varphi_i(T)\, I$ is nilpotent on $\mathfrak{B}_i$, then $\mathfrak{L}$ is nilpotent.

The converse part of the theorem follows from Engel's theorem; namely, it suffices to show that the restriction $\mathfrak{L}_i$ of $\mathfrak{L}$ to each $\mathfrak{B}_i$ is a nilpotent Lie algebra; and in $\mathfrak{L}_i$, $\operatorname{ad} T = \operatorname{ad}(T - \varphi_i(T)\, I)$ is nilpotent since $T - \varphi_i(T)\, I$ is nilpotent as endomorphism of $\mathfrak{B}_i$; Cor. 2 above now yields that $\mathfrak{L}_i$ is a nilpotent Lie algebra.

For the direct part of the theorem, we first prove $\mathfrak{B}_\varphi\, \mathfrak{L} \subseteqq \mathfrak{B}_\varphi$. Thus let $v \in \mathfrak{B}$, $S, T \in \mathfrak{L}$; we wish to show that if $v(T - \lambda\, I)^m = 0$ for some m, some $\lambda \in \mathfrak{F}$, then $v\, S(T - \lambda\, I)^n = 0$ for some n. By induction,

$$(7) \quad v S(T - \lambda I)^r = v(T - \lambda I)^r S + \sum_{j=0}^{r-1} v(T - \lambda I)^{r-j-1} [ST] (T - \lambda I)^j .$$

Now if $d = \dim. \mathfrak{B}$, then the polynomial in T of lowest degree annihilating v is a power of $X - \lambda$, X indeterminate, and has degree at most d; thus it divides $(X - \lambda)^d$, and $v(T - \lambda\, I)^d = 0$. Since $\mathfrak{L}$ is nilpotent, we have $S(\operatorname{ad} T)^k = 0$ for some $k \geqq 0$. We argue by induction on k to show our assertion; for $k = 0$, it is trivial. Hence we assume $k \geqq 1$, so that $[ST]\, (\operatorname{ad} T)^{k-1} = 0$, and by induction $[ST]$ maps into itself the space belonging to the characteristic root λ of T, call it $\mathfrak{B}_\lambda$. Taking $r \geqq 2d$ in (7), with $v \in \mathfrak{B}_\lambda$, all terms on the right are zero except possibly those with $r - j - 1 < d$, for which $j > r - d - 1 \geqq d - 1$, i.e., $j \geqq d$. Since $v(T - \lambda\, I)^{r-j-1}\, [ST] \in \mathfrak{B}_\lambda$ for these values of j, we have $v(T - \lambda\, I)^{r-j-1}\, [ST]\, (T - \lambda\, I)^j = 0$ here as well. Hence if φ is as in the theorem, we have $\mathfrak{B}_\varphi\, S = \left(\bigcap_{T \in \mathfrak{L}} \mathfrak{B}_{\varphi(T)} \right) S \subseteqq \bigcap_T \mathfrak{B}_{\varphi(T)} = \mathfrak{B}_\varphi$. (Here $\mathfrak{B}_{\varphi(T)}$ refers to the subspace belonging to the characteristic root $\varphi(T)$ of T.)

Next we note that if $T \in \mathfrak{L}$, and if $T = T_s + T_n$ is the Jordan decomposition of T, with T_s semisimple (being represented by a diagonal matrix) and with T_n nilpotent, $[T_s T_n] = 0$, then $\operatorname{ad} T_s$ and $\operatorname{ad} T_n$, acting in $\mathfrak{E}(\mathfrak{B})$, are in turn the semisimple and nilpotent parts of $\operatorname{ad} T$ (cf. the proof of Lemma II.1.1). Moreover, both of these may be expressed as polynomials in $\operatorname{ad} T$ (cf. [71, p. 71]), hence induce mappings of $\mathfrak{L}$ into $\mathfrak{L}$. On $\mathfrak{L}$, $\operatorname{ad} T_s$ induces the nilpotent mapping $\operatorname{ad} T - \operatorname{ad} T_n$; hence $[\mathfrak{L}, T_s] = 0$, so that $[S, T_s] = 0 = [S_s, T_s]$ for all $S, T \in \mathfrak{L}$. Let $\mathfrak{L}_s = \{T_s \mid T \in \mathfrak{L}\}$; from the above, $\mathfrak{L}_s$ is a commutative set of diagonalizable transformations of $\mathfrak{B}$, from which we see that $\mathfrak{B} = \sum \oplus \mathfrak{B}'_{\varphi_i}$, where the φ_i are distinct $\mathfrak{F}$-valued functions on $\mathfrak{L}_s$, and where $\mathfrak{V}'_{\varphi_i} = \{v \mid vT_s = \varphi_i(T_s)\, v \text{ for all } T \in \mathfrak{L}\}$. Moreover, if φ^* is an $\mathfrak{F}$-valued function on $\mathfrak{L}_s$, and if $v \in \mathfrak{V}$ satisfies $vT_s = \varphi^*(T_s)\, v$ for all $T_s \in \mathfrak{L}_s$, then $v \in \mathfrak{V}'_{\varphi_i}$ for some i.

Since $[\mathfrak{L}, \mathfrak{L}_s] = 0$, we have $\mathfrak{V}'_{\varphi_i}\mathfrak{L} \subseteqq \mathfrak{V}'_{\varphi_i}$ for all i, and if $T \in \mathfrak{L}$, $(T - \varphi_i(T_s)\, I) \mid \mathfrak{V}'_{\varphi_i} = (T - T_s) \mid \mathfrak{V}'_{\varphi_i} = T_n \mid \mathfrak{V}'_{\varphi_i}$ is nilpotent. It remains only to show that if φ is an $\mathfrak{F}$-valued function on $\mathfrak{L}$, and if $v \in \mathfrak{V}_\varphi$, then $v \in \mathfrak{V}'_{\varphi_i}$ for some i as above. But if $T \in \mathfrak{L}$, we see from the above that T_s and T_n map $\mathfrak{V}_\varphi$ into $\mathfrak{V}_\varphi$, where $T = \varphi(T)\, I + N$, N nilpotent. Thus $T_s \mid \mathfrak{V}_\varphi = \varphi(T)\, I$, so that $vT_s = \varphi(T)\, v$ for all $T \in \mathfrak{L}$. By the remarks above, this completes the proof.

The functions φ with $\mathfrak{V}_\varphi \neq 0$ are called the *weights* of $\mathfrak{L}$.

Corollary. If $\mathfrak{V}$ and $\mathfrak{L}$ are as above, and if $\mathfrak{V}$ is indecomposable relative to $\mathfrak{L}$, then every $T \in \mathfrak{L}$ has only one characteristic root.

§ 6. Cartan subalgebras

Let $\mathfrak{L}$ be a Lie algebra over the field $\mathfrak{F}$. A *Cartan subalgebra* $\mathfrak{H}$ of $\mathfrak{L}$ is a nilpotent subalgebra which is its own normalizer in $\mathfrak{L}$, i.e., $[x\,\mathfrak{H}] \subseteqq \mathfrak{H}$ implies $x \in \mathfrak{H}$. A powerful tool for studying the structure of a Lie algebra $\mathfrak{L}$ over an algebraically closed field has been the study of the adjoint representation on $\mathfrak{L}$ of a Cartan subalgebra $\mathfrak{H}$. Since the image of $\mathfrak{H}$ in such a representation is a nilpotent Lie algebra of linear transformations, the considerations of § 5 apply, where the weights φ may be regarded as functions on $\mathfrak{H}$ via the homomorphism $h \to \operatorname{ad} h$. In this setting, one refers to these functions on $\mathfrak{H}$ as the *roots* of $\mathfrak{L}$ relative to $\mathfrak{H}$. Since we are here concerned with the modular case, where the trace criteria of Cartan fail (cf. Chap. V), and since these criteria constitute one of the most striking applications of the Cartan subalgebra, this notion cannot contribute positively to our study in so many ways as in the case of characteristic zero. On the other hand, it will be central to the study of an important class of semi-simple algebras in the next chapters. For the classical proof of existence of Cartan subalgebras for the case where $\mathfrak{F}$ is sufficiently large cf. [53, 64, 72, 128, 234]. The result in question is stated below as Th. I.6.1.

An element $x \in \mathfrak{L}$ is called *regular* if the multiplicity of zero as characteristic root of $\operatorname{ad} x$ is minimal among all elements of $\mathfrak{L}$; if $\mathfrak{L} \neq 0$, then from $[x\,x] = 0$ it is clear that this multiplicity is at least one. If $\mathfrak{L} = 0$, then $\mathfrak{L}$ is a Cartan subalgebra of itself.

Theorem I.6.1 (Cartan). Let the field $\mathfrak{F}$ have more elements than the dimension over $\mathfrak{F}$ of the finite-dimensional Lie algebra $\mathfrak{L}$, and let x be a regular element of $\mathfrak{L}$. Let $\mathfrak{H}$ be the subspace of $\mathfrak{L}$ belonging to the characteristic root 0 of $\operatorname{ad} x$, i.e., $\mathfrak{H} = \{h \mid h \in \mathfrak{L} \cdot h(\operatorname{ad} x)^m = 0 \text{ for some } m\}$. Then $\mathfrak{H}$ is a Cartan subalgebra of $\mathfrak{L}$.

A property of Cartan subalgebras which is formally stronger than the self-normalizing property is

Lemma I.6.1. Let $\mathfrak{L}$ be a Lie algebra over $\mathfrak{F}$, $\mathfrak{H}$ a Cartan subalgebra, and let $x \in \mathfrak{L}$ be such that $x(\operatorname{ad} h)^k = 0$ holds for each $h \in \mathfrak{H}$ and sufficiently large k. Then $x \in \mathfrak{H}$.

For if $\mathfrak{M}$ is the set of all such $x \in \mathfrak{L}$, then $\mathfrak{M}$ is a subspace, and it follows as in the proof of Th. 5.2 that the nilpotent Lie algebra ad $\mathfrak{H}$ of linear transformations of $\mathfrak{L}$ maps $\mathfrak{M}$ into itself. Moreover, the restrictions to $\mathfrak{M}$ of all ad h, $h \in \mathfrak{H}$, are nilpotent, and $\mathfrak{H}$ is an invariant subspace. Hence ad $\mathfrak{H}$ induces a Lie algebra $\mathfrak{N}$ of nilpotent transformations in the quotient space $\mathfrak{M}/\mathfrak{H}$. Applying Engel's theorem, we see that if $\mathfrak{M} \neq \mathfrak{H}$ there is $x \in \mathfrak{M}$, $x \notin \mathfrak{H}$, with $[x\,\mathfrak{H}] \subseteqq \mathfrak{H}$. But $\mathfrak{H}$ is its own normalizer, so $\mathfrak{M} = \mathfrak{H}$, which proves the lemma.

§ 7. Semisimplicity. The Killing form

Of the several equivalent notions of semisimplicity for Lie algebras over fields of characteristic zero, we take as our definition in characteristic $p > 0$ the weakest one: $\mathfrak{L}$ is called *semisimple* if the only ideal $\mathfrak{J}$ in $\mathfrak{L}$ with $[\mathfrak{J}\,\mathfrak{J}] = 0$ is $\mathfrak{J} = 0$. This is equivalent to saying that the only *solvable* ideal in $\mathfrak{L}$ is the zero ideal, where a Lie algebra $\mathfrak{S}$ is defined to be solvable if its *derived series* $\mathfrak{S}^{(0)} = \mathfrak{S}$, $\mathfrak{S}^{(1)} = [\mathfrak{S}\,\mathfrak{S}], \ldots,$ $\mathfrak{S}^{(k)} = [\mathfrak{S}^{(k-1)}\,\mathfrak{S}^{(k-1)}], \ldots$ terminates in zero. If $\mathfrak{L}$ is an arbitrary finite-dimensional Lie algebra over a field $\mathfrak{F}$, and if $\mathfrak{S}_1$, $\mathfrak{S}_2$ are solvable ideals in $\mathfrak{L}$, then so is $\mathfrak{S}_1 + \mathfrak{S}_2$, as one sees at once from the second isomorphism theorem: $(\mathfrak{S}_1 + \mathfrak{S}_2)/\mathfrak{S}_2 \cong \mathfrak{S}_1/(\mathfrak{S}_1 \cap \mathfrak{S}_2)$; as a homomorph of $\mathfrak{S}_1$, this algebra $\mathfrak{B}$ satisfies $\mathfrak{B}^{(k)} = 0$ if $\mathfrak{S}_1^{(k)} = 0$, i.e., $(\mathfrak{S}_1 + \mathfrak{S}_2)^{(k)} \subseteqq \mathfrak{S}_2$; now if $\mathfrak{S}_2^{(j)} = 0$, then $(\mathfrak{S}_1 + \mathfrak{S}_2)^{(k+j)} \subseteqq \mathfrak{S}_2^{(j)} = 0$. It follows that if $\mathfrak{S}$ is a maximal solvable ideal in $\mathfrak{L}$, then $\mathfrak{S}$ contains all other solvable ideals, so that $\mathfrak{L}$ has a unique maximal solvable ideal $\mathfrak{S}$, called the *radical* of $\mathfrak{L}$. Furthermore, $\mathfrak{L}/\mathfrak{S}$ is readily seen to be semisimple.

A Lie algebra $\mathfrak{L}$ is *simple* if the only ideals of $\mathfrak{L}$ are $\mathfrak{L}$ and 0, and if $[\mathfrak{L}\,\mathfrak{L}] \neq 0$. Thus the radical of $\mathfrak{L}$ is zero, so that $\mathfrak{L}$ is semisimple and $\mathfrak{L}^{(1)} = [\mathfrak{L}\,\mathfrak{L}] = \mathfrak{L}$. More generally, let $\mathfrak{L}$ be a Lie algebra over a field $\mathfrak{F}$ such that $\mathfrak{L}$ is the direct sum of ideals $\mathfrak{L} = \mathfrak{L}_1 \oplus \cdots \oplus \mathfrak{L}_m$, with each $\mathfrak{L}_i$ being a simple Lie algebra. Then the only ideals of $\mathfrak{L}$ are of the form $\mathfrak{L}_{i_1} + \cdots + \mathfrak{L}_{i_s}$, $1 \leqq i_1 < \cdots < i_s \leqq m$, and $(\mathfrak{L}_{i_1} + \cdots + \mathfrak{L}_{i_s})^{(1)} = \mathfrak{L}_{i_1}^{(1)} + \cdots + \mathfrak{L}_{i_s}^{(1)} = \mathfrak{L}_{i_1} + \cdots + \mathfrak{L}_{i_s}$, so that the only abelian ideal is zero, and $\mathfrak{L}$ is semisimple. One sees the assertion about the ideals of $\mathfrak{L}$ as follows: Let $\mathfrak{B}$ be an ideal of $\mathfrak{L}$, $\mathfrak{B}_i$ the projection of $\mathfrak{B}$ on $\mathfrak{L}_i$. Then clearly $\mathfrak{B} \subseteqq \sum_{j=1}^{s} \mathfrak{L}_{i_j}$, where $\mathfrak{B}_i = 0$ if $i \neq i_j$, $1 \leqq j \leqq s$. On the other hand, $[\mathfrak{L}_i\,\mathfrak{B}_i] = [\mathfrak{L}_i\,\mathfrak{B}] \subseteqq \mathfrak{L}_i \cap \mathfrak{B} \subseteqq \mathfrak{B}_i$, so that $\mathfrak{B}_i$ is an ideal in $\mathfrak{L}_i$; thus if $\mathfrak{B}_i \neq 0$ we have $\mathfrak{B}_i = \mathfrak{L}_i$, $\mathfrak{L}_i = [\mathfrak{L}_i \mathfrak{B}_i] = [\mathfrak{L}_i \mathfrak{B}] \subseteqq \mathfrak{B}$, which yields the assertion.

Let $\mathfrak{L}$ be a Lie algebra over the field $\mathfrak{F}$, and let $\varrho: x \to x^\varrho$ be a representation of $\mathfrak{L}$ in the finite-dimensional vector space $\mathfrak{B}$ over $\mathfrak{F}$. Then the *trace form* $(x, y) = \mathrm{Tr}(x^\varrho y^\varrho)$ (where $x^\varrho y^\varrho$ denotes the ordinary composite of the endomorphisms x^ϱ, y^ϱ of $\mathfrak{B}$) is a symmetric bilinear form on $\mathfrak{L}$ with values in $\mathfrak{F}$. It is moreover *associative* (or *invariant*), i.e., $([x\,y], z) = (x, [y\,z])$ for all x, y, z. In particular, when ϱ is the adjoint representation the trace form $(x, y) = \mathrm{Tr}((\mathrm{ad}\, x)(\mathrm{ad}\, y))$ is called the *Killing form* of $\mathfrak{L}$, in honor of that pioneer in the structure theory [250].

Theorem I.7.1. Let $\mathfrak{J}$ be an abelian ideal in $\mathfrak{L}$. Then $\mathfrak{J}$ is contained in the radical of the Killing form. Thus if the Killing form of $\mathfrak{L}$ is non-singular, $\mathfrak{L}$ is semisimple.

For the proof, we choose a basis $v_1, \ldots, v_n$ for $\mathfrak{L}$ whose first s vectors form a basis for $\mathfrak{J}$. Then if $y \in \mathfrak{L}$, $v_i(\mathrm{ad}\, y)$ is a linear combination of $v_1, \ldots, v_s$, $1 \leqq i \leqq s$. If $x \in \mathfrak{J}$, $v_i(\mathrm{ad}\, x) = 0$, $1 \leqq i \leqq s$, and $v_i(\mathrm{ad}\, x)$ is a combination of $v_1, \ldots, v_s$ in any case. Hence $v_i(\mathrm{ad}\, x)(\mathrm{ad}\, y) = 0$ for $i \leqq s$, while for $i > s$, $v_i(\mathrm{ad}\, x)(\mathrm{ad}\, y)$ is a combination of $v_1, \ldots, v_s$. It follows that $\mathrm{Tr}((\mathrm{ad}\, x)(\mathrm{ad}\, y)) = 0$ for $x \in \mathfrak{J}$, $y \in \mathfrak{L}$, and this is our assertion.

Lemma I.7.1. Let $\mathfrak{J}$ be an ideal in $\mathfrak{L}$. Then the Killing form of $\mathfrak{J}$ is the restriction to $\mathfrak{J}$ of the Killing form of $\mathfrak{L}$.

For if the basis $v_1, \ldots, v_n$ for $\mathfrak{L}$ is chosen as in the proof of the above theorem, then for $x, y \in \mathfrak{J}$, $v_i(\mathrm{ad}\, x)(\mathrm{ad}\, y)$ is a combination of $v_1, \ldots, v_s$. Thus $\mathrm{Tr}((\mathrm{ad}\, x)(\mathrm{ad}\, y))$ is the sum of the coefficients for $1 \leqq i \leqq s$ of the v_i in $v_i(\mathrm{ad}\, x)(\mathrm{ad}\, y)$; but this is just the value of the Killing form of $\mathfrak{J}$ at the pair x, y.

Theorem I.7.2 (Cartan [53]–Dieudonné [105]): Let $\mathfrak{L}$ be a semisimple Lie algebra over a field $\mathfrak{F}$, and let (x, y) be a non-singular, symmetric, associative bilinear form on $\mathfrak{L}$. Then $\mathfrak{L}$ is a direct sum of ideals which are simple Lie algebras, and which are orthogonal with respect to the form, hence non-singular with respect to (x, y).

We may assume $\mathfrak{L} \neq 0$ and that the result is valid for algebras of lower dimension. If $\mathfrak{J}$ is a minimal non-zero ideal in $\mathfrak{L}$, then $\mathfrak{J}^\perp$, the orthogonal space to $\mathfrak{J}$ with respect to the form, is an ideal by the associativity of the form. If $\mathfrak{J}^\perp \cap \mathfrak{J} = \mathfrak{J}$, then for $x \in \mathfrak{L}$, $y, z \in \mathfrak{J}$, we have $(x, [y\,z]) = ([x\,y], z) = 0$. Thus $[\mathfrak{J}\,\mathfrak{J}] = 0$ by non-singularity of the form, and this contradicts semisimplicity. Hence $\mathfrak{J}^\perp \cap \mathfrak{J} = 0$ by minimality of $\mathfrak{J}$, and from this we see that: a) the form is non-singular on $\mathfrak{J}$; b) $\mathfrak{L} = \mathfrak{J} \oplus \mathfrak{J}^\perp$; c) $[\mathfrak{J}\,\mathfrak{J}^\perp] = 0$; d) the form is non-singular on $\mathfrak{J}^\perp$. By b) and c) the ideals of $\mathfrak{L}$ contained in $\mathfrak{J}$ (or in $\mathfrak{J}^\perp$) are just the ideals of $\mathfrak{J}$ (or of $\mathfrak{J}^\perp$). It follows that: e) $\mathfrak{J}$ is a simple Lie algebra;

f) $\mathfrak{J}^{\perp}$ is semi-simple. By d), f) and induction, $\mathfrak{J}^{\perp}$ is a direct sum of ideals which are simple Lie algebras and which are orthogonal. But these are ideals in $\mathfrak{L}$, and $\mathfrak{L}$ is the direct sum of $\mathfrak{J}$ and these. This completes the proof.

The lemma and the two theorems of this section yield the

Corollary. If $\mathfrak{L}$ is a Lie algebra over $\mathfrak{F}$ with non-singular Killing form, then $\mathfrak{L}$ is the direct sum of ideals in $\mathfrak{L}$, each of which is a simple Lie algebra with non-singular Killing form.

Thus we see that each of these propositions implies the next:

(A) $\mathfrak{L}$ has non-singular Killing form.

(B) $\mathfrak{L}$ is a direct sum of simple Lie algebras.

(C) $\mathfrak{L}$ is semisimple.

Whereas these are equivalent for the non-modular case [47, 64, 72, 234], we shall see in Chapter V that all reverse implications fail in the modular case. It is the equivalence of (B) and (C) in the classical case which reduces the classification of semisimple algebras to that of simple algebras, and the Killing form is an essential tool in carrying out the classification. The classification of simple Lie algebras over an algebraically closed modular field is an open problem, whose solution must include a rather complicated list of algebras (cf. Chap. V). On the other hand, the condition (A), or slightly weaker forms of it, leads to a classification theory, parallel to the classical work of KILLING [250] and CARTAN [53], for Lie algebras over algebraically closed fields of characteristic $p > 3$ (Chap. II).

§ 8. Trace forms, derivations, and restrictedness

The assumption of the existence of a non-degenerate trace form on a modular Lie algebra $\mathfrak{L}$ allows the introduction into $\mathfrak{L}$ of the structure of a restricted Lie algebra. This has been shown by Block [34]:

Lemma I.8.1 (BLOCK). Let $\mathfrak{L}$ be a Lie subalgebra of $\mathfrak{E}(\mathfrak{V})$, $\mathfrak{V}$ a finite-dimensional vector space over a field $\mathfrak{F}$, and suppose that the trace form $(x, y) = \mathrm{Tr}(x\,y)$ of $\mathfrak{E}(\mathfrak{V})$ is non-singular on $\mathfrak{L}$. Let $u \in \mathfrak{E}(\mathfrak{V})$, $[u\,\mathfrak{L}] \subseteqq \mathfrak{L}$; then there is $v \in \mathfrak{L}$ with $[u\,x] = [v\,x]$ for all $x \in \mathfrak{L}$.

One need only take $v \in \mathfrak{L}$ such that $(v, x) = \mathrm{Tr}(u\,x)$ for all $x \in \mathfrak{L}$; then if $x, y \in \mathfrak{L}$, $([v\,x], y) = (v, [x\,y]) = (u, [x\,y]) = ([u\,x], y)$, and the lemma follows by non-singularity of (x, y) on $\mathfrak{L}$.

Theorem I.8.1 (BLOCK). Let $\mathfrak{L}$ be a Lie subalgebra of $\mathfrak{E}(\mathfrak{V})$, $\mathfrak{V}$ a finite-dimensional vector space over a field $\mathfrak{F}$ of prime characteristic p. Suppose the form $(x, y) = \mathrm{Tr}(x\,y)$ is non-singular on $\mathfrak{L}$. Then an operation $x \to x^{[p]}$ can be introduced in $\mathfrak{L}$, relative to which $\mathfrak{L}$ becomes a restricted Lie algebra.

For by the lemma and the identity $[y, x^p] = y(\mathrm{ad}\, x)^p$, there is for each $x \in \mathfrak{L}$ an element $v = v(x)$ in $\mathfrak{L}$ such that $[y, x^p] = [y\, v]$ for all $y \in \mathfrak{L}$. The result now follows by Th. 3.3.

Corollary. Let $\mathfrak{L}$ be a semisimple modular Lie algebra, and let ϱ be a representation of $\mathfrak{L}$ with non-singular trace form. Then there is a unique mapping $x \to x^{[p]}$ of $\mathfrak{L}$ into $\mathfrak{L}$ relative to which $\mathfrak{L}$ is a restricted Lie algebra.

Corollary. Let $\mathfrak{L}$ be a modular Lie algebra with non-singular Killing form. Then $\mathfrak{L}$ admits a unique operation $x \to x^{[p]}$ relative to which $\mathfrak{L}$ becomes a restricted Lie algebra.

Corollary (Zassenhaus). Let $\mathfrak{L}$ be a Lie algebra with non-singular Killing form. Then every derivation of $\mathfrak{L}$ is inner.

The first of these follows from the theorem, using the fact that ϱ is faithful and, for the uniqueness, the fact that a Lie algebra with center zero admits at most one structure of restricted Lie algebra. The second is then immediate, using Th. 7.1. For the third, we have seen that $\mathrm{ad}(\mathfrak{L})$ is an ideal in the Lie algebra $\mathfrak{D}(\mathfrak{L})$ of derivations of $\mathfrak{L}$; the assumption is that $\mathrm{ad}(\mathfrak{L})$ satisfies the hypotheses of the lemma in $\mathfrak{E}(\mathfrak{L})$. Then if $D \in \mathfrak{D}(\mathfrak{L})$, there is $d \in \mathfrak{L}$ with $[(\mathrm{ad}\, x), D] = [(\mathrm{ad}\, x), (\mathrm{ad}\, d)]$ for all $x \in \mathfrak{L}$, i.e., with $\mathrm{ad}(xD) = \mathrm{ad}[x\, d]$. The corollary follows.

From the first corollary and the results of § 7 it follows that a semisimple modular Lie algebra with a non-singular trace form is restricted, and is the direct sum of simple ideals which are restricted. More generally, one easily proves:

Theorem I.8.2. Let $\mathfrak{L}$ be a restricted Lie algebra, and let $\mathfrak{L}$ be the direct sum of ideals $\mathfrak{L}_1, \ldots, \mathfrak{L}_n$, each of which has center zero. Then the $\mathfrak{L}_i$ are restricted ideals.

§ 9. Extension of the base ring

Let $\mathfrak{L}$ be a Lie algebra over the ring $\mathfrak{F}$, and let $\mathfrak{K}$ be an extension ring of $\mathfrak{F}$. An algebra over $\mathfrak{K}$ obtained from $\mathfrak{L}$ by *extension of the base ring* may be defined as a linear algebra $\mathfrak{A}$ over $\mathfrak{K}$, together with an $\mathfrak{F}$-algebra homomorphism φ: $\mathfrak{L} \to \mathfrak{A}$, such that if $\mathfrak{B}$ is a second $\mathfrak{K}$-algebra and ψ an $\mathfrak{F}$-algebra homomorphism $\mathfrak{L} \to \mathfrak{B}$, there is a unique $\mathfrak{K}$-algebra homomorphism η: $\mathfrak{A} \to \mathfrak{B}$ making the following diagram commutative:

$$\begin{array}{ccc} \mathfrak{L} & \xrightarrow{\varphi} & \mathfrak{A} \\ & {\scriptstyle\psi}\searrow \quad \swarrow{\scriptstyle\eta} & \\ & \mathfrak{B} & \end{array}$$

It follows that $\mathfrak{A}$ is unique to within a $\mathfrak{K}$-algebra isomorphism compatible with the mapping φ of $\mathfrak{L}$ into $\mathfrak{A}$. The existence of such an algebra is seen by taking the tensor product $\mathfrak{L} \otimes_{\mathfrak{F}} \mathfrak{K}$, which, if $\mathfrak{L}$ is a free $\mathfrak{F}$-module, may be regarded as a free $\mathfrak{K}$-module of dimension equal to that of $\mathfrak{L}$ over $\mathfrak{F}$. The map $\varphi: x \to x \otimes 1$ is $\mathfrak{F}$-linear from $\mathfrak{L}$ into $\mathfrak{L} \otimes_{\mathfrak{F}} \mathfrak{K}$, its image spans $\mathfrak{L} \otimes_{\mathfrak{F}} \mathfrak{K}$ over $\mathfrak{K}$, and there is a unique structure of $\mathfrak{K}$-algebra on $\mathfrak{L} \otimes_{\mathfrak{F}} \mathfrak{K}$ in which $(x \otimes \xi)(y \otimes \eta) = [x\, y] \otimes \xi \eta$ for $x, y \in \mathfrak{L}$; $\xi, \eta \in \mathfrak{K}$. With this structure, the pair $(\mathfrak{L} \otimes_{\mathfrak{F}} \mathfrak{K}, \varphi)$ satisfies the condition required of $(\mathfrak{A}, \varphi)$ in the definition above. We shall normally denote the algebra $\mathfrak{L} \otimes_{\mathfrak{F}} \mathfrak{K}$ by $\mathfrak{L}_{\mathfrak{K}}$, and when φ is one-one, we regard $\mathfrak{L}$ as an $\mathfrak{F}$-subalgebra of $\mathfrak{L}_{\mathfrak{K}}$ via the embedding φ. One sees at once that $\mathfrak{L}_{\mathfrak{K}}$ is a Lie algebra.

If $e_1, \ldots, e_n$ is a basis for $\mathfrak{L}$ over $\mathfrak{F}$, this set is again a basis for $\mathfrak{L}_{\mathfrak{K}}$ over $\mathfrak{K}$. Since the conditions that $\mathfrak{L}$ be, respectively, abelian, nilpotent or solvable can be expressed as the vanishing of all Lie monomials of a certain type in these basis elements, it is clear that $\mathfrak{L}$ is respectively abelian, nilpotent or solvable if and only if $\mathfrak{L}_{\mathfrak{K}}$ is. We now assume that $\mathfrak{F}$ and $\mathfrak{K}$ are fields. If $\mathfrak{A}$ is an ideal in $\mathfrak{L}$, then $\mathfrak{A}_{\mathfrak{K}}$ is an ideal in $\mathfrak{L}_{\mathfrak{K}}$, so that if $\mathfrak{L}_{\mathfrak{K}}$ has no abelian ideals other than zero neither does $\mathfrak{L}$, and if $\mathfrak{L}_{\mathfrak{K}}$ has no proper ideals neither does $\mathfrak{L}$. The converses of these propositions are false (Chap. V). The Killing form of $\mathfrak{L}_{\mathfrak{K}}$ is the $\mathfrak{K}$-bilinear extension of that of $\mathfrak{L}$, and is thus non-singular if and only if $\mathfrak{L}$ has non-singular Killing form.

More generally, one may extend the base ring of an arbitrary linear algebra $\mathfrak{A}$ over $\mathfrak{F}$ to obtain an algebra $\mathfrak{A}_{\mathfrak{K}}$ over $\mathfrak{K}$, or of a module $\mathfrak{V}$ over $\mathfrak{F}$ and of an $\mathfrak{F}$-module $\mathfrak{M}$ of endomorphisms of $\mathfrak{V}$ to obtain $\mathfrak{M}_{\mathfrak{K}}$, a $\mathfrak{K}$-module of $\mathfrak{K}$-endomorphisms of $\mathfrak{V}_{\mathfrak{K}}$. In the latter case, if $\mathfrak{M}$ is a Lie algebra of endomorphisms of $\mathfrak{V}$, $\mathfrak{M}_{\mathfrak{K}}$ is a Lie algebra of endomorphisms of $\mathfrak{V}_{\mathfrak{K}}$, and if $\varphi: \mathfrak{L} \to \mathfrak{M}$ is an $\mathfrak{F}$-Lie homomorphism (i.e., if $\mathfrak{V}$ is an *$\mathfrak{L}$-module*), then φ extends uniquely to a $\mathfrak{K}$-Lie homomorphism $\varphi_{\mathfrak{K}}: \mathfrak{L}_{\mathfrak{K}} \to \mathfrak{M}_{\mathfrak{K}}$ (that is, a unique structure of $\mathfrak{L}_{\mathfrak{K}}$-module on $\mathfrak{V}_{\mathfrak{K}}$ is determined). If φ is a Lie isomorphism, then so is $\varphi_{\mathfrak{K}}$, and if φ is an automorphism of $\mathfrak{L}$, so is $\varphi_{\mathfrak{K}}$ (of $\mathfrak{L}_{\mathfrak{K}}$). Analogous assertions hold when $\mathfrak{L}$ and $\mathfrak{M}$ are replaced by arbitrary linear algebras $\mathfrak{A}$, $\mathfrak{B}$ over $\mathfrak{F}$. *Now let $\mathfrak{F}$ and $\mathfrak{K}$ be fields, an assumption we make hereafter.* If $\mathfrak{A}$ is a finite-dimensional linear algebra over $\mathfrak{F}$, $\mathfrak{D}(\mathfrak{A})$ the Lie algebra of derivations of $\mathfrak{A}$, then a linear transformation D is in $\mathfrak{D}(\mathfrak{A})$ if and only if its matrix (relative to a fixed basis for $\mathfrak{A}$) is a solution to a certain system of homogeneous linear equations with coefficients in $\mathfrak{F}$. By comparison of dimensions it follows that $\mathfrak{D}(\mathfrak{A}_{\mathfrak{K}}) = (\mathfrak{D}(\mathfrak{A}))_{\mathfrak{K}}$ [211]. If $\mathfrak{V}$ is finite-dimensional over $\mathfrak{F}$, and if $\mathfrak{M}$ is a Lie algebra of linear transformations of $\mathfrak{V}$ with non-singular trace form, then $\mathfrak{M}_{\mathfrak{K}}$ (acting in $\mathfrak{V}_{\mathfrak{K}}$) has non-singular trace form.

If the finite-dimensional algebra $\mathfrak{A}$ has no proper ideals (i.e., subspaces invariant under all left and right multiplications), the set of linear transformations of $\mathfrak{A}$ commuting with all left and right multiplications (the *centroid* of $\mathfrak{A}$) is an $\mathfrak{F}$-division algebra $\mathfrak{Z}$, by Schur's lemma. If, in addition, $\mathfrak{A} = \mathfrak{A}\mathfrak{A}$, then for $Y, Z \in \mathfrak{Z}$, it follows from $(a\,b)\,YZ = (aY)\,(b\,Z) = (a\,b)\,Z\,Y$ that $\mathfrak{Z}$ is a field. In particular this is the case if $\mathfrak{A} = \mathfrak{L}$, a simple Lie algebra over $\mathfrak{F}$, in which case $\mathfrak{Z}$ is a finite extension field of $\mathfrak{F}$. Subject to the above conditions, $\mathfrak{A}$ may be regarded as an algebra $\mathfrak{A}^*$ over $\mathfrak{Z}$ (*not* by extending the base field, but simply by taking advantage of the action of $\mathfrak{Z}$ on $\mathfrak{A}$). In this setting, it is clear that the centroid of $\mathfrak{A}^*$ is $\mathfrak{Z}$, and that $\mathfrak{A}^*$ has no proper $\mathfrak{Z}$-subspaces invariant under left and right multiplications. It follows that the enveloping associative $\mathfrak{Z}$-algebra of the multiplications of $\mathfrak{A}^*$ is the full algebra of $\mathfrak{Z}$-linear transformations of $\mathfrak{A}^*$, hence that $\mathfrak{A}^*_{\mathfrak{K}}$ has a similar property (with $\mathfrak{Z}$ replaced by $\mathfrak{K}$) for every field extension $\mathfrak{K}$ of $\mathfrak{Z}$. One says that $\mathfrak{A}^*$ is a *central simple* (or *normal simple*) algebra over $\mathfrak{Z}$; if $\mathfrak{Z} = \mathfrak{F}$, then $\mathfrak{A}$ is *central* (or *normal*) *simple*. In particular, if $\mathfrak{L}$ is a simple Lie algebra over $\mathfrak{F}$, then $\mathfrak{L}$ is a normal simple Lie algebra over $\mathfrak{Z}$, so that the study of simple Lie algebras over $\mathfrak{F}$ may be reduced to the study of normal simple Lie algebras over finite extensions $\mathfrak{K}$ of $\mathfrak{F}$ (cf. [234, p. 292]).

If $\mathfrak{L}$ is a restricted Lie algebra over $\mathfrak{F}$, then the basis $e_1, \ldots, e_n$ of $\mathfrak{L}$ has $(\mathrm{ad}\, e_i)^p = \mathrm{ad}\, f_i$ for each i, and for some $f_i \in \mathfrak{L}$. Acting in $\mathfrak{L}_{\mathfrak{K}}$, we again have $(\mathrm{ad}\, e_i)^p = \mathrm{ad}\, f_i$, from which it follows by Th. 3.3 that $\mathfrak{L}_{\mathfrak{K}}$ is a restricted Lie algebra over $\mathfrak{K}$ under a p-power operation which extends the given one on $\mathfrak{L}$. From the formulas for the p-th power in associative algebras one sees that if $\mathfrak{M}$ is a Lie algebra of linear transformations of $\mathfrak{V}$ closed under p-th powers, then $\mathfrak{M}_{\mathfrak{K}}$, acting in $\mathfrak{V}_{\mathfrak{K}}$, is also closed under p-th powers. It also follows that the unique $\mathfrak{K}$-extension of an $\mathfrak{F}$-homomorphism of restricted Lie algebras is a restricted $\mathfrak{K}$-homomorphism, and similarly for restricted representations.

If $\mathfrak{L}$ is a Lie algebra over $\mathfrak{F}$, and if $\mathfrak{H}$ is a Cartan subalgebra of $\mathfrak{L}$, then $\mathfrak{H}_{\mathfrak{K}}$ is a Cartan subalgebra of $\mathfrak{L}_{\mathfrak{K}}$; for $\mathfrak{H}_{\mathfrak{K}}$ is a nilpotent subalgebra, and the normalizer in $\mathfrak{L}_{\mathfrak{K}}$ of a subalgebra of the form $\mathfrak{M}_{\mathfrak{K}}$, where $\mathfrak{M}$ is a subalgebra of $\mathfrak{L}$, is $\mathfrak{N}_{\mathfrak{K}}$, where $\mathfrak{N}$ is the normalizer of $\mathfrak{M}$ in $\mathfrak{L}$; since $\mathfrak{H}$ is its own normalizer in $\mathfrak{L}$, it follows that $\mathfrak{H}_{\mathfrak{K}}$ is its own normalizer in $\mathfrak{L}_{\mathfrak{K}}$.

Chapter II

Classical Semisimple Lie Algebras

§ 1. The Cartan decomposition

Let $\mathfrak{F}$ be a field of characteristic $p \neq 0$; let $\mathfrak{L}$ be a finite-dimensional Lie algebra over $\mathfrak{F}$, and let ϱ be a representation of $\mathfrak{L}$ in a finite-dimensional vector space $\mathfrak{V}$ over $\mathfrak{F}$ such that the form $(x, y) = \mathrm{Tr}(x^\varrho y^\varrho)$ is non-singular on $\mathfrak{L}$. We further assume that $\mathfrak{L}$ has a Cartan subalgebra $\mathfrak{H}$ such that for all $h \in \mathfrak{H}$, all characteristic roots of h^ϱ are in $\mathfrak{F}$, an assumption which is satisfied trivially if $\mathfrak{F}$ is algebraically closed.

The non-singularity of the form implies at once that ϱ is an isomorphism of $\mathfrak{L}$ onto $\mathfrak{L}^\varrho$. It now follows that each ad h, $h \in \mathfrak{H}$, has all its characteristic roots in $\mathfrak{F}$, by the following:

Lemma II.1.1. Let T be an endomorphism of a finite-dimensional vector space $\mathfrak{V}$ over the field $\mathfrak{F}$, and let the distinct characteristic roots of T be $\alpha_1, \ldots, \alpha_n$. Then the characteristic roots of ad T in $\mathfrak{E}(\mathfrak{V})$ are the differences $\alpha_i - \alpha_j$, $1 \leqq i, j \leqq n$.

By enlarging the field, we may assume all $\alpha_i \in \mathfrak{F}$; thus T is a zero of a polynomial

$$q(X) = (X - \alpha_1)^{\nu_1} \ldots (X - \alpha_n)^{\nu_n} \in \mathfrak{F}[X].$$

Let v_{ij}, $1 \leqq j \leqq m_i$, be a basis for the subspace of $\mathfrak{V}$ annihilated by $(T - \alpha_i I)^{\nu_i}$, $1 \leqq i \leqq n$, so that the totality of the v_{ij} form a basis $\{w_k\}$ for $\mathfrak{V}$. Now if $U \in \mathfrak{E}(\mathfrak{V})$, the effect of applying $(\mathrm{ad}\, T - \beta + \alpha)^\nu$ to U is seen by induction to be

$$(1) \qquad \sum_{s=0}^{\nu} (-1)^{\nu - s} \binom{\nu}{s} (T - \alpha I)^{\nu - s} U (T - \beta I)^s.$$

From this it follows that if w_k, w_r are vectors from our basis, annihilated by $(T - \alpha I)^\lambda$, $(T - \beta I)^\mu$ respectively, then the matrix unit $U = E_{kr}$ is annihilated by $(\mathrm{ad}\, T - \beta + \alpha)^{\lambda + \mu - 1}$. Since these matrix units form a basis for $\mathfrak{E}(\mathfrak{V})$, the lemma follows.

Lemma II.1.2. Let $\mathfrak{L}$ be a Lie algebra over $\mathfrak{F}$, $\mathfrak{H}$ a Cartan subalgebra such that each ad h, $h \in \mathfrak{H}$, has all its characteristic roots in $\mathfrak{F}$.

Let ϱ be a representation of $\mathfrak{L}$, and let $\varphi \neq 0$ be a root of $\mathfrak{H}$ in $\mathfrak{L}$, denoting the corresponding root-space by $\mathfrak{L}_\varphi$. Then if $h \in \mathfrak{H}$, $x \in \mathfrak{L}_\varphi$, $\operatorname{Tr}(h^\varrho x^\varrho) = 0$.

Writing (x, y) for $\operatorname{Tr}(x^\varrho y^\varrho)$, we have an associative form on $\mathfrak{L}$. Since $\varphi \neq 0$, there is an element $h_1 \in \mathfrak{H}$ such that $\varphi(h_1) \neq 0$. Now $\operatorname{ad} h_1$ maps $\mathfrak{L}_\varphi$ into itself, and has there the single characteristic root $\varphi(h_1)$; hence there is an element $y \in \mathfrak{L}_\varphi$ such that $x = y(\operatorname{ad} h_1)^k$, where k may be taken so large that $\mathfrak{H}^{k+1} = 0$.

Thus $(h, x) = (h, y(\operatorname{ad} h_1)^k) = (-1)^k (h(\operatorname{ad} h_1)^k, y) = 0$.

Corollary. If the form $(x, y) = \operatorname{Tr}(x^\varrho y^\varrho)$ of Lemma 2 is non-singular, then its restriction to $\mathfrak{H}$ is non-singular.

This follows by Th. I.5.2 and Lemma 2.

Lemma II.1.3 (Zassenhaus). Under the hypotheses at the beginning of this section (i.e., (x, y) non-singular, all characteristic roots of all h^ϱ in $\mathfrak{F}$), $\mathfrak{H}$ is abelian.

For suppose $\mathfrak{H}^{k-1} \neq 0$, $\mathfrak{H}^k = 0$, where $k \geqq 3$; then $\mathfrak{H}^{k-1}$ is central in $\mathfrak{H}$. Let $0 \neq z \in \mathfrak{H}^{k-1} \subseteqq [\mathfrak{H}\mathfrak{H}]$, and decompose $\mathfrak{V}$ into weight spaces $\mathfrak{V}_\varphi$ relative to $\mathfrak{H}$. Let $\mathfrak{U} = \mathfrak{V}_\varphi$ for some φ, and let $h \in \mathfrak{H}$; let $d = \dim. \mathfrak{U}$. Then since z^ϱ and h^ϱ commute, there is a basis for $\mathfrak{U}$ relative to which they both have triangular matrices; thus $\operatorname{Tr}_\mathfrak{U}(z^\varrho h^\varrho) = d\varphi(z)\varphi(h)$. But $z \in [\mathfrak{H}\mathfrak{H}]$ implies $\operatorname{Tr}_\mathfrak{U}(z^\varrho) = 0$, i.e., $d\varphi(z) = 0$. Thus $\operatorname{Tr}_\mathfrak{U}(z^\varrho h^\varrho) = 0$ for all $h \in \mathfrak{H}$, and $(z, h) = 0$ for all $h \in \mathfrak{H}$ since (z, h) is the sum of the $\operatorname{Tr}_\mathfrak{U}(z^\varrho h^\varrho)$ as $\mathfrak{U}$ runs over all $\mathfrak{V}_\varphi$. This contradicts the last corollary.

By Th. I.8.1, $\mathfrak{L}^\varrho$, hence $\mathfrak{L}$, admits an operation $x \to x^{[p]}$ relative to which it is a restricted Lie algebra. If $h \in \mathfrak{H}$, then for all $x \in \mathfrak{H}$, $[x, h^{[p]}] = x(\operatorname{ad} h)^p = 0$ by Lemma 3; thus $h^{[p]} \in \mathfrak{H}$.

Lemma II.1.4 (Jacobson, Block). If $h \in \mathfrak{H}$, $(\operatorname{ad} h)^p = 0$, then $(x, h) = 0$ for all $x \in \mathfrak{H} \cap [\mathfrak{L}\mathfrak{L}]$. Hence if $\mathfrak{L} = [\mathfrak{L}\mathfrak{L}]$ the mapping $h \to h^{[p]}$ is a one-one p-semilinear transformation of $\mathfrak{H}$ into $\mathfrak{H}$.

For let $\mathfrak{V} = \mathfrak{V}_0 \supset \mathfrak{V}_1 \supset \cdots \supset \mathfrak{V}_s = 0$ be a composition series of invariant subspaces of $\mathfrak{V}$ relative to $\mathfrak{L}$, i.e., $\mathfrak{L}$ has an irreducible representation ϱ_{i+1} in $\mathfrak{V}_i/\mathfrak{V}_{i+1}$; then for $z \in \mathfrak{H}$, all characteristic roots of z^{ϱ_i} are in $\mathfrak{F}$, and $(x, y) = \sum_i \operatorname{Tr}(x^{\varrho_i} y^{\varrho_i})$ for all x, y. Now for all $y \in \mathfrak{L}$, $[y^{\varrho_i}, (h^{\varrho_i})^p] = (y(\operatorname{ad} h)^p)^{\varrho_i} = 0$, so that $(h^{\varrho_i})^p$ centralizes $\mathfrak{L}^{\varrho_i}$. Taking λ to be a characteristic root of $(h^{\varrho_i})^p$, we see that $(h^{\varrho_i})^p = \lambda I$ on $\mathfrak{V}_{i-1}/\mathfrak{V}_i$. Thus the only characteristic root of h^{ϱ_i} is μ, where $\mu^p = \lambda$. If $x \in \mathfrak{H} \cap [\mathfrak{L}\mathfrak{L}]$, then $\operatorname{Tr}(x^{\varrho_i}) = 0$ since $x \in [\mathfrak{L}\mathfrak{L}]$, and x^{ϱ_i} and h^{ϱ_i} can both be taken in triangular form. Hence $\operatorname{Tr}(h^{\varrho_i} x^{\varrho_i}) = \mu \operatorname{Tr}(x^{\varrho_i}) = 0$, so that $(h, x) = 0$ for all $x \in \mathfrak{H} \cap [\mathfrak{L}\mathfrak{L}]$. In particular, $(h, x) = 0$ for all $x \in \mathfrak{H}$ when $\mathfrak{L} = [\mathfrak{L}\mathfrak{L}]$, so that then $h = 0$ by the corollary to Lemma 2, and $h^{[p]} = 0$ implies $h = 0$. Since $\mathfrak{H}$ is abelian $h \to h^{[p]}$ is p-semilinear on $\mathfrak{H}$, and is thus one-one if $\mathfrak{L} = [\mathfrak{L}\mathfrak{L}]$.

Lemma II.1.5. Let $\mathfrak{F}$ be a perfect field of prime characteristic p; let $\mathfrak{V}$ be a finite-dimensional vector space over $\mathfrak{F}$, and T a p-semilinear transformation of $\mathfrak{V}$ into $\mathfrak{V}$. Then T is one-one if and only if T maps $\mathfrak{V}$ onto $\mathfrak{V}$.

This follows at once by an argument on dimension.

Theorem II.1.1. Let $\mathfrak{L}$ be a finite-dimensional Lie algebra over a perfect field $\mathfrak{F}$ of prime characteristic p; let ϱ be a representation of $\mathfrak{L}$ in a vector space $\mathfrak{V}$, having non-singular trace form. Suppose that $[\mathfrak{L}\,\mathfrak{L}] = \mathfrak{L}$, and that $\mathfrak{L}$ has a Cartan subalgebra $\mathfrak{H}$ such that each $h^\varrho\,(h \in \mathfrak{H})$ has all its characteristic roots in $\mathfrak{F}$. Then $\mathfrak{H}$ is abelian, the restriction to $\mathfrak{H}$ of the trace form is non-singular, and an operation $x \to x^{[p]}$ can be introduced into $\mathfrak{L}$ which makes $\mathfrak{L}$ into a restricted Lie algebra. Each such operation maps $\mathfrak{H}$ into $\mathfrak{H}$, and is a p-semilinear isomorphism of $\mathfrak{H}$ onto $\mathfrak{H}$.

This theorem is a summary of earlier remarks and lemmas.

Theorem II.1.2. Let $\mathfrak{L}$, $\mathfrak{H}$, ϱ be as in Th. 1. Let φ be a root of $\mathfrak{H}$ in $\mathfrak{L}$. Then for $h \in \mathfrak{H}$, $x \in \mathfrak{L}_\varphi$, we have $[x\,h] = \varphi(h)\,x$.

Namely, if d is the dimension of $\mathfrak{L}_\varphi$, then for $x \in \mathfrak{L}_\varphi$, $y \in \mathfrak{H}$, $x(\mathrm{ad}\, y - \varphi(y)\, I)^d = 0$. Choose k so that $p^k \geqq d$; then $x(\mathrm{ad}\, y)^{p^k} - \varphi(y)^{p^k} x = 0$. Now by Th. 1, if $h \in \mathfrak{H}$, then $h = y^{([p]^k)} \big(= (\ldots ((y^{[p]})^{[p]}) \ldots {}^{[p]})\big)$ for some $y \in \mathfrak{H}$, and we have for this y, $x(\mathrm{ad}\, h) - \varphi(y)^{p^k} x = 0$. Thus $\varphi(y)^{p^k}$ is a characteristic root of $\mathrm{ad}\, h$ in $\mathfrak{L}$, where $\mathrm{ad}\, h$ has the single characteristic root $\varphi(h)$. Hence $\varphi(y)^{p^k} = \varphi(h)$, and the assertion follows.

Lemma II.1.6. Let $\mathfrak{L}$ be as in Th. 1. Let α and β be roots of $\mathfrak{L}$ with respect to $\mathfrak{H}$. Then $[\mathfrak{L}_\alpha\, \mathfrak{L}_\beta] \subseteqq \mathfrak{L}_{\alpha+\beta}$.

For let $x \in \mathfrak{L}_\alpha$, $y \in \mathfrak{L}_\beta$, $h \in \mathfrak{H}$. Then by Th. 2 and the Jacobi identity, $[[x\,y]\,h] = [[x\,h]\,y] + [x[y\,h]] = (\alpha + \beta)\,(h)\,[x\,y]$. It will be noted that $[\mathfrak{L}_\alpha\, \mathfrak{L}_\beta] = 0$ if $\alpha + \beta$ is not a root.

Lemma II.1.7. With the notations above, let α and β be roots of $\mathfrak{L}$ with respect to $\mathfrak{H}$; then $(\mathfrak{L}_\alpha, \mathfrak{L}_\beta) = 0$ unless $\alpha + \beta = 0$. Hence if α is a root, so is $-\alpha$.

For if either α or β is zero, the assertion holds by Lemmas 2 and I.6.1. If $\alpha \neq 0$ and $\beta \neq -\alpha$, let $h \in \mathfrak{H}$ with $\alpha(h) \neq 0$; let $x \in \mathfrak{L}_\alpha$, $y \in \mathfrak{L}_\beta$; then $(x, y) = -\alpha(h)^{-1}([h\,x], y)$ by Th. 2; but $([h\,x], y) = (h, [x\,y]) = 0$ by Lemmas 2 and 6. This proves the first assertion; the second follows from the first and the non-singularity of the form.

Lemma II.1.8. With the hypotheses of Th. 1, the roots of $\mathfrak{H}$ in $\mathfrak{L}$ are linear functions on $\mathfrak{H}$.

This follows at once from Th. 2.

Lemma II.1.9. With the same hypotheses, let α be a root of $\mathfrak{H}$ in $\mathfrak{L}$. Then there is a unique $h_\alpha \in \mathfrak{H}$ with $(h, h_\alpha) = \alpha(h)$ for all $h \in \mathfrak{H}$.

This follows by Lemma 8 and the corollary to Lemma 2.

Lemma II.1.10. Let $e_\alpha \in \mathfrak{L}_\alpha$, $e_- \in \mathfrak{L}_{-\alpha}$. Then $[e_{-\alpha} e_\alpha] = (e_{-\alpha}, e_\alpha) h_\alpha$.

For if $h \in \mathfrak{H}$, we have $([e_{-\alpha} e_\alpha], h) = (e_{-\alpha}, [e_\alpha h]) = \alpha(h)(e_{-\alpha}, e_\alpha) = ((e_{-\alpha}, e_\alpha) h_\alpha, h)$, the middle step by Th. 2 and the last by definition of h_α. The assertion now follows.

Lemma II.1.11. If $\alpha \neq 0$ is a root, then $[\mathfrak{L}_{-\alpha} \mathfrak{L}_\alpha]$ is one-dimensional.

This follows at once from Lemmas 7 and 10.

§ 2. Split 3-dimensional algebras and applications

If $\mathfrak{F}$ is an arbitrary field, let $\mathfrak{L}$ be the 3-dimensional Lie algebra over $\mathfrak{F}$ with basis e, f, h satisfying $[e f] = h$, $[e h] = e$, $[f h] = -f$. We first study the representations of $\mathfrak{L}$. Proofs are given only in the modular case; that of characteristic zero requires only minor changes and may in any case be found in standard references, notably [234].

The following is easily proved by induction.

Lemma II.2.1. Let ϱ be a representation of $\mathfrak{L}$ in $\mathfrak{B}$. Let $\beta \in \mathfrak{F}$, $v \in \mathfrak{B}$, $v(h^\varrho) = \beta v$. Then

$$v(e^\varrho)^i h^\varrho = \beta v(e^\varrho)^i + i v(e^\varrho)^i; \tag{2}$$

$$v(e^\varrho)^i f^\varrho = i \beta v(e^\varrho)^{i-1} + \binom{i}{2} v(e^\varrho)^{i-1} + v f^\varrho (e^\varrho)^i; \tag{3}$$

for all $i \geqq 0$, where for $i = 0$ it is to be understood that the first two terms on the right side of (3) are zero.

Lemma II.2.2. Let $\mathfrak{L}, \mathfrak{B}, \varrho$ be as above. Let $\beta \in \mathfrak{F}$ be a characteristic root of h^ϱ such that $\beta - 1$ is not a characteristic root of h^ϱ. Suppose that $\beta, \beta + 1, \ldots, \beta + j$ are characteristic roots of h^ϱ, while $\beta + j + 1$ is not. Then $2\beta = -j$, and if $0 \neq v \in \mathfrak{B}$, $v h^\varrho = \beta v$, then $v(e^\varrho)^i \neq 0$, $0 \leqq i \leqq j$.

If the characteristic is 2, then $j = 0$, since $\beta + 1 = \beta - 1$ is not a characteristic root, and the assertion holds. Thus let $\mathfrak{F}$ have prime characteristic $p \geqq 3$. By (2), if v is as in the statement, $v(e^\varrho)^{j+1}$, if not zero, belongs to the characteristic root $\beta + j + 1$ of h^ϱ; hence $v(e^\varrho)^{j+1} = 0$. Also, $j < p - 1$; thus we have an integer k, $0 \leqq k \leqq j < p - 1$, with $v(e^\varrho)^k \neq 0$, $v(e^\varrho)^{k+1} = 0$. By analogy with (2), $v f^\varrho h^\varrho = (\beta - 1) v f^\varrho$, from which $v f^\varrho = 0$ by assumption. Applying (3) with $i = k + 1$ yields $2\beta = -k$, since $0 < k + 1 \leqq p - 1$ and since $v(e^\varrho)^k \neq 0$. Thus the lemma will be proved once we show $k = j$.

Suppose $k < j$; then let $0 \neq u \in \mathfrak{B}$, $u h^\varrho = (\beta + k + 1) u$. By induction, $u(f^\varrho)^m h^\varrho = (\beta + k - m + 1) u(f^\varrho)^m$, $m \geqq 0$, from which $u(f^\varrho)^{k+2} = 0$. Now let m be minimal with $u(f^\varrho)^m (e^\varrho)^m = 0$, so that $1 \leqq m \leqq k + 2$. Then $u(f^\varrho)^{m-1} \neq 0$, and for $n \geqq 0$,

$$u(f^\varrho)^{m-1} (e^\varrho)^n h^\varrho = (\beta + k - m + n + 2) u(f^\varrho)^{m-1} (e^\varrho)^n. \tag{4}$$

From (4) it follows that $u(f^\varrho)^{m-1}(e^\varrho)^n = 0$ for some $n < p$, so that if we assume n minimal with this property, we have $m - 1 < n < p$. Applying (3) with v replaced by $u(f^\varrho)^{m-1}$ yields

$$\begin{aligned} 0 &= u(f^\varrho)^{m-1}(e^\varrho)^n f^\varrho \\ &= \left(n(\beta + k - m + 2) + \binom{n}{2}\right) u(f^\varrho)^{m-1}(e^\varrho)^{n-1} + u(f^\varrho)^m (e^\varrho)^n. \end{aligned}$$

The second term is zero since $n \geqq m$, so that the vanishing of the first gives $2(\beta + k - m + 2) = -(n-1)$ by $0 < n < p$. Substituting $2\beta = -k$ gives

(5) $$k \equiv 2m - n - 3 \pmod{p}.$$

Now $k < j < p - 1$ gives $k + 2 \leqq p - 1$, so that $k + 2 - m \equiv m - n - 1$ and $0 \leqq k + 2 - m < p - 1$; furthermore, $m - 1 < n < p$ gives $0 < n + 1 - m < p + 1 - m \leqq p$, or $0 > m - n - 1 > -p$. Thus the congruence (5) can only result from the equality

(6) $$p + m - n - 1 = k + 2 - m,$$

from which $n = p + 2m - k - 3 > p + m - k - 3$. Thus $u(f^\varrho)^{m-1} \times (e^\varrho)^{p+m-k-3} \neq 0$; but application of (2) shows that this vector belongs to the characteristic root $\beta - 1$ of h^ϱ, a contradiction. Hence $k = j$, and the lemma is proved.

Lemma II.2.3. Let $\mathfrak{L}, \mathfrak{V}, \varrho$ be as above, and let j be an element of the prime field which is a characteristic root of h^ϱ. Then so is $-j$.

We again prove the lemma for prime characteristics $p > 2$, it being trivial for $p = 2$. We may also assume that not all elements of the prime field are characteristic roots of h^ϱ. Thus we have an integer ν, $0 \leqq \nu < p - 1$, such that $j, j-1, \ldots, j - \nu$ are characteristic roots while $j - \nu - 1$ is not. In Lemma 2, let $\beta = j - \nu$. Then we have $2\beta = -r$, where $\beta, \beta + 1, \ldots, \beta + r$ are characteristic roots, while $\beta + r + 1$ is not. That is, $2(j - \nu) \equiv -r \pmod{p}$, and clearly $r \geqq \nu$, so that $r \geqq r - \nu \geqq 0$. Let $0 \neq v \in \mathfrak{V}$, $v h^\varrho = \beta v$; by Lemma 2, $v(e^\varrho)^k \neq 0$, $0 \leqq k \leqq r$, so $v(e^\varrho)^{r-\nu} \neq 0$. But $v(e^\varrho)^{r-\nu} h^\varrho = (\beta + r - \nu) \times v(e^\varrho)^{r-\nu} = -(\beta + \nu) v(e^\varrho)^{r-\nu}$, and $\beta + \nu = j$. Thus $-j$ is a characteristic root.

Lemma II.2.4 (Jacobson). Let $\mathfrak{L}, \mathfrak{V}, \varrho, \mathfrak{F}$ be as in Th. 1.1; let α be a root of $\mathfrak{H}$ in $\mathfrak{L}$, and let h_α be as in Lemma 1.9. Suppose $\alpha(h_\alpha) \neq 0$, and that $p \neq 2, 3$. Then not all integral multiples of α are roots.

For let $p > 3$, and assume that all $k\alpha$, $1 \leqq k \leqq p - 1$, are roots; let $\mathfrak{M} = \mathfrak{F} h_\alpha + \sum_{k=1}^{p-1} \mathfrak{L}_{k\alpha}$. Let $0 \neq f \in \mathfrak{L}_{-\alpha}$, and let $e \in \mathfrak{L}_\alpha$ be such that $(e, f) = -\alpha(h_\alpha)^{-1}$. Let $h = \alpha(h_\alpha)^{-1} h_\alpha$. From Lemma 1.10 and Th. 1.2 we have $[e f] = h$, $[e h] = e$, $[f h] = -f$. By Lemmas 1.6 and 1.10,

$\mathfrak{M}$ is a subalgebra of $\mathfrak{L}$, hence is mapped into itself by all $\operatorname{ad} x$, for x in the three-dimensional algebra $\mathfrak{T}$ spanned by e, f, h.

Now let $2 \leqq k \leqq p-2$, and let $0 \neq x \in \mathfrak{L}_{k\alpha}$, $[x f] = 0$. By Lemma 1.6, $x(\operatorname{ad} e)^{p-k} \in \mathfrak{L}_0 = \mathfrak{H}$, and is a multiple of h_α by Lemma 1.10. Since $p-k \geqq 2$, and since $h_\alpha(\operatorname{ad} e)^2 = -\alpha(h_\alpha)[e\, e] = 0$, it follows by associativity of the form that $(x(\operatorname{ad} e)^{p-k}, h_\alpha) = 0$. From $(h_\alpha, h_\alpha) = \alpha(h_\alpha) \neq 0$, we have $x(\operatorname{ad} e)^{p-k} = 0$. Thus if j is the first non-negative integer with $x(\operatorname{ad} e)^{j+1} = 0$, the space $\mathfrak{W}$ spanned by $x(\operatorname{ad} e)^i$, $0 \leqq i \leqq j$, is stable under the adjoint representation of $\mathfrak{T}$ in $\mathfrak{M}$, and $j < p-k \leqq \leqq p-2$. The characteristic roots of $\operatorname{ad} h$ in $\mathfrak{W}$ are $k, k+1, \ldots, k+j$, but not $k-1$. Lemma 2 now yields $2k \equiv -j \pmod{p}$.

Since $h_{r\alpha} = r h_\alpha$, we see that the hypotheses of the lemma are satisfied by each root $\beta = r\alpha \neq 0$. Thus if $0 \neq y \in \mathfrak{L}_{s\beta}$, $2 \leqq s \leqq \leqq p-2$, and if $0 \neq z \in \mathfrak{L}_{-\beta}$ satisfies $[y\, z] = 0$, we have $2s \equiv -t \pmod{p}$, where $t < p-s$. Applying these considerations with $\beta = -2\alpha$, $s = \frac{p+1}{2}$, we have $s\beta = -\alpha$, $-\beta = 2\alpha$. Taking $y = f$, $z = x$ above, we see that $[x f] = 0$ implies $p+1 \equiv -t \pmod{p}$, where $0 \leqq t < \frac{p-1}{2}$. This is clearly absurd, from which we conclude that the mapping $x \to [x f]$ is one-one from $\mathfrak{L}_{2\alpha}$ into $\mathfrak{L}_\alpha$.

Thus we have $\dim \mathfrak{L}_{2\alpha} \leqq \dim \mathfrak{L}_\alpha$, and in general $\dim \mathfrak{L}_{2^k\alpha} \leqq \leqq \dim \mathfrak{L}_{2\alpha} \leqq \dim \mathfrak{L}_\alpha$, $k \geqq 1$. With $k = p-1$, we find $\dim \mathfrak{L}_{2\alpha} = \dim \mathfrak{L}_\alpha$, from which $[\mathfrak{L}_{2\alpha} f] = \mathfrak{L}_\alpha$. Thus $e = [x f]$, some $x \in \mathfrak{L}_{2\alpha}$, and $0 \neq (e, f) = ([x f], f) = (x, [f f]) = 0$, in contradiction to the choice of e. This establishes the lemma.

Lemma II.2.5. Let the hypotheses be as in Lemma 4. Then 2α is not a root.

For if $\mathfrak{T}$ is the subalgebra of the proof of Lemma 4, let $\mathfrak{W} = \mathfrak{F} f + + \mathfrak{F} h + \sum_{k=1}^{r} \mathfrak{L}_{k\alpha}$, where $\alpha, 2\alpha, \ldots, r\alpha$ are roots, but $(r+1)\alpha$ is not ($r \leqq p-3$ by Lemma 4). Then $\mathfrak{W}$ is invariant under the adjoint representation of $\mathfrak{T}$ in $\mathfrak{L}$, and in $\mathfrak{W}$, $\operatorname{ad} h$ has the characteristic roots $-1, 0, 1, 2, \ldots, r$, but not -2. If $r \geqq 2$, this contradicts Lemma 3.

Lemma II.2.6. Let the hypotheses be as in Lemma 4. Then $\mathfrak{L}_\alpha$ has dimension one.

Namely, let $\mathfrak{T}$ be as above; if $\dim \mathfrak{L}_\alpha > 1$, there is an $x \neq 0$ in $\mathfrak{L}_\alpha$ with $(x, f) = 0$, hence with $[x f] = 0$ by Lemma 1.10. Now $[x\, e] = 0$ since $\mathfrak{L}_{2\alpha} = 0$, and the Jacobi identity yields $[x\, h] = 0$; but $[x\, h] = x$ since $\alpha(h) = 1$. The lemma follows.

Lemma II.2.7. Let the hypotheses be as in Lemma 4. Then the only integral multiples of α which are roots are $\pm\alpha$ and 0.

For by Lemma 5, 2α is not a root; thus if $k\alpha$, $1 < k < p-1$, is a root, we have $2 < k < p - 2$, and we may assume $2 < k \leqq \frac{p-1}{2}$. Also, $k < \frac{p-1}{2}$, since $2\left(\frac{p-1}{2}\alpha\right) = -\alpha$ is a root, which is impossible by Lemma 5. Now let k be minimal in the interval $\left(2, \frac{p-1}{2}\right)$ with $k\alpha$ a root. Let $\mathfrak{T}$ be as above, and let $\mathfrak{W} = \mathfrak{L}_{k\alpha} + \mathfrak{L}_{(k+1)\alpha} + \cdots + \mathfrak{L}_{(k+r)\alpha}$, where $k\alpha, (k+1)\alpha, \ldots, (k+r)\alpha$ are roots, $(k+r+1)\alpha$ not a root; then $k + r < \frac{p-1}{2}$ and $\mathfrak{W}$ is a representation-space for $\mathfrak{T}$, in which $\operatorname{ad} h$ has characteristic roots $k, \ldots, k+r$. By Lemma 3, $-k = k + j$ for some $j \leqq r$, or $j \equiv -2k (\operatorname{mod} p)$; but $0 < j + k \leqq \leqq r + k < \frac{p-1}{2}$, and $0 < k < \frac{p-1}{2}$, so that $0 < j + 2k < p - 1$, a contradiction.

Lemma II.2.8. Let $\mathfrak{L}, \mathfrak{B}, \varrho, \mathfrak{F}$ be as above. Let α and β be non-zero roots of $\mathfrak{H}$ in $\mathfrak{L}$, and let $h_\alpha, h_\beta \in \mathfrak{H}$ be as before. Let $\beta(h_\beta) \neq 0$ and let $p \neq 2, 3$. Let $0 \neq e_\gamma \in \mathfrak{L}_\gamma$, $\gamma = \alpha, \pm\beta$. Then if $[e_\alpha e_\beta] = 0 = [e_\alpha e_{-\beta}]$, we have $(h_\alpha, h_\beta) = 0$.

For since $\mathfrak{L}_\beta, \mathfrak{L}_{-\beta}$ are one-dimensional, we have $(e_{-\beta}, e_\beta) \neq 0$, hence $[e_{-\beta} e_\beta] = (e_{-\beta}, e_\beta) h_\beta \neq 0$. But $[e_\alpha [e_{-\beta} e_\beta]] = 0$ by the Jacobi identity, from which $(e_{-\beta}, e_\beta)(h_\alpha, h_\beta) e_\alpha = 0$. The lemma follows.

Corollary. Under the conditions above on α, β, suppose that neither of $\alpha + \beta, \alpha - \beta$ is a root. Then $(h_\alpha, h_\beta) = 0$.

Lemma II.2.9 (Kaplansky [247]). Let $\mathfrak{L}, \mathfrak{B}, \varrho, \mathfrak{F}$ be as above, and assume further that the center of $\mathfrak{L}$ is zero. Then if $p \neq 2, 3$, and if $\alpha \neq 0$ is a root of $\mathfrak{H}$ in $\mathfrak{L}$, we have $\alpha(h_\alpha) \neq 0$.

For suppose $\alpha(h_\alpha) = 0$. If $\beta(h_\alpha) = 0$ for all roots β, then h_α is in the center of $\mathfrak{L}$, which is impossible. Hence there is a root β with $\beta(h_\alpha) = (h_\alpha, h_\beta) \neq 0$. Now all $\beta + k\alpha$, $0 \leqq k \leqq p - 1$, are roots; to see this, it suffices to show that $\beta - \alpha$ is a root, since $\beta - \alpha$ then satisfies the same conditions as does β. Thus let $0 \neq e_\gamma \in \mathfrak{L}_\gamma$, $\gamma = \beta, \pm\alpha$, where $(e_\alpha, e_{-\alpha}) = 1$. If $\beta - \alpha$ is not a root, then $[e_\beta e_{-\alpha}] = 0$, from which it follows, using $\alpha(h_\alpha) = 0$, that $e_\beta (\operatorname{ad} e_\alpha)^j (\operatorname{ad} e_{-\alpha}) = -j\beta(h_\alpha) \times \times e_\beta (\operatorname{ad} e_\alpha)^{j-1}$ for all $j > 0$. Moreover $e_\beta (\operatorname{ad} e_\alpha)^{p-1} \in \mathfrak{L}_{\beta-\alpha} = 0$, from which we have by the preceding relation that $e_\beta (\operatorname{ad} e_\alpha)^j = 0$ for $j = p-2, p-3, \ldots, 1, 0$, i.e., $e_\beta = 0$, which is absurd.

Now $\beta(h_\beta) \neq 0$; for if $\beta(h_\beta) = 0$, then $(\beta + \alpha)(h_{\beta+\alpha}) = 2(h_\alpha, h_\beta) \neq 0$, so that $2(\alpha + \beta)$ is not a root (Lemma 5). On the other hand, $(\beta + 2\alpha)(h_\beta) = 2(h_\alpha, h_\beta) \neq 0$, so that $(\beta + 2\alpha) + \beta$ is a root by the above. Our assertion follows.

Next note that $\gamma = \alpha + 2\beta, \alpha - 2\beta$ are not roots; for in each case $\gamma(h_\alpha) \neq 0$, so that $\gamma \pm \alpha$ is a root, i.e., $2(\alpha + \beta)$ is a root. But $(\alpha + \beta)(h_\alpha) \neq 0$, from which $(\alpha + \beta)(h_{\alpha+\beta}) \neq 0$ by the preceding paragraph; Lemma 5 now yields the assertion.

Let $\mathfrak{T}$ be the three-dimensional subalgebra $\mathfrak{L}_\beta + \mathfrak{L}_{-\beta} + \mathfrak{F} h_\beta$ as before, and let h be a multiple of h_β with $\beta(h) = 1$. Let $\mathfrak{W}$ be the subspace $\mathfrak{L}_{\alpha-\beta} + \mathfrak{L}_\alpha + \mathfrak{L}_{\alpha+\beta}$ of $\mathfrak{L}$, a subspace which is a representation-space for $\mathfrak{T}$ (adjoint representation), and in which ad h has the characteristic roots $\alpha(h) - 1$, $\alpha(h)$, $\alpha(h) + 1$, and only these. By Lemma 2, $2(\alpha(h) - 1) = -2$, or $2\alpha(h) = 0$, a contradiction to $(h_\alpha, h_\beta) \neq 0$. Thus the lemma is proved.

Lemma II.2.10. Let the hypotheses be as in Lemma 9, and let α and β be roots, $\beta \neq 0$. Then not all of $\alpha, \alpha+\beta, \alpha+2\beta, \alpha+3\beta, \alpha+4\beta$ are roots. Thus not all $\alpha + k\beta$ are roots.

Namely, if one of the above sequence is zero, then by Lemma 7 either $\alpha = 0$, $\alpha = -\beta$, or $\alpha = \beta\,(p = 5)$, so that either $\alpha + 2\beta$, $\alpha + 3\beta$, or $\alpha + \beta$ is 2β, which is not a root. If none is zero, and if all are roots, then $(\alpha + 2\beta) \pm \alpha$ cannot be roots by Lemma 5, nor can $(\alpha + 4\beta) \pm (\alpha + 2\beta)$. By the corollary to Lemma 8, $(h_{\alpha+2\beta}, h_\alpha) = 0 = (h_{\alpha+4\beta}, h_{\alpha+2\beta})$, from which $4(h_\beta, h_{\alpha+2\beta}) = 0$, and $0 = (h_{\alpha+2\beta}, h_\alpha + 2h_\beta) = (\alpha + 2\beta)(h_{\alpha+2\beta})$, in contradiction to Lemma 9.

§ 3. Classical Lie algebras

Let $\mathfrak{F}$ be a field of characteristic not 2, 3. A Lie algebra $\mathfrak{L}$ over $\mathfrak{F}$ is called *classical* if:

i) the center of $\mathfrak{L}$ is zero;

ii) $[\mathfrak{L}\,\mathfrak{L}] = \mathfrak{L}$;

iii) $\mathfrak{L}$ has an abelian Cartan subalgebra $\mathfrak{H}$ (called a *classical Cartan subalgebra*), relative to which:

a) $\mathfrak{L} = \sum \mathfrak{L}_\alpha$, where $[x\,h] = \alpha(h)\,x$ for all $x \in \mathfrak{L}_\alpha$, $h \in \mathfrak{H}$;

b) if $\alpha \neq 0$ is a root, $[\mathfrak{L}_\alpha\,\mathfrak{L}_{-\alpha}]$ is one-dimensional;

c) if α and β are roots, and if $\beta \neq 0$, then not all $\alpha + k\beta$ are roots.

Thus §§ 1, 2 show that if $\mathfrak{L}$ is a Lie algebra over a perfect field of prime characteristic $\neq 2, 3$, satisfying i) and ii) and having a representation ϱ with non-singular trace form and a Cartan subalgebra $\mathfrak{H}$ such that each h^ϱ $(h \in \mathfrak{H})$ has all its characteristic roots in the ground field, then $\mathfrak{L}$ is classical with $\mathfrak{H}$ as classical Cartan subalgebra. If we observe that the only use of the assumptions that the associative form is a trace form and that $\mathfrak{F}$ is perfect was made in proving Lemmas 1.4 and 1.5 and Th. 1.2, we note that the relevant parts of these are trivial if we assume:

(*) For each $h \in \mathfrak{H}$, ad h is semi-simple and has characteristic roots in $\mathfrak{F}$ (i.e., has a diagonal matrix).

Thus we see that if $\mathfrak{F}$ is arbitrary of prime characteristic $\neq 2, 3$; and if $\mathfrak{L}$ satisfies i), ii), has a Cartan subalgebra $\mathfrak{H}$ satisfying (*), and carries a non-singular associative symmetric bilinear form, then $\mathfrak{L}$ is classical relative to $\mathfrak{H}$. In particular, $\mathfrak{L}$ is classical relative to every Cartan subalgebra if $\mathfrak{F}$ is algebraically closed and $\mathfrak{L}$ has non-singular Killing form (Th. 1.7.1, corollary to Th. 1.7.2, and results of §§ 1, 2). In fact, under the assumption of non-singular Killing form, it is enough to assume either (*) or that $\mathfrak{F}$ is perfect and that $\mathfrak{L}$ has a Cartan subalgebra $\mathfrak{H}$ such that every ad h ($h \in \mathfrak{H}$) has its characteristic roots in $\mathfrak{F}$, in order to assure that $\mathfrak{L}$ is classical.

It may be remarked here that every semisimple Lie algebra of characteristic zero, having a Cartan subalgebra $\mathfrak{H}$ with all characteristic roots of ad h ($h \in \mathfrak{H}$) in the ground field, is classical (see, e.g., Chap. IV of [234]).

Another procedure for the construction of classical Lie algebras is that of CHEVALLEY [76], who has shown that if $\mathfrak{L}_C$ is a semisimple Lie algebra over the complex field, then $\mathfrak{L}_C$ has a basis consisting of a basis for a Cartan subalgebra $\mathfrak{H}_C$ and of root-vectors, such that the additive subgroup of $\mathfrak{L}_C$ generated by this basis is closed under the Lie product in $\mathfrak{L}_C$. Thus one has a Lie algebra $\mathfrak{L}_Z$ over the integers, and if $\mathfrak{F}$ is an arbitrary field, $\mathfrak{L} = \mathfrak{F} \otimes_Z \mathfrak{L}_Z$ becomes a Lie algebra over $\mathfrak{F}$. When $\mathfrak{F}$ has characteristic $\neq 2, 3$, then either $\mathfrak{L}$ is classical relative to $\mathfrak{H} = \mathfrak{F} \otimes_Z \mathfrak{H}_Z$, $\mathfrak{H}_Z = \mathfrak{H}_C \cap \mathfrak{L}_Z$, or $\mathfrak{L}_C$ has a simple ideal isomorphic to the n by n complex matrices of trace zero for some n divisible by the characteristic of $\mathfrak{F}$. In the latter case, $\mathfrak{L}$ has a center $\mathfrak{Z}$ of dimension equal to the number of such ideals in $\mathfrak{L}_C$, and $\mathfrak{L}/\mathfrak{Z}$ is classical relative to $\mathfrak{H}/\mathfrak{Z}$ (STEINBERG [381]). It will be a consequence of later results that all classical Lie algebras can be obtained by this procedure.

Now let $\mathfrak{L}$ be a classical Lie algebra, $\mathfrak{H}$ a classical Cartan subalgebra, to be fixed in the sequel. Thus roots are always roots of $\mathfrak{H}$ in $\mathfrak{L}$. From iii), a) and c), and from $[\mathfrak{L}_\alpha \mathfrak{L}_\beta] \subseteq \mathfrak{L}_{\alpha+\beta}$, it follows that if $\alpha \neq 0$ is a root and if $e \in \mathfrak{L}_\alpha$, then $(\text{ad } e)^{p-1} = 0$ (in characteristic zero, that ad e is nilpotent). From iii), a), the roots are linear functions on $\mathfrak{H}$. We now indicate how some of the other results of the preceding two sections can be modified so as to apply in the new setting.

Lemma II.3.1. If $\alpha \neq 0$ is a root, then $\alpha([\mathfrak{L}_\alpha \mathfrak{L}_{-\alpha}]) \neq 0$.

For by iii), $[\mathfrak{L}_\alpha \mathfrak{L}_{-\alpha}] = \mathfrak{F} h$, $0 \neq h \in \mathfrak{H}$, and $\beta(h) \neq 0$ for some root β, by i). We may write $h = [e_\alpha e_{-\alpha}]$, $e_\gamma \in \mathfrak{L}_\gamma$. Now let $\beta, \beta - \alpha, \ldots, \beta - r\alpha$ be roots, while $\beta - (r+1)\alpha$ is not; let $0 \neq y \in \mathfrak{L}_{\beta - r\alpha}$, and let $\mathfrak{W}$ be the subspace of $\mathfrak{L}$ spanned by the $y(\text{ad } e_\alpha)^j$, $0 \leqq j < p - 1$. Then each of these generators is a root-vector, and for each $j \geqq 1$ one proves by induction that

$$y(\text{ad } e_\alpha)^j (\text{ad } e_{-\alpha}) = j\left(\beta - r\alpha + \frac{j-1}{2}\alpha\right)(h)\, y(\text{ad } e_\alpha)^{j-1}.$$

In particular if j ($1 \leqq j \leqq p - 1$) is minimal with $y(\text{ad } e_\alpha)^j = 0$, we find $\left(\beta - r\alpha + \frac{j-1}{2}\alpha\right)(h) = 0$. Thus $\alpha(h) \neq 0$.

The proofs of Lemmas 2.5, 6, 7 now go over without essential change. We state them as one, incorporating Lemma 1 above:

Lemma II.3.2. If $\alpha \neq 0$ is a root, then $\mathfrak{L}_\alpha$ is one-dimensional, and only $0, \pm\alpha$ are roots among the integral multiples of α.

Namely, in Lemma 2.5, we choose $\mathfrak{T}$ as the subalgebra spanned by $h \neq 0$ in $[\mathfrak{L}_\alpha\,\mathfrak{L}_{-\alpha}]$, so normalized (Lemma 1) that $\alpha(h) = 1$, and by $e_\gamma \in \mathfrak{L}_\gamma$, $\gamma = \pm\alpha$, such that $[e_\alpha\,e_{-\alpha}] = h$. In Lemma 2.6, with $f = e_{-\alpha}$, the mapping $x \to [x\,f]$ maps $\mathfrak{L}_\alpha$ onto the one-dimensional space $[\mathfrak{L}_\alpha\,\mathfrak{L}_{-\alpha}]$, so that if $\dim \mathfrak{L}_\alpha > 1$, there is an $x \in \mathfrak{L}_\alpha$, $x \neq 0$, with $[x\,f] = 0$; the proof now goes as before. With $\mathfrak{T}$ as above, the proof of Lemma 2.7 goes through without change.

Lemma II.3.3. Let α and β be roots, $\beta \neq 0$. Let $0 \neq e_\alpha \in \mathfrak{L}_\alpha$, $0 \neq e_\beta \in \mathfrak{L}_\beta$, $0 \neq e_{-\beta} \in \mathfrak{L}_{-\beta}$, and let $[e_\alpha\,e_\beta] = 0 = [e_\alpha\,e_{-\beta}]$. Then $\alpha([\mathfrak{L}_\beta\,\mathfrak{L}_{-\beta}]) = 0$.

By Lemma 2 and iii), b), $[e_\beta\,e_{-\beta}]$ is a basis for $[\mathfrak{L}_\beta\,\mathfrak{L}_{-\beta}]$. But $\alpha([e_\beta\,e_{-\beta}])\,e_\alpha = [e_\alpha[e_\beta\,e_{-\beta}]] = 0$ by assumption and the Jacobi identity.

Lemma II.3.4. Let α and β be roots, $\beta \neq 0$. Then not all of $\alpha, \alpha+\beta$, $\alpha+2\beta, \alpha+3\beta, \alpha+4\beta$ are roots.

As in the proof of Lemma 2.10, one has in the contrary case, $\alpha([\mathfrak{L}_{\alpha+2\beta}\,\mathfrak{L}_{-\alpha-2\beta}]) = 0 = (\alpha+4\beta)([\mathfrak{L}_{\alpha+2\beta}\,\mathfrak{L}_{-\alpha-2\beta}])$ (using Lemma 3), hence $(\alpha+2\beta)([\mathfrak{L}_{\alpha+2\beta}\,\mathfrak{L}_{-\alpha-2\beta}]) = 0$, in contradiction to Lemma 1.

§ 4. Strings of roots and Cartan integers

We assume that $\mathfrak{L}$ is a classical Lie algebra, with classical Cartan subalgebra $\mathfrak{H}$. By "roots" we mean roots of $\mathfrak{H}$ in $\mathfrak{L}$.

Lemma II.4.1. Let $\alpha, \beta, \alpha+\beta$ be non-zero roots. Then $[\mathfrak{L}_\alpha\,\mathfrak{L}_\beta] = \mathfrak{L}_{\alpha+\beta}$.

By Lemma 3.2 we may assume α is not a multiple of β, and it suffices to show $[\mathfrak{L}_\alpha\,\mathfrak{L}_\beta] \neq 0$. Each $\mathfrak{L}_{\alpha+k\beta}$ which is non-zero has dimension one, and we may assume $\alpha - r\beta, \ldots, \alpha-\beta, \alpha, \alpha+\beta, \ldots, \alpha+q\beta$ are roots, while $\alpha-(r+1)\beta$ and $\alpha+(q+1)\beta$ are not roots; here q and r are non-negative integers, and $q > 0$ by assumption. Let $\mathfrak{T}$ be the three-dimensional algebra $\mathfrak{L}_\beta + [\mathfrak{L}_\beta\,\mathfrak{L}_{-\beta}] + \mathfrak{L}_{-\beta}$, where $h \in [\mathfrak{L}_\beta\,\mathfrak{L}_{-\beta}]$ has $\beta(h) = 1$, and where $e \in \mathfrak{L}_\beta$, $f \in \mathfrak{L}_{-\beta}$ have $[e\,f] = h$. Let $\mathfrak{W} = \mathfrak{L}_{\alpha-r\beta} + \cdots + \mathfrak{L}_{\alpha+q\beta}$; then $\mathfrak{W}$ is invariant under the adjoint representation of $\mathfrak{T}$, and the characteristic roots of $\operatorname{ad} h$ in $\mathfrak{W}$ are $\alpha(h) - r, \ldots, \alpha(h)+q$, but not $\alpha(h)-r-1$, $\alpha(h)+q+1$. By Lemma 2.2, if $0 \neq w \in \mathfrak{L}_{\alpha-r\beta}$, then $w(\operatorname{ad} e)^i \neq 0$, $0 \leqq i \leqq r+q$. In particular, $w(\operatorname{ad} e)^r$ is a basis for $\mathfrak{L}_\alpha$, and $0 \neq w(\operatorname{ad} e)^{r+1}$ in $[\mathfrak{L}_\alpha\,\mathfrak{L}_\beta]$.

Lemma II.4.2. Let α and β be roots, $\beta \neq 0$. Let r and q be the smallest non-negative integers with $\alpha-(r+1)\beta$ not a root and $\alpha+(q+1)\beta$ not a root. Let $h_\beta \in [\mathfrak{L}_\beta\,\mathfrak{L}_{-\beta}]$. Then $2\alpha(h_\beta) = (r-q)\,\beta(h_\beta)$.

For let $\mathfrak{T}$ and $\mathfrak{W}$ be as in the proof of Lemma 1, admitting now the possibility that some $\alpha + k\beta = 0$. It suffices to show $2\alpha(h) = r - q$.

But this is immediate from Lemma 2.2, since this lemma yields $2(\alpha(h) - r) = -(r + q)$.

The rational integer $r - q$ of Lemma 2 will be written $A_{\alpha,\beta}$, and will be called the *Cartan integer* of the ordered pair of roots α, β $(\beta \neq 0)$. By Lemma 3.4, we have $-3 \leqq A_{\alpha,\beta} \leqq 3$, and $A_{\alpha,\beta} \leqq -2$ implies that $\alpha - \beta$ is not a root. Clearly $A_{\alpha,-\beta} = -A_{\alpha,\beta} = A_{-\alpha,\beta}$, and by Lemma 3.2, $A_{0,\beta} = 0$. It will be noted that $A_{\alpha,\beta} < 0$ implies that $\alpha + \beta$ is a root or, by Lemma 1, that $[\mathfrak{L}_\alpha \mathfrak{L}_\beta] \neq 0$ (since $\beta \neq 0$, this holds even for $\alpha = 0$, $\alpha = -\beta$).

Lemma II.4.3. If α, β are non-zero roots, then $A_{\alpha,\beta} = 0$ implies $A_{\beta,\alpha} = 0$.

We may assume $A_{\alpha,\beta} = 0$, $A_{\beta,\alpha} < 0$. Then $\beta + \alpha$ is a root, so that $A_{\alpha,\beta} = 0$ yields $\alpha - \beta$, hence $\beta - \alpha$, as a root. It follows that $A_{\beta,\alpha} = -1$, and that $\beta + 2\alpha$ is a root. Now $(\beta + 2\alpha) \pm \beta$ are not roots, so that $A_{\beta+2\alpha,\beta} = 0$. Taking $h_\beta \neq 0$ in $[\mathfrak{L}_\beta \mathfrak{L}_{-\beta}]$, Lemma 2 yields $\alpha(h_\beta) = 0 = (\beta + 2\alpha)(h_\beta)$, or $\beta(h_\beta) = 0$, in contradiction to Lemma 3.1.

Lemma II.4.4. If $\alpha, \beta, \alpha + \beta$ and γ are roots, $\gamma \neq 0$, then $A_{\alpha,\gamma} + A_{\beta,\gamma} = A_{\alpha+\beta,\gamma}$.

By Lemma 2, the two sides are congruent modulo the characteristic. Since $-6 \leqq A_{\alpha,\gamma} + A_{\beta,\gamma} \leqq 6$ and $-3 \leqq A_{\alpha+\beta,\gamma} \leqq 3$, we have the result except for characteristics 5 and 7, in which cases it may be obtained by exhaustion (cf. [303]).

Lemma II.4.5. Let $\alpha_1, \ldots, \alpha_k$ be non-zero roots; let A_{ij} denote the Cartan integer of α_i, α_j, and suppose the matrix (A_{ij}) is non-singular. Then every sequence of roots of the form $\beta, \beta - \alpha_{i_1}, \beta - \alpha_{i_1} - \alpha_{i_2}, \ldots, 1 \leqq i_j \leqq k$, terminates in a root γ such that no $\gamma - \alpha_i$ is a root.

For assuming the existence of an infinite sequence of roots $\beta = \beta_0$, $\beta_1 = \beta - \alpha_{i_1}, \ldots$ of the above type, we see by Lemma 4 that $A_{\beta_m,\alpha_r} = A_{\beta,\alpha_r} - \sum_{j=1}^{k} n_j^{(m)} A_{jr}$, $1 \leqq r \leqq k$, where $n_j^{(m)} \geqq 0$ is the number of $i_s = j$ for $1 \leqq s \leqq m$. Thus $m = \sum_j n_j^{(m)}$. Since the totality of roots is finite, we have $\beta_m = \beta_q$ for some m and some $q > m$, from which $\sum_j (n_j^{(q)} - n_j^{(m)}) A_{jr} = 0$, $1 \leqq r \leqq k$, and $\sum_j (n_j^{(q)} - n_j^{(m)}) = q - m > 0$. This contradicts the non-singularity of (A_{ij}).

§ 5. Fundamental root systems

The notations and conventions are as in § 4. Since $[\mathfrak{L}\,\mathfrak{L}] = \mathfrak{L}$ and $[\mathfrak{L}_\alpha \mathfrak{L}_\beta] \subsetneqq \mathfrak{L}_{\alpha+\beta}$, $[\mathfrak{H}\,\mathfrak{H}] = 0$, it follows that $\mathfrak{H}$ is the sum of the one-dimensional spaces $[\mathfrak{L}_\beta \mathfrak{L}_{-\beta}]$, $\beta \neq 0$ a root. Thus if we choose $h_\beta \neq 0$

in $[\mathfrak{L}_\beta\, \mathfrak{L}_{-\beta}]$ for each $\beta \neq 0$, we see that a basis for $\mathfrak{H}$ can be chosen from among these h_β. Let the corresponding values of β be $\beta_1, \ldots, \beta_r$. With this modification, we follow CURTIS [94], denoting for each root α the rational integer A_{α, β_j} by $b_j(\alpha)$, and writing $\alpha < \beta$ for two roots α, β if the first integer j $(1 \leqq j \leqq r)$ for which $b_j(\alpha) \neq b_j(\beta)$ has $b_j(\alpha) < < b_j(\beta)$.

Lemma II.5.1. The relation $\alpha < \beta$ is a total ordering of the roots such that if α, β and $\alpha + \beta$ are roots, $\beta > 0$, then $\alpha < \alpha + \beta$, and if $\alpha < \beta$, then $-\beta < -\alpha$.

For we have seen in § 4 that for each j, $b_j(\alpha) = -b_j(-\alpha)$, and that $b_j(\alpha + \beta) = b_j(\alpha) + b_j(\beta)$ if $\alpha, \beta, \alpha + \beta$ are roots. From these observations and the obvious transitivity, it only remains to prove trichotomy, and for this to prove that if $b_j(\alpha) = b_j(\beta)$ for all j, then $\alpha = \beta$. Letting $0 \neq h_j \in [\mathfrak{L}_{\beta_j}\, \mathfrak{L}_{-\beta_j}]$, we see from Lemma 4.2 that $b_j(\alpha) = b_j(\beta)$ implies $2\alpha(h_j) = b_j(\alpha)\, \beta_j(h_j) = 2\beta(h_j)$, hence that $\alpha(h_j) = \beta(h_j)$ for all j. Since the h_j are a basis for $\mathfrak{H}$, we have $\alpha = \beta$.

Following DYNKIN [128], we call a positive root α (i.e., $0 < \alpha$) *simple* if there are no two positive roots β, γ with $\alpha = \beta + \gamma$. Such roots evidently exist; e.g., take α to be the smallest positive root. Let Π be the totality of simple roots, $\Pi = \{\alpha_1, \ldots, \alpha_m\}$. Now it is clear that if $\alpha_i \neq \alpha_j$ are members of Π, then $\alpha_i - \alpha_j$ is not a root. Thus, writing $A_{\alpha_i, \alpha_j} = A_{ij}$, we have $A_{ij} \leqq 0$ for $i \neq j$, while $A_{ii} = 2$.

Lemma II.5.2. Let α be any positive root. Then there is a sequence of roots of the form $\beta_1 = \alpha_{i_1}, \beta_2 = \alpha_{i_1} + \alpha_{i_2}, \ldots, \beta_k = \beta_{k-1} + \alpha_{i_k} = \alpha$, where $1 \leqq i_j \leqq m$.

For since there are only finitely many roots, and since the least positive root is simple, we may assume that α is not a simple root, and that the conclusion holds for all positive roots $\beta < \alpha$; then it suffices to deduce from these assumptions the assertion for α. We have $\alpha = \beta + \gamma$, where β, γ are positive roots less than α. Thus we have chains $\beta_1 = \alpha_{i_1}$, $\beta_2 = \beta_1 + \alpha_{i_2}$, $\ldots$, $\beta_{s-1} + \alpha_{i_s} = \beta_s = \beta$; $\gamma_1 = \alpha_{j_1}$, $\gamma_2 = \gamma_1 + \alpha_{j_2}, \ldots, \gamma_t = \gamma_{t-1} + \alpha_{j_t} = \gamma$, all members being roots. Now all these members are positive roots, so by Lemma 4.1, $\mathfrak{L}_\beta$ has a basis $e_\beta = [\ldots[[e_{i_1}\, e_{i_2}]\, e_{i_3}] \ldots e_{i_s}]$ and $\mathfrak{L}_\gamma$ a basis $e_\gamma = [\ldots[[e_{j_1}\, e_{j_2}]\, e_{j_3}] \ldots e_{j_t}]$, where e_i is a basis for $\mathfrak{L}_{\alpha_i}$. By the same lemma, $[e_\beta\, e_\gamma]$ is a basis for $\mathfrak{L}_\alpha$. Now we show that for some indexing $\{k_1, \ldots, k_{s+t}\}$ of the full set of indices $\{i_1, \ldots, i_s\} + \{j_1, \ldots, j_t\}$ we have $[\ldots[e_{k_1}\, e_{k_2}] \ldots e_{k_{s+t}}] \neq 0$. Then $\alpha_{k_1}, \alpha_{k_1} + \alpha_{k_2}, \ldots, \alpha_{k_1} + \cdots + \alpha_{k_{s+t}} = \alpha$ are roots, as required. We proceed by induction on γ in the linearly ordered and finite set of positive roots. If γ is minimal positive then γ is simple, $e_\gamma = e_{j_1}$, and $0 \neq [[\ldots[e_{i_1}\, e_{i_2}] \ldots e_{i_s}]\, e_{j_1}] \in \mathfrak{L}_\alpha$. Now assume the assertion whenever α is written $\alpha = \beta' + \gamma'$, where β', γ' are positive roots, $\gamma' < \gamma$; we may assume γ is not simple. Letting e_β, e_γ be as above, the Jacobi identity and the fact that $[e_\beta\, e_\gamma] \neq 0$ yield that either

$$\text{a)}\qquad \big[[[\ldots[e_{i_1}\, e_{i_2}]\ldots e_{i_s}]\, e_{j_t}],\ [\ldots[e_{j_1}\, e_{j_2}]\ldots e_{j_{t-1}}]\big] \neq 0,$$

or

$$\text{b)}\qquad \big[[[\ldots[e_{i_1}\, e_{i_2}]\ldots e_{i_s}],\ [\ldots[e_{j_1}\, e_{j_2}]\ldots e_{j_{t-1}}]],\ e_{j_t}\big] \neq 0.$$

In case a), we have $\alpha = \beta' + \gamma'$, where $\beta' = \beta + \alpha_{j_t}$, $\gamma' = \gamma - \alpha_{j_t}$ are positive roots; in case b), $\alpha' = \alpha - \alpha_{j_t}$ is a root, and $\alpha = \alpha' + \alpha_{j_t}$. In either case the assertion follows by inductive hypothesis.

It follows from Lemma 2 and the fact that the roots span the dual space of $\mathfrak{H}$ (otherwise we should have $0 \neq h \in \mathfrak{H}$ with $\alpha(h) = 0$ for all roots α, from which h is in the center of $\mathfrak{L}$) that the simple roots span the dual space of $\mathfrak{H}$. (In contrast to the situation treated by Curtis, the simple roots may fail to be linearly independent.) We shall say that a set $\alpha_1, \ldots, \alpha_m$ of roots is a *fundamental system* of roots (relative to $\mathfrak{H}$) if:

a) $\alpha_i - \alpha_j$ is not a root for $i \neq j$;

b) if α is a root then one of the following holds:

i) α is a member of a sequence of roots of the form $\alpha_{i_1}, \alpha_{i_1} + \alpha_{i_2}, \ldots$;

ii) $-\alpha$ is a member of such a sequence;

iii) $\alpha = 0$;

c) Every diagonal minor of the matrix (A_{ij}) is positive.

Since c) implies that (A_{ij}) is non-singular, it is clear by Lemma 4.4 that only one of i), ii), iii) can hold for a given root α, and that the number of times the root α_j occurs in a sequence $\alpha_{i_1}, \alpha_{i_1} + \alpha_{i_2}, \ldots, \alpha_{i_1} + \cdots + \alpha_{i_m} = \alpha$ of roots is independent of the manner of formation of this sequence. (It also follows that a) is redundant, but we choose to emphasize a) by making it part of the definition anyway.) By Lemma 2 and trichotomy it is clear that the simple roots in the ordering introduced at the beginning of this section satisfy a) and b). We now prove that c) is also satisfied, and thereby show the existence of fundamental systems. Here we may assume every diagonal minor smaller than k by k is positive, and that the k-rowed diagonal minor corresponding to $\alpha_1, \ldots, \alpha_k$ is not positive. Being equal to 2, every one-rowed diagonal minor is positive, so we may assume $k > 1$. The desired conclusion will now follow from the more general principle of Lemma 4, which we shall use again in classifying fundamental systems of roots. First we prove

Lemma II.5.3 (Mills). Let (a_{ij}) be a k by k real matrix such that: $a_{ij} \leqq 0$ for all $i \neq j$; every proper diagonal minor is positive; $\det(a_{ij}) \leqq 0$; then if $\lambda_1, \ldots, \lambda_k$ are non-negative real numbers satisfying $\sum_i \lambda_i\, a_{ij} \geqq 0$ for all j, we have $\sum_i \lambda_i\, a_{ij} = 0$ for all j.

For suppose some inequality is strict, say $\sum_i \lambda_i a_{i1} > 0$. Then $\lambda_1 > 0$, $a_{11} > 0$, and replacing a_{11} by $a_{11} - \varepsilon$ for ε sufficiently small and positive we have a new matrix (b_{ij}) with $\sum_i \lambda_i b_{ij} \geqq 0$, all j, with proper diagonal minors positive, with $b_{ij} = a_{ij} \leqq 0$ for $i \neq j$, and with $\det(b_{ij}) = \det(a_{ij}) - \varepsilon M_{11}$, where M_{11} is the diagonal minor of (a_{ij}) obtained by deleting the first row and column. Thus $\det(b_{ij}) < 0$. We show that this is impossible; more generally, we show that if $b_{ij} \leqq 0$ for $i \neq j$; if every proper diagonal minor of (b_{ij}) is non-negative; and if $\lambda_1, \ldots, \lambda_k$ are non-negative real numbers, not all zero, such that $\sum_i \lambda_i b_{ij} \geqq 0$ for all j, then $\det(b_{ij}) \geqq 0$. For $k = 1$, we have $\lambda_1 b_{11} \geqq 0$, $\lambda_1 > 0$, from which $b_{11} \geqq 0$. Thus we may assume $k > 1$ and the assertion proved for smaller matrices. We may further assume some $\sum_i \lambda_i b_{ij} > 0$, since otherwise (b_{ij}) is singular. Thus the assertion holds for k by k matrices satisfying the hypotheses when the number r of strict inequalities is zero. Now let $r \geqq 1$, and assume the assertion proved for k by k matrices satisfying a set of k similar inequalities, fewer then r of which are strict. Then we may assume $\sum_i \lambda_i b_{ij} > 0$, $1 \leqq j \leqq r$; $\sum_i \lambda_i b_{ij} = 0$, $r < j \leqq k$. Thus $\lambda_1 > 0$, and letting $c_{ij} = b_{ij}$, $(i, j) \neq (1, 1)$, $c_{11} = -\lambda_1^{-1} \sum_{i=2}^{k} \lambda_i b_{i1}$, we have $b_{11} > c_{11} \geqq 0$. Now consider the matrix (c_{ij}).

If all proper diagonal minors of (c_{ij}) are non-negative, we note first that $\det(b_{ij}) - \det(c_{ij}) = (b_{11} - c_{11}) M_{11} \geqq 0$. We further have $\sum_i \lambda_{ij} c_{ij} = \sum_i \lambda_i b_{ij}$ for $2 \leqq j \leqq k$, $\sum_i \lambda_i c_{i1} = 0$; thus (c_{ij}) satisfies the hypotheses for (b_{ij}) with only $r - 1$ of the associated inequalities being strict, and hence by induction $\det(c_{ij}) \geqq 0$, from which $\det(b_{ij}) \geqq 0$.

On the other hand, if some proper diagonal minor of (c_{ij}) is negative, let q be the minimal size for such a minor, and let (d_{ij}) be a q-rowed diagonal submatrix of negative determinant. Then all proper diagonal minors of (d_{ij}) are non-negative, $q < k$, and the fact that $\lambda_i c_{ij} \leqq 0$ for $i \neq j$ implies that those λ_i with indices used in forming (d_{ij}) satisfy inequalities of the form $\sum_i \mu_i d_{ij} \geqq 0$, $1 \leqq j \leqq q$. Thus $\det(d_{ij}) \geqq 0$ by induction, and the proof of Lemma 3 is complete.

Lemma II.5.4 (Mills). Let $\alpha_1, \ldots, \alpha_k$ be non-zero roots. Suppose there is a k by k real matrix (a_{ij}) such that: $A_{ij} \leqq a_{ij} \leqq 0$, $i \neq j$; $a_{ii} = 2$ $(= A_{ii})$, $1 \leqq i \leqq k$; every proper diagonal minor of (a_{ij}) is positive, while $\det(a_{ij}) \leqq 0$. Then $(A_{ij}) = (a_{ij})$ and $\det(A_{ij}) = 0$. There are positive integers m_i such that $\sum_i m_i A_{ij} = 0$, $1 \leqq j \leqq k$; for each such solution, there is a sequence of roots $\alpha_{i_1}, \alpha_{i_1} + \alpha_{i_2}, \ldots,$

$\alpha_{i_1} + \cdots + \alpha_{i_r} = \sum m_i \alpha_i$, $1 \leqq i_j \leqq k$, and $\sum m_i \alpha_i = 0$; furthermore, the roots $\alpha_1, \ldots, \alpha_k$ are linearly dependent.

That c) of our definition of fundamental system holds for the set of simple roots is seen from the lemma by taking $a_{ij} = A_{ij}$ to begin with; then the existence of a sequence $\alpha_{i_1}, \ldots, \alpha_{i_1} + \cdots + \alpha_{i_r} = 0$ and the positivity of all roots in the sequence contradicts trichotomy. To prove the lemma, let $\beta_0 = 0$, $\beta_1 = \alpha_{i_1}$, $\beta_2 = \alpha_{i_1} + \alpha_{i_2}, \ldots, \beta_r = \alpha_{i_1} + \cdots + \alpha_{i_r}$ be any sequence of roots as above; we show that for some j, $\beta_r + \alpha_j$ is a root. For otherwise we have $A_{\beta_r, \alpha_j} \geqq 0$ for all j, from which by Lemma 4.4, $\sum_{i=1}^{k} s_i^{(r)} A_{ij} \geqq 0$ for all j, where $s_i^{(r)}$ is the number of times the root α_i has been added in forming the sequence leading to β_r. Then by Lemma 3, $\sum_i s_i^{(r)} a_{ij} = 0 = \sum_i s_i^{(r)} A_{ij}$, all j; that is, $A_{\beta_r, \alpha_j} = 0$ for all j. Now if $r = 0$, $\beta_0 + \alpha_1 = \alpha_1$ is a root, while if $r > 0$, let $j = i_r$; then $\beta_r - \alpha_j$ is a root and $A_{\beta_r, \alpha_j} = 0$, from which $\beta_r + \alpha_j$ is a root. As in the proof of Lemma 4.5 it follows that there are non-negative integers m_i, not all zero, such that $\sum_i m_i A_{ij} = 0$, $1 \leqq j \leqq k$, thus that $\sum_i m_i a_{ij} \geqq 0$, $1 \leqq j \leqq k$, and that $\det(A_{ij}) = 0$. Lemma 3 applies as before to give $\sum_i m_i a_{ij} = 0$, all j. By the fact that all proper diagonal minors of (a_{ij}) are non-zero, we must have all $m_i > 0$, so that $a_{ij} > A_{ij}$ for some values of i, j would give $\sum_i m_i a_{ij} > \sum_i m_i A_{ij} = 0$ for this value of j. Thus $(A_{ij}) = (a_{ij})$.

To complete the proof, it suffices to construct the sequence of roots associated with every positive integral solution of $\sum m_i A_{ij} = 0$; for then the root $\alpha = \sum m_i \alpha_i$ will have its double $(= \sum (2m_i) \alpha_i)$ as a root as well, so must be zero. Since we can find a solution where not all m_i are divisible by the characteristic, it follows that the α_i are dependent. Thus let (m_i) be a positive integral solution. We claim that a sequence of roots of the type considered can be so chosen that if $r \; (= \sum_i s_i^{(r)}) \leqq \sum_i m_i$, then $s_i^{(r)} \leqq m_i$ for all i; in particular, since there is a root β_r in our sequence for $r = \sum m_i$, we have for this root $s_i^{(r)} = m_i$ for all i, hence $\beta_r = \sum m_i \alpha_i$. For $r = 0$ or 1, the assertion is evident. Thus let $1 < r \leqq \sum m_i$, and assume $\beta_{r-1} = \alpha_{i_1} + \cdots + \alpha_{i_{r-1}}$ has been found in our sequence satisfying $s_i^{(r-1)} \leqq m_i$ for all i. If $A_{\beta_{r-1}, \alpha_j} < 0$ for some j, then $\beta_{r-1} + \alpha_j$ is a root; if then $s_j^{(r-1)} = m_j$, we have $0 > A_{\beta_{r-1}, \alpha_j} = \sum_i s_i^{(r-1)} A_{ij} \geqq \sum_i m_i A_{ij} = 0$, a contradiction; thus in this case $\beta_r = \beta_{r-1} + \alpha_j$ has $s_j^{(r)} \leqq m_j$, therefore $s_i^{(r)} \leqq m_i$ for all i. If $A_{\beta_{r-1}, \alpha_j} \geqq 0$ for all j, then by Lemma 3 $A_{\beta_{r-1}, \alpha_j} = 0$ for all j, and $\sum_i s_i^{(r-1)} A_{ij} = 0$, all j. Now (A_{ij}) has rank $k - 1$, from which it follows that $(s_1^{(r-1)}, \ldots, s_k^{(r-1)})$ is a positive rational multiple of (m_i), with sum $r - 1 < \sum m_i$; hence $s_i^{(r-1)} < m_i$ for all i. From this and the fact that $\beta_{r-1} - \alpha_{i_{r-1}}$ is a root, it follows that we can take $\beta_r = \beta_{r-1} + \alpha_{i_{r-1}}$. This completes the proof.

By our remarks following the definition of fundamental system of roots, we see that if a root α is a member of a sequence $\alpha_{i_1}, \ldots, \alpha_{i_1} + \cdots + \alpha_{i_m} = \alpha$ of roots, where the α_i are taken from a given fundamental system, then the length m of this sequence is uniquely determined. We shall say that α has *level* m, writing $L(\alpha) = m$. We set $L(\alpha) = -m$ if $L(-\alpha) = m > 0$, and $L(0) = 0$. We also see that $s_i(\alpha)$, the number of i_j equal to i in the sequence above, is uniquely determined, and that $m = L(\alpha) = \sum_i s_i(\alpha)$. For $L(\alpha) < 0$, we set $s_i(\alpha) = -s_i(-\alpha)$, and $s_i(0) = 0$; then for all α_j, $A_{\alpha,\alpha_j} = \sum_i s_i(\alpha) A_{ij}$. From the non-singularity of (A_{ij}) and from $A_{\alpha,\alpha_j} + A_{\beta,\alpha_j} = A_{\alpha+\beta,\alpha_j}$ if $\alpha, \beta, \alpha + \beta$ are roots, we have $s_i(\alpha + \beta) = s_i(\alpha) + s_i(\beta)$ under these circumstances. In particular, we have the

Lemma II.5.5. If α, β, and $\alpha + \beta$ are roots, then $L(\alpha + \beta) = L(\alpha) + L(\beta)$.

§ 6. Semisimplicity and simplicity

Let $\mathfrak{L}$ and $\mathfrak{H}$ be classical as in the preceding sections, and let $\Sigma = \{\alpha_1, \ldots, \alpha_n\}$ be a fundamental system of roots relative to $\mathfrak{H}$. We say that two roots $\alpha, \beta \in \Sigma$ *belong to the same component* if there are roots $\gamma_1, \ldots, \gamma_k$ from Σ such that $\alpha = \gamma_1$, $\beta = \gamma_k$, and $A_{\gamma_i,\gamma_{i+1}} \neq 0$, $1 \leqq i < k$. This relation is an equivalence, and the equivalence classes are the *components* of Σ. We say that Σ is *connected* if Σ consists of a single component.

Lemma II.6.1. Let α be a root, and let $\alpha_{i_1}, \alpha_{i_1} + \alpha_{i_2}, \ldots, \alpha_{i_1} + \cdots + \alpha_{i_m} = \alpha$ be a sequence of roots, all $\alpha_{i_j} \in \Sigma$. Then all α_{i_j} belong to a single component of Σ.

We argue by induction on the level of α, which is by hypothesis positive. If $L(\alpha) = 1$, then $m = 1$, $\alpha = \alpha_{i_1}$, and we are done. Thus we may assume $L(\alpha) > 1$, and that $\alpha_{i_1}, \ldots, \alpha_{i_{m-1}}$ all belong to a single component Σ' of Σ. Let $\alpha_{i_m} = \alpha_i \in \Sigma$; if $\alpha_i \notin \Sigma'$, then $A_{ij} = 0 = A_{ji}$ for all $\alpha_j \in \Sigma'$; thus $A_{\alpha-\alpha_i,\alpha_i} = 0$ by Lemma 4.4, from which $\alpha - 2\alpha_i$ is a root, $L(\alpha - 2\alpha_i) = L(\alpha) - 2 \neq 0$ since $2\alpha_i$ is not a root. Since $L(\alpha) \geqq 2$, $L(\alpha - 2\alpha_i) > 0$, so that there is a component Σ'' of Σ and a sequence of roots $\beta_{i_1}, \beta_{i_1} + \beta_{i_2}, \ldots, \beta_{i_1} + \cdots + \beta_{i_{m-2}} = \alpha - 2\alpha_i$, all $\beta_{i_j} \in \Sigma''$; hence $\beta_{i_1}, \beta_{i_1} + \beta_{i_2}, \ldots, \beta_{i_1} + \cdots + \beta_{i_{m-2}} +$

$+\alpha_i = \alpha - \alpha_i$ is a sequence yielding $\alpha - \alpha_i$. By the uniqueness of the multiplicity of each $\alpha_k \in \Sigma$ in such a sequence, we have $\Sigma'' = \Sigma'$, $\alpha_i \in \Sigma'$, and the lemma is proved.

Lemma II.6.2. Let Σ' be a component of Σ, and let $\mathfrak{S}$ be the subalgebra of $\mathfrak{L}$ generated by the subspaces $\mathfrak{L}_\alpha$, $\mathfrak{L}_{-\alpha}$, where $\alpha \in \Sigma'$. Then $\mathfrak{S}$ is an ideal in $\mathfrak{L}$, and $\mathfrak{S}$ is classical relative to the classical Cartan subalgebra $\mathfrak{H} \cap \mathfrak{S}$; the non-zero roots of $\mathfrak{S}$ relative to $\mathfrak{H} \cap \mathfrak{S}$ are the restrictions of those roots β of $\mathfrak{L}$ relative to $\mathfrak{H}$ which are members of root-sequences $\alpha_{i_1}, \alpha_{i_1} + \alpha_{i_2}, \ldots, \alpha_{i_1} + \cdots + \alpha_{i_m} = \beta$, where all $\alpha_{i_j} \in \Sigma'$, and of the negatives of such roots β. The restrictions of the roots $\alpha \in \Sigma'$ constitute a fundamental system of roots relative to $\mathfrak{S} \cap \mathfrak{H}$.

First we show that $\mathfrak{S}$ is the linear subspace $\mathfrak{T}$ of $\mathfrak{L}$ spanned by the $\mathfrak{L}_\beta$, $\mathfrak{L}_{-\beta}$, and $[\mathfrak{L}_\alpha \mathfrak{L}_{-\alpha}]$, where $\alpha \in \Sigma'$, and where β is as in the statement of the lemma. By Lemma 4.1, $\mathfrak{T} \subseteq \mathfrak{S}$. Thus it suffices to show that $\mathfrak{T}$ is a subalgebra; we show in fact that $\mathfrak{T}$ is an ideal. To do so, it suffices to show that $\mathfrak{T}$ is mapped into itself by every $\mathrm{ad}(e_\gamma)$, $\pm\gamma \in \Sigma$, $0 \neq e_\gamma \in \mathfrak{L}_\gamma$; for these elements generate $\mathfrak{L}$. First suppose $\gamma \notin \Sigma'$, $-\gamma \notin \Sigma'$, where one of $\pm\gamma \in \Sigma$. Then for each β as above it follows from $[\mathfrak{L}_\alpha e_\gamma] = 0 = [\mathfrak{L}_\alpha e_{-\gamma}]$ for all $\alpha \in \Sigma'$ that $[\mathfrak{L}_{\pm\beta} e_{\pm\gamma}] = 0$, as well as that $[[\mathfrak{L}_\alpha \mathfrak{L}_{-\alpha}] e_{\pm\gamma}] = 0$, thus that $[\mathfrak{T} e_\gamma] = 0$. Next suppose that $\gamma \in \Sigma'$, and let β be a root as above, $L(\beta) > 0$. If $\beta + \gamma$ is a root, then $\beta + \gamma$ is a member of an admissible root sequence $\ldots, \beta, \beta + \gamma$ as above, so that $[\mathfrak{L}_\beta e_\gamma] = \mathfrak{L}_{\beta+\gamma} \subseteq \mathfrak{T}$; if $\beta + \gamma$ is not a root, then $[\mathfrak{L}_\beta e_\gamma] = 0$. Finally, let $-\gamma \in \Sigma'$, β as before; we may assume that $\beta + \gamma$ is a root. Then either $\gamma = -\beta$, $[\mathfrak{L}_\gamma \mathfrak{L}_\beta] = [\mathfrak{L}_\gamma \mathfrak{L}_{-\gamma}] \subseteq \mathfrak{T}$, or $L(\beta + \gamma) = L(\beta) - 1 > 0$. We thus have a Σ-sequence of roots $\ldots, \beta + \gamma, \beta$; it follows from the uniqueness of multiplicities of summands from Σ that $\mathfrak{L}_{\beta+\gamma} \subseteq \mathfrak{T}$. That $\mathfrak{T}$ is an ideal is now immediate.

Thus $\mathfrak{S}$ is the sum of the $[\mathfrak{L}_\alpha \mathfrak{L}_{-\alpha}]$, $\alpha \in \Sigma'$, and of certain root-spaces for non-zero roots β. Hence $\mathfrak{S} \cap \mathfrak{H}$ is the span of these $[\mathfrak{L}_\alpha \mathfrak{L}_{-\alpha}]$, and none of the roots β can vanish on $\mathfrak{H} \cap \mathfrak{S}$ since this would imply $A_{\beta,\alpha} = 0$ for all $\alpha \in \Sigma'$, and would combine with Lemma 4.4 to contradict the positivity of the diagonal minor corresponding to Σ' in the matrix (A_{ij}) for Σ. Thus $\mathfrak{H} \cap \mathfrak{S}$ is a commutative Cartan subalgebra of $\mathfrak{S}$. Now let $\beta \neq 0$ be a root, $\mathfrak{L}_\beta \subseteq \mathfrak{S}$, and let $\delta \neq 0$ be any root; then either $\mathfrak{L}_\delta \subseteq \mathfrak{S}$ or $\pm\delta$ is a member of a Σ''-sequence of roots, where Σ'' is another component of Σ. In the latter case, it follows that $[\mathfrak{L}_\beta \mathfrak{L}_\delta] = 0 = [\mathfrak{L}_\beta \mathfrak{L}_{-\delta}]$; hence the root β is uniquely determined by its effect on $\mathfrak{H} \cap \mathfrak{S}$, since $\mathfrak{H} = \sum_{\gamma \neq 0} [\mathfrak{L}_\gamma \mathfrak{L}_{-\gamma}]$. Therefore all roots β with $\mathfrak{L}_\beta \subseteq \mathfrak{S}$ take on distinct values on $\mathfrak{H} \cap \mathfrak{S}$, and if $z \in \mathfrak{S}$ is central in $\mathfrak{S}$, then $z \in \mathfrak{H} \cap \mathfrak{S}$, $[z\, \mathfrak{L}_\delta] = 0$ for all δ as above, so that $\gamma(z) = 0$ for all roots γ, and $z = 0$. From these remarks we see that i), ii) and iii) a), b) of the definition (§ 3) are satisfied by $\mathfrak{S}$ relative to $\mathfrak{S} \cap \mathfrak{H}$. It also follows that $\mathfrak{H}$ is the direct sum of $\mathfrak{S} \cap \mathfrak{H}$ and $\mathfrak{H}' = \sum_{\delta} [\mathfrak{L}_\delta \, \mathfrak{L}_{-\delta}]$, where δ runs over all roots $\delta \neq 0$ such that $\mathfrak{L}_\delta \subsetneqq \mathfrak{S}$. Thus the restriction mapping from $\mathfrak{H}$ to $\mathfrak{S} \cap \mathfrak{H}$ of those elements of $\mathfrak{H}^*$ vanishing on $\mathfrak{H}'$ is a vector-space isomorphism; from this one sees that iii) c) is satisfied by $\mathfrak{S}$ relative to $\mathfrak{S} \cap \mathfrak{H}$, as well as that the restrictions to $\mathfrak{S} \cap \mathfrak{H}$ of the roots $\alpha \in \Sigma'$ constitute a fundamental system of roots with the same matrix of Cartan integers as when these functions are considered as roots of $\mathfrak{H}$.

Theorem II.6.1. Let $\mathfrak{L}$, $\mathfrak{H}$, Σ be as above, and let $\Sigma = \Sigma_1 \cup \cdots \cup \Sigma_r$ be the decomposition of Σ into components. Let $\mathfrak{S}_i$ be the subalgebra of $\mathfrak{L}$ generated by the $\mathfrak{L}_\alpha$, $\mathfrak{L}_{-\alpha}$, where $\alpha \in \Sigma_i$. Then $\mathfrak{S}_i$ is an ideal in $\mathfrak{L}$, a simple classical Lie algebra, and the restriction of Σ_i to the classical Cartan subalgebra $\mathfrak{H} \cap \mathfrak{S}_i$ is a fundamental system of roots with the same matrix of Cartan integers as that of Σ_i. $\mathfrak{L}$ is the direct sum of the $\mathfrak{S}_i$. Thus $\mathfrak{L}$ is simple if and only if Σ is connected.

Except for simplicity, the assertions about the individual $\mathfrak{S}_i$ have been proved in the proof of Lemma 2. From the proofs of Lemmas 1 and 2 we see that each root-space $\mathfrak{L}_\beta (\beta \neq 0)$ and each $[\mathfrak{L}_\beta \, \mathfrak{L}_{-\beta}]$ is contained in some $\mathfrak{S}_i$, hence that $\mathfrak{L}$ is contained in the sum of the $\mathfrak{S}_i$; we have also seen that $[\mathfrak{S}_i \, \mathfrak{S}_j] = 0$ if $i \neq j$ and that the center of $\mathfrak{S}_i$ is zero, from which it follows that the sum of the $\mathfrak{S}_i$ is direct. The remaining conclusions will be established once we show that the connectedness of Σ implies the simplicity of $\mathfrak{L}$. Thus assume $\Sigma = \{\alpha_1, \ldots, \alpha_n\}$ is connected, and let $\mathfrak{J}$ be an ideal in $\mathfrak{L}$, $\mathfrak{J} \neq 0$. Let $0 \neq x \in \mathfrak{J}$, and let $x = h + \sum_{\beta} x_\beta$, $x_\beta \in \mathfrak{L}_\beta$; we further assume that the number of non-zero terms in this representation is minimized by x among all non-zero elements of $\mathfrak{J}$. Now if $h (\in \mathfrak{H}) \neq 0$, we have $x = h$; for otherwise, if $x_\beta \neq 0$, we choose $h' \in \mathfrak{H}$ such that $\beta(h') \neq 0$ and have $[x\, h'] = \sum_{\beta} \beta(h')\, x_\beta \neq 0$ in $\mathfrak{J}$, with fewer non-zero terms than has x. If $h = 0$, and if $x_\beta \neq 0$, $x_\gamma \neq 0$ for $\beta \neq \gamma$, we choose $h \in \mathfrak{H}$ with $\beta(h) \neq \gamma(h)$, and we have $\beta(h)\, x - [x\, h] \in \mathfrak{J}$, $\beta(h)\, x - [x\, h] = \sum_{\delta} (\beta(h) - \delta(h))\, x_\delta \neq 0$, with fewer non-zero terms than has x. Thus either $x = h \in \mathfrak{H}$ or $x = x_\beta$, $\beta \neq 0$. If $x = h$, then $\alpha_i(h) \neq 0$ for some i (otherwise h is central), from which $\mathfrak{J}$ contains $[\mathfrak{L}_{\alpha_i} h] = \mathfrak{L}_{\alpha_i}$, as well as $\mathfrak{L}_{-\alpha_i}$, hence also $[\mathfrak{L}_{\alpha_i} \mathfrak{L}_{-\alpha_i}]$. If $\alpha_j \in \Sigma$ has $A_{ij} \neq 0$, then α_j does not vanish on $[\mathfrak{L}_{\alpha_i} \mathfrak{L}_{-\alpha_i}]$, so that $\mathfrak{J}$ contains $\mathfrak{L}_{\pm\alpha_j}$. Using the connectedness of Σ, we see that all $\mathfrak{L}_{\pm\alpha_j}$, $\alpha_j \in \Sigma$, are contained in $\mathfrak{J}$, hence that $\mathfrak{J} = \mathfrak{L}$. If $x = x_\beta$, then $\mathfrak{J}$ contains $[\mathfrak{L}_\beta \, \mathfrak{L}_{-\beta}] \neq 0$, a one-dimensional subspace of $\mathfrak{H}$, so that $\mathfrak{J} = \mathfrak{L}$ as above.

§ 7. Determination of the fundamental systems

We consider certain integral n by n matrices (a_{ij}), with all $a_{ii} = 2$, denoted and defined as follows:

$$
(7)\quad \begin{cases}
A_n (n \geqq 1): & a_{ij} = -1 \text{ if } |i-j| = 1,\ a_{ij} = 0 \text{ otherwise;} \\
B_n (n \geqq 2): & a_{ij} = -1 \text{ if } |i-j| = 1, \text{ except that} \\
& \quad a_{n-1,n} = -2;\ a_{ij} = 0 \text{ otherwise;} \\
C_n (n \geqq 3): & a_{ij} = -1 \text{ if } |i-j| = 1, \text{ except that} \\
& \quad a_{n,n-1} = -2;\ a_{ij} = 0 \text{ otherwise;} \\
D_n (n \geqq 4): & a_{ij} = -1 \text{ if } |i-j| = 1,\ 1 \leqq i,j \leqq n-1; \\
& \quad a_{n-2\ n} = -1 = a_{n,n-2},\ a_{ij} = 0 \\
& \quad \text{otherwise;} \\
E_n (n = 6,7,8): & a_{ij} = -1 \text{ if } |i-j| = 1,\ 1 \leqq i,j \leqq n-1; \\
& \quad a_{n-3,n} = -1 = a_{n,n-3},\ a_{ij} = 0 \\
& \quad \text{otherwise;} \\
F_n (n = 4): & a_{ij} = -1 \text{ if } |i-j| = 1, \text{ except that} \\
& \quad a_{23} = -2;\ a_{ij} = 0 \text{ otherwise;} \\
G_n (n = 2): & a_{12} = -3,\ a_{21} = -1.
\end{cases}
$$

In each case, every diagonal submatrix consists of diagonal blocks which are matrices in the list (7), and of zeros elsewhere. Expansion by minors gives by induction the following determinants: A_n: $n+1$; B_n, C_n, E_7: 2; D_n: 4; E_6: 3; E_8, F_4, G_2: 1. Hence all diagonal minors in the matrices (7) are positive.

We also consider a second list of n by n integral matrices (a_{ij}), with $a_{ii} = 2$, denoted and defined as follows:

$$
(8)\quad \begin{cases}
I_2 (n = 2): & a_{12} = -2 = a_{21}; \\
G_3 (n = 3): & a_{12} = -1 = a_{21},\ a_{13} = 0 = a_{31};\ \text{either} \\
& \quad a_{23} = -3,\ a_{32} = -1, \text{ or } a_{32} = -3, \\
& \quad a_{23} = -1; \\
H_n (n \geqq 3): & a_{ij} = -1 \text{ if } |i-j| = 1,\ a_{1n} = -1 = a_{n1}; \\
& \quad a_{ij} = 0 \text{ otherwise;} \\
BB_n (n \geqq 3): & a_{ij} = -1 \text{ if } |i-j| = 1, \text{ except that} \\
& \quad a_{21} = -2 = a_{n-1,n};\ a_{ij} = 0 \text{ otherwise;} \\
CB_n (n \geqq 3): & a_{ij} = -1 \text{ if } |i-j| = 1, \text{ except that} \\
& \quad a_{12} = -2 = a_{n-1,n};\ a_{ij} = 0 \text{ otherwise;}
\end{cases}
$$

$$(8)\quad \begin{cases} CC_n(n \geqq 3): & a_{ij} = -1 \text{ if } |i-j| = 1, \text{ except that} \\ & a_{12} = -2 = a_{n,n-1};\ a_{ij} = 0 \text{ otherwise;} \\ BD_n(n \geqq 4): & a_{ij} = -1 \text{ if } |i-j| = 1,\ 1 \leqq i,j \leqq n-1, \\ & \text{except that } a_{21} = -2;\ a_{n,n-2} = -1 \\ & = a_{n-2,n};\ a_{ij} = 0 \text{ otherwise;} \\ CD_n(n \geqq 4): & a_{ij} = -1 \text{ if } |i-j| = 1,\ 1 \leqq i,j \leqq n-1 \\ & \text{except that } a_{12} = -2;\ a_{n,n-2} = -1, \\ & = a_{n-2,n};\ a_{ij} = 0 \text{ otherwise;} \\ F_5(n=5): & a_{ij} = -1 \text{ if } |i-j| = 1, \text{ except that} \\ & a_{34} = -2, \text{ or except that } a_{43} = -2; \\ & a_{ij} = 0 \text{ otherwise;} \\ DD_n(n \geqq 5): & a_{ij} = -1 \text{ if } |i-j| = 1,\ 2 \leqq i,j \leqq n-1; \\ & a_{13} = -1 = a_{31} = a_{n-2,n} = a_{n,n-2}; \\ & a_{ij} = 0 \text{ otherwise;} \\ J_7(n=7): & a_{ij} = -1 \text{ if } |i-j| = 1, \text{ except that } a_{56} = 0 \\ & = a_{65};\ a_{36} = -1 = a_{63};\ a_{ij} = 0 \\ & \text{otherwise;} \\ J_8(n=8): & a_{ij} = -1 \text{ if } |i-j| = 1,\ 1 \leqq i,j \leqq 7; \\ & a_{48} = -1 = a_{84};\ a_{ij} = 0 \text{ otherwise;} \\ E_9(n=9): & a_{ij} = -1 \text{ if } |i-j| = 1,\ 1 \leqq i,j \leqq 8; \\ & a_{69} = -1 = a_{96};\ a_{ij} = 0 \text{ otherwise.} \end{cases}$$

For the list (8), every proper diagonal submatrix consists of diagonal blocks, each similar by a permutation matrix to one of the list (7), and of zeros outside the diagonal blocks; hence all proper diagonal minors are positive. Expansion by minors using the values of the determinants of (7) shows that every matrix (8) is singular.

Lemma II.7.1. Let $\mathfrak{L}, \mathfrak{H}, \Sigma$ be as in earlier sections. Then there is no subset $\{\alpha_1, \ldots, \alpha_n\}$ of Σ such that for all $i \neq j$, $A_{ij} \leqq a_{ij}$, where the n by n matrix (a_{ij}) is in the list (8).

The lemma follows at once from Lemma 5.4, since that lemma yields $A_{ij} = a_{ij}$, all i, j, hence that the diagonal minor corresponding to $\alpha_1, \ldots, \alpha_n$ in the matrix (A_{ij}) of Σ is zero. This contradicts the definition of fundamental system.

Theorem II.7.1. If a fundamental system of roots Σ is connected, then upon suitable relabeling of its members Σ has one of the matrices (7).

For let $\alpha_1 \neq \alpha_2$ in Σ. Then if $A_{12} \leqq -2$ and $A_{21} \leqq -2$, we apply Lemma 1 to get a contradiction with $(a_{ij}) = I_2$. Hence if $A_{ij} \neq 0$ $(i \neq j)$, we have one of $A_{ij}, A_{ji} = -1$, and the other is among $-1, -2, -3$. If Σ consists of only one root, its matrix is evidently A_1; if Σ consists of two roots, these remarks and connectedness show that its matrix is one of A_2, B_2, G_2. Now suppose $A_{ij} = -3$ for some i, j. Then if Σ consists of at least 3 roots, there is a third root α_k with either $A_{ki} \neq 0$ or $A_{kj} \neq 0$; letting $k = 1$, $i = 2$, $j = 3$ in the former case, and $k = 1$, $j = 2$, $i = 3$ in the latter, we compare with G_3 and apply Lemma 1 to get a contradiction. Thus if $A_{ij} = -3$ for some i, j, Σ has the matrix G_2, and we may henceforth assume $A_{ij} \geqq -2$ for all i, j. Now let $A_{ij} = -2 = A_{qr}$, where $(q, r) \neq (i, j)$; that $(q, r) = (j, i)$ is impossible by the above. If only three distinct indices are involved, we may compare the matrix of the three roots with one of BB_3, CB_3, CC_3 to obtain a contradiction; if all four indices are distinct, then the connectedness of Σ enables us to connect one of the pairs $(\alpha_i, \alpha_q), (\alpha_i, \alpha_r), (\alpha_j, \alpha_q), (\alpha_j, \alpha_r)$ by a sequence of distinct roots $\alpha_{i_1}, \ldots, \alpha_{i_m}$ from Σ where only α_{i_1} and α_{i_m} are among the four given roots, and where $A_{i_k, i_{k+1}} < 0$, $1 \leqq k \leqq m$. We then compare the matrix of $\alpha_i, \alpha_j, \alpha_{i_2}, \ldots, \alpha_{i_{m-1}}, \alpha_q, \alpha_r$ with BB_{m+2}, CB_{m+2}, or CC_{m+2} and apply Lemma 1 to get a contradiction. Hence there is at most one pair (i, j) with $A_{ij} = -2$.

Assuming that such a pair is present, say $A_{ij} = -2$, we next suppose there are distinct indices p, q, r, s with $A_{pq}A_{pr}A_{ps} \neq 0$; an argument like that above yields a subset of Σ satisfying the condition $A_{km} \leqq a_{km}$ for all k, m, where (a_{km}) is a matrix BD_n or CD_n. Thus no such set of indices can be present, and for each index q there are at most two indices $r \neq q$ with $A_{qr} \neq 0$; for such indices $A_{qr} = -1$ unless $(q, r) = (i, j)$. Now suppose $A_{ik} \neq 0$ for some $k \neq i, j$, while $A_{jm} \neq 0$, $m \neq i, j$. Then $k = m$ is impossible by comparison with H_3; it follows that if Σ consists of 3 roots, the matrix of Σ (upon relabeling) is one of B_3, C_3. If $k \neq m$ above, and if either $A_{kr} \neq 0$ (or $A_{mr} \neq 0$) for some $r \neq i, k$ (or $r \neq j, m$), comparison with H_4 shows that $r \neq m$ (or $r \neq k$), and previous remarks show that r is a new index. The subset $\alpha_i, \alpha_j, \alpha_k, \alpha_m, \alpha_r$ now may be compared with F_5 of (8) to show the assumptions to be contradictory. Thus the assumption $A_{ik} \neq 0$, $A_{jm} \neq 0$ for k, m as above leads to the conclusion that our matrix is F_4. Finally, if we assume that $A_{ik} \neq 0$ implies $k = i$ or $k = j$, and if $A_{jk} \neq 0$ for some $k \neq i, j$, then there is only one such k and $A_{jk} = -1 = A_{kj}$. Repeating, $A_{km} \neq 0$ for $m \neq k, j$ implies that m is new, and $A_{km} = -1 = A_{mk}$; continuing, we see that Σ is C_n. On the other hand, if $A_{jk} \neq 0$ implies $k = i$ or j and if $A_{ik} \neq 0$ for some $k \neq i, j$, a similar argument shows that Σ is B_n.

We may thus assume $A_{ij} = -1$ whenever $A_{ij} \neq 0$, $i \neq j$. Comparison with H_n, DD_n, J_7, J_8, E_9 then yields as above that Σ is one of A_n, D_n, E_n of (7).

Theorem II.7.2. Assume that the fundamental system Σ is connected. Then Σ is linearly independent unless the matrix of Σ is A_n, where $n \equiv -1 \pmod{p}$, p being the characteristic. In the latter case, if the roots are labeled consistently with the matrix (a_{ij}) of (7), the roots $\alpha_1, \ldots, \alpha_{n-1}$ are linearly independent, and $\alpha_n = \alpha_1 + 2\alpha_2 + \cdots + (n-1)\alpha_{n-1}$.

For letting $h_i \in [\mathfrak{L}_{\alpha_i}, \mathfrak{L}_{-\alpha_i}]$ be such that $\alpha_i(h_i) = 2$, we see by Lemma 4.2 that $\alpha_j(h_i) = A_{ji}$ for all i, j. Thus the matrix $(\alpha_j(h_i))$ has non-zero determinant except for the case cited above, this by Th. 1 and the values of the determinants of (7). It follows that the α_j are linearly independent (hence a basis for $\mathfrak{H}^*$), as well as that the h_i are linearly independent (hence a basis for $\mathfrak{H}$) except in this case. In the exceptional case, the $(n-1)$-rowed diagonal minor corresponding to $\alpha_1, \ldots, \alpha_{n-1}$ of $(\alpha_j(h_i))$ has non-zero determinant, whereas that of $(\alpha_j(h_i))$ is zero. For $2 \leqq i \leqq n-1$, $(\alpha_1 + 2\alpha_2 + \cdots + n\alpha_n)(h_i) = -(i-1) + 2i - (i+1) = 0$, and $(\alpha_1 + 2\alpha_2 + \cdots + n\alpha_n)(h_1) = 2 - 2 = 0$; also $(\alpha_1 + 2\alpha_2 + \cdots + n\alpha_n)(h_n) = 2n - (n-1) = n + 1 = 0$, so that $\alpha_1 + 2\alpha_2 + \cdots + n\alpha_n$ vanishes on the generators $h_1, \ldots, h_n$ (see proof of Lemma 6.2) for the linear space $\mathfrak{H}$, hence is zero. The assertion follows. Similarly, evaluation of all roots α_i at $h_1 + 2h_2 + \cdots + n\,h_n$ shows that $h_n = h_1 + 2h_2 + \cdots + (n-1)\,h_{n-1}$. Thus $h_1, \ldots, h_{n-1}$ form a basis for $\mathfrak{H}$.

§ 8. Existence of isomorphisms

Theorem II.8.1. Let $\mathfrak{L}$ and $\mathfrak{M}$ be classical Lie algebras over $\mathfrak{F}$ with classical Cartan subalgebras $\mathfrak{H}$ and $\mathfrak{K}$, respectively. Let Σ and T be fundamental systems of roots relative to these Cartan subalgebras, $\Sigma = \{\alpha_1, \ldots, \alpha_n\}$, $T = \{\beta_1, \ldots, \beta_n\}$, and suppose that $A_{\alpha_i, \alpha_j} = A_{\beta_i, \beta_j}$ for all i, j. Then $\mathfrak{L}$ and $\mathfrak{M}$ are isomorphic; in fact, if we choose $0 \neq e_{\pm\alpha_i} \in \mathfrak{L}_{\pm\alpha_i}$, $0 \neq e_{\pm\beta_i} \in \mathfrak{M}_{\pm\beta_i}$, such that $\alpha_i([e_{\alpha_i}, e_{-\alpha_i}]) = 2 = \beta_i([e_{\beta_i}, e_{-\beta_i}])$, there is a unique isomorphism of $\mathfrak{L}$ onto $\mathfrak{M}$ sending $e_{\pm\alpha_i}$ onto $e_{\pm\beta_i}$.

We first show there is an isomorphism of the dual space $\mathfrak{H}^*$ of $\mathfrak{H}$ onto $\mathfrak{K}^*$ which maps roots onto roots and preserves levels. In view of the results of § 7, the only relations of linear dependence among roots in Σ (or T) come in components whose matrix is A_r, $r \equiv -1 \pmod{p}$, and here the r-th root is a uniquely determined combination of the remaining $r-1$ roots, these being linearly independent. It follows that there is an isomorphism τ of $\mathfrak{H}^*$ onto $\mathfrak{K}^*$ sending α_i onto β_i for each i. Thus it suffices to show that if α is a root relative to $\mathfrak{H}$, $L(\alpha) > 0$, then $\tau(\alpha)$ is a root relative to $\mathfrak{K}$, and $L(\tau(\alpha)) = L(\alpha)$, $A_{\tau(\alpha), \beta_i} = A_{\alpha, \alpha_i}$, all i. We proceed by induction on $L(\alpha)$, which we may assume to exceed one. Thus let $L(\alpha) > 1$, so that $\alpha = \gamma + \alpha_i$ for some α_i. Assume j is the smallest positive integer such that $\alpha - (j+1)\alpha_i$ is not a root. Then $1 \leqq j \leqq 3$, and since $\alpha \neq \alpha_i, 2\alpha_i, 3\alpha_i$, $L(\alpha) > L(\alpha - j\alpha_i) > 0$, $L(\alpha - j\alpha_i) = L(\alpha) - j$. Hence $\delta = \tau(\alpha - j\alpha_i)$ is a root, by induction, $L(\delta) = L(\alpha) - j$, and $A_{\delta,\beta_k} = A_{\alpha - j\alpha_i, \alpha_k} = A_{\alpha,\alpha_k} - jA_{\alpha_i,\alpha_k}$ for all k, $1 \leqq k \leqq n$. In particular, $A_{\delta,\beta_i} = A_{\alpha - j\alpha_i,\alpha_i} \leqq -j$, so that $\delta + j\beta_i = \tau(\alpha)$ is a root β, and $L(\beta) = L(\delta) + j = L(\alpha)$. Moreover, $A_{\beta,\beta_k} = A_{\delta,\beta_k} + jA_{\beta_i,\beta_k} = A_{\alpha,\alpha_k} - jA_{\alpha_i,\alpha_k} + jA_{\alpha_i,\alpha_k} = A_{\alpha,\alpha_k}$ for all k, and the assertion is proved.

The relations of dependence among the $h_i \in [\mathfrak{L}_{\alpha_i}\, \mathfrak{L}_{-\alpha_i}]$, $\alpha_i(h_i) = 2$, show that we have an isomorphism of $\mathfrak{H}$ onto $\mathfrak{K}$ mapping h_i onto $k_i \in \mathfrak{K}$ (k_i being similarly defined for the β_i). Now let $e_{\pm\alpha_i} \in \mathfrak{L}_{\pm\alpha_i}$, $e_{\pm\beta_i} \in \mathfrak{M}_{\pm\beta_i}$ be such that $[e_{\alpha_i}\, e_{-\alpha_i}] = h_i$, $[e_{\beta_i}\, e_{-\beta_i}] = k_i$. Denoting by $\mathfrak{L}_r$ the subspace of $\mathfrak{L}$ spanned by all $\mathfrak{L}_\alpha$, $0 \leqq |L(\alpha)| \leqq r$, and by $\mathfrak{M}_r$ the analogous subspace of $\mathfrak{M}$, we see that there is a linear isomorphism σ_1 of $\mathfrak{L}_1$ onto $\mathfrak{M}_1$ sending h_i onto k_i, $e_{\pm\alpha_i}$ onto $e_{\pm\beta_i}$. Furthermore, σ_1 has the property that if $w_\alpha \in \mathfrak{L}_\alpha$, $w_\gamma \in \mathfrak{L}_\gamma$, $(\alpha, \gamma \neq 0)$, h, $h' \in \mathfrak{H}$ are in $\mathfrak{L}_1$, then $\sigma_1([h\,h'])$ $= [\sigma_1(h), \sigma_1(h')]$, $\sigma_1([w_\alpha\, h]) = \alpha(h)\,\sigma_1(w_\alpha) = (\tau(\alpha))\,(\sigma_1(h))\,\sigma_1(w_\alpha) = [\sigma_1(w_\alpha), \sigma_1(h)]$, and that either $[w_\alpha\, w_\gamma] \notin \mathfrak{L}_1$ or $\sigma_1([w_\alpha\, w_\gamma]) = [\sigma_1(w_\alpha)\,\sigma_1(w_\gamma)]$. To see the last, consider the case $L(\alpha) = 1$, the case $L(\alpha) = -1$ being obtained by analogy: thus $w_\alpha = \lambda\, e_{\alpha_i}$, $w_\gamma = \mu\, e_{\alpha_j}$ or $\mu\, e_{-\alpha_j}$. In the former case $[w_\alpha\, w_\gamma] \in \mathfrak{L}_1$ only if $\alpha_i + \alpha_j$ is not a root, in which case $[w_\alpha\, w_\gamma] = 0$ $= [\sigma_1(w_\alpha)\,\sigma_1(w_\gamma)]$. If $w_\gamma = \mu\, e_{-\alpha_j}$, then $[w_\alpha\, w_\gamma] = 0 = [\sigma_1(w_\alpha)\,\sigma_1(w_\gamma)]$ unless $i = j$, and then $\sigma_1([w_\alpha\, w_\gamma]) = \lambda\,\mu\,\sigma_1(h_i) = \lambda\,\mu\, k_i = [\sigma_1(w_\alpha)\,\sigma_1(w_\gamma)]$. Thus we have shown that for $r = 1$, if w_α, w_γ are root-vectors contained in $\mathfrak{L}_r$, such that $[w_\alpha\, w_\gamma] \in \mathfrak{L}_r$, then $\sigma_r([w_\alpha\, w_\gamma]) = [\sigma_r(w_\alpha)\,\sigma_r(w_\gamma)]$, where σ_r is a linear isomorphism of $\mathfrak{L}_r$ onto $\mathfrak{M}_r$ mapping $\mathfrak{L}_\alpha$ onto $\mathfrak{M}_{\tau(\alpha)}$. We prove by induction the existence of such a mapping σ_r for all $r > 0$; for large r, σ_r will be the desired isomorphism. The uniqueness follows at once since the $e_{\pm\alpha_i}$ generate $\mathfrak{L}$.

Thus let $r \geqq 1$, and suppose that $\sigma_r\colon \mathfrak{L}_r \to \mathfrak{M}_r$ has been defined with the above properties. If α is a root of $\mathfrak{L}$, $L(\alpha) = r + 1$, we have $\alpha = \gamma + \alpha_i$ for some i, where γ is a root, $L(\gamma) = r$; hence if $0 \neq e_\gamma \in \mathfrak{L}_\gamma$, $e_\alpha = [e_\gamma\, e_{\alpha_i}]$ is a basis for $\mathfrak{L}_\alpha$, and we set $\sigma(e_\alpha) = [\sigma_r(e_\gamma), e_{\beta_i}] \neq 0$ in $\mathfrak{M}_{\tau(\alpha)}$ since $\tau(\alpha) = \tau(\gamma) + \beta_i$ is a root. Similarly, we set $\sigma([e_{-\gamma}\, e_{-\alpha_i}])$ $= [\sigma_r(e_{-\gamma}), e_{-\beta_i}]$. We define σ_{r+1} to be the unique linear mapping of $\mathfrak{L}_{r+1}$ onto $\mathfrak{M}_{r+1}$ agreeing with σ_r on $\mathfrak{L}_r$ and with σ on the vectors $e_{\pm\alpha}$ chosen as above. Writing σ for σ_{r+1}, it remains only to show that if w_α, w_β are root-vectors in $\mathfrak{L}_{r+1}$ such that $[w_\alpha\, w_\beta] \in \mathfrak{L}_{r+1}$, then $\sigma([w_\alpha\, w_\beta]) = [\sigma(w_\alpha)\,\sigma(w_\beta)]$. We may further assume that one of α, β, $\alpha + \beta$, if all are roots, has level $\pm(r+1)$, and that if $\alpha + \beta$ is not a root one of α, β has level $\pm(r+1)$; in the former case, we may assume $|L(\alpha+\beta)| \leqq r+1$. If (say) $\beta = 0$, $w_\beta = h \in \mathfrak{H}$, $[w_\alpha w_\beta] = \alpha(h) w_\alpha$, from which since $(\tau(\alpha))(\sigma(h)) = \alpha(h)$ we see that $\sigma([w_\alpha w_\beta]) = [\sigma(w_\alpha)\sigma(w_\beta)]$. Thus we may assume $\alpha, \beta \neq 0$. Now suppose $L(\alpha) = r+1$, $L(\beta) > 0$; if $\alpha+\beta$ is a root, then $L(\alpha+\beta) > r+1$, so that we may assume that $\alpha+\beta$ is not a root. Then $\tau(\alpha)+\tau(\beta)$ is not a root, so that $[\sigma(w_\alpha)\sigma(w_\beta)] = 0 = \sigma([w_\alpha w_\beta])$. Thus we may assume that if $|L(\alpha)|$ or $|L(\beta)|$ is $r+1$, then $L(\alpha)L(\beta) < 0$.

Now let $L(\alpha) = r+1$ (the case $L(\alpha) = -(r+1)$ is analogous), $L(\beta) < 0$, and we may assume that $\alpha+\beta$ is a root. If $L(\beta) \geqq -r$, then $0 < L(\alpha+\beta) \leqq r$, so that $\mathfrak{L}_\beta, \mathfrak{L}_{\alpha+\beta} \subseteqq \mathfrak{L}_r$. Then $w_\alpha = \lambda[e_\gamma e_{\alpha_i}]$ as above, so that $[w_\alpha w_\beta] = \lambda[[e_\gamma w_\beta] e_{\alpha_i}] + \lambda[e_\gamma[e_{\alpha_i} w_\beta]]$. But now $\mathfrak{L}_\delta \subseteqq \mathfrak{L}_r$ for $\delta = \beta, \gamma, \beta+\gamma, \alpha_i, \beta+\alpha_i, \beta+\gamma+\alpha_i$, so that by induction we see that $\sigma([w_\alpha w_\beta]) = \lambda[[\sigma(e_\gamma)\sigma(w_\beta)]\sigma(e_{\alpha_i})] + \lambda[\sigma(e_\alpha)[\sigma(e_{\alpha_i}), \sigma(w_\beta)]] = \lambda[[\sigma(e_\gamma)\sigma(e_{\alpha_i})]\sigma(w_\beta)] = [\sigma(w_\alpha)\sigma(w_\beta)]$. If $L(\beta) = -(r+1)$, then $\alpha+\beta$ is not a root unless $\beta = -\alpha$, in which case $w_{-\alpha} = \mu[e_{-\gamma} e_{-\alpha_i}]$; substitution and use of the induction hypotheses as above again gives

$$\sigma([w_\alpha w_\beta]) = [\sigma(w_\alpha)\sigma(w_\beta)].$$

We finally consider the case $L(\alpha)L(\beta) > 0$, $|L(\alpha+\beta)| = r+1$, for which it will suffice by analogy to assume $L(\alpha), L(\beta) > 0$. Let δ, α_i be the previously chosen roots for $\alpha+\beta$, i.e., $\alpha+\beta = \delta+\alpha_i$, and $e_{\alpha+\beta} = [e_\delta e_{\alpha_i}]$ has had its image under σ defined as $[\sigma(e_\delta)\sigma(e_{\alpha_i})]$. Now we have $[w_\alpha w_\beta] = \lambda e_{\alpha+\beta}$, $0 \neq \lambda \in \mathfrak{F}$, and $[[w_\alpha w_\beta] e_{-\alpha_i}] = \lambda[e_{\alpha+\beta} e_{-\alpha_i}] \neq 0$ in $\mathfrak{L}_\delta$. Then $\sigma([w_\alpha w_\beta]) = \lambda[\sigma(e_\delta)\sigma(e_{\alpha_i})]$, and by the preceding, $\sigma([[w_\alpha w_\beta] e_{-\alpha_i}]) = \lambda[\sigma(e_{\alpha+\beta})\sigma(e_{-\alpha_i})]$. We also have $[\sigma(w_\alpha)\sigma(w_\beta)] = \mu\sigma(e_{\alpha+\beta})$, some $\mu \neq 0$, and $\mu[\sigma(e_{\alpha+\beta})\sigma(e_{-\alpha_i})] = [[\sigma(w_\alpha)\sigma(e_{-\alpha_i})]\sigma(w_\beta)] + [\sigma(w_\alpha)[\sigma(w_\beta)\sigma(e_{-\alpha_i})]]$, which by induction is equal to $\sigma([[w_\alpha e_{-\alpha_i}] w_\beta]) + \sigma([w_\alpha[w_\beta e_{-\alpha_i}]]) = \sigma([[w_\alpha w_\beta] e_{-\alpha_i}])$. Hence $\mu = \lambda$, which completes the inductive step, and with it the proof of the theorem.

§ 9. The Weyl group

If we have classical Lie algebras $\mathfrak{L}_1, \mathfrak{L}_2$ over fields $\mathfrak{F}_1, \mathfrak{F}_2$, with classical Cartan subalgebras $\mathfrak{H}_1, \mathfrak{H}_2$ and fundamental systems of roots $\Sigma_1 = \{\alpha_1, \ldots, \alpha_n\}$ and $\Sigma_2 = \{\beta_1, \ldots, \beta_n\}$ such that $A_{\beta_i, \beta_j} = A_{\alpha_i, \alpha_j}$ for all i, j, then the argument of the proof of Th. 8.1 can be used to prove that there is a one-one mapping τ of the set of roots relative to $\mathfrak{H}_1$ onto that relative to $\mathfrak{H}_2$, such that: $\tau(\alpha_i) = \beta_i$ for all i; $\tau(-\alpha) = -\tau(\alpha)$ for all roots α relative to $\mathfrak{H}_1$; $A_{\tau(\alpha), \beta_i} = A_{\alpha, \alpha_i}$ for all roots α and all i; $L(\tau(\alpha)) = L(\alpha)$ for all roots α; $\tau(\alpha+\alpha') = \tau(\alpha)+\tau(\alpha')$ if α, α', and $\alpha+\alpha'$ are roots relative to $\mathfrak{H}_1$, and $\tau(\alpha)+\tau(\alpha')$ is a root if and only if $\alpha+\alpha'$ is a root. For $L(\alpha) > 0$, one proceeds by induction, where one may assume $L(\alpha) > 1$, $1 \leqq k \leqq 3$, $\gamma = \alpha - k\alpha_i$ a root, $\gamma - \alpha_i$

not a root, to define $\tau(\alpha) = \tau(\gamma) + k\beta_i$; for $L(\alpha) \leqq 0$, one defines $\tau(\alpha) = -\tau(-\alpha)$; that the mapping has the properties claimed is then verified exactly as in the cited proof (see, e.g., [357]).

With $\mathfrak{L}, \mathfrak{H}, \Sigma$ as above, we denote by $\mathfrak{W}$ the group of linear transformations of the dual space $\mathfrak{H}^*$ generated by the mappings $S_i : \varphi S_i = \varphi - \varphi(h_i)\alpha_i$ $(\alpha_i \in \Sigma)$, where $h_i \in [\mathfrak{L}_{\alpha_i} \mathfrak{L}_{-\alpha_i}]$, $\alpha_i(h_i) = 2$. If α is a root, then $\alpha(h_i)\alpha_i = A_{\alpha,\alpha_i}\alpha_i$ by Lemma 4.2, so that $\alpha S_i = \alpha - A_{\alpha,\alpha_i}\alpha_i$, which is always a root. Thus $\mathfrak{W}$ permutes the roots, and since Σ contains a basis for $\mathfrak{H}^*$, each $S \in \mathfrak{W}$ is determined by its effect on Σ. It follows that $\mathfrak{W}$ is a finite group, which we shall call the *Weyl group* defined by $\mathfrak{L}, \mathfrak{H}, \Sigma$, and sometimes write $\mathfrak{W}(\mathfrak{L}, \mathfrak{H}, \Sigma)$. (It is in fact independent of Σ, as will be seen below.) Now let $\mathfrak{L}_1, \mathfrak{H}_1, \Sigma_1$ and $\mathfrak{L}_2, \mathfrak{H}_2, \Sigma_2$ be as in the preceding paragraph; let $S_i^{(1)}$, and $S_i^{(2)}$ be the corresponding generators of the Weyl groups $\mathfrak{W}_1 = \mathfrak{W}(\mathfrak{L}_1, \mathfrak{H}_1, \Sigma_1)$ and $\mathfrak{W}_2 = \mathfrak{W}(\mathfrak{L}_2, \mathfrak{H}_2, \Sigma_2)$. From our first paragraph it follows that $\tau(\alpha) S_i^{(2)} = \tau(\alpha S_i^{(1)})$ for all roots α relative to $\mathfrak{H}_1$ and for $1 \leqq i \leqq n$. Thus $\tau(\alpha S_{i_1}^{(1)} \ldots S_{i_k}^{(1)}) = \tau(\alpha) S_{i_1}^{(2)} \ldots S_{i_k}^{(2)}$ for all roots α and all sets of indices $i_1, \ldots, i_k$; it follows that there a group-isomorphism $S \to S'$ of $\mathfrak{W}_1$ onto $\mathfrak{W}_2$ such that $S_i^{(1)\prime} = S_i^{(2)}$ for all i, and such that if α is a root relative to $\mathfrak{H}_1$, then $\tau(\alpha S) = \tau(\alpha) S'$ for all $S \in \mathfrak{W}_1$. From the fact that both τ and elements of the Weyl group are one-one additive mappings on roots, we see that if α, β are roots relative to $\mathfrak{H}_1$, then $A_{\tau(\alpha),\tau(\beta)} = A_{\alpha,\beta}$ and $A_{\alpha S,\beta S} = A_{\alpha,\beta}$ for all $S \in \mathfrak{W}_1$.

§ 10. Existence of the classical algebras

In case the ground field is the complex field $\boldsymbol{C}$, our "classical Lie algebras" are the semisimple Lie algebras; this follows by the results of § 6 and the well-known fact that every semisimple Lie algebra over an algebraically closed field of characteristic zero satisfies our axioms (e.g., see [64, 128, 234]). Now it is known, and has been established in various ways by KILLING [250], CARTAN [53], WITT [416], and HARISH-CHANDRA [166] (see also [234, Chap. 7]), that for each matrix of rational integers which decomposes into matrices $A_n, B_n, C_n, D_n, G_2, F_4, E_6, E_7, E_8$ of § 7, there is a semisimple Lie algebra $\mathfrak{L}_C$ over $\boldsymbol{C}$ and a fundamental system of roots relative to a (necessarily classical) Cartan subalgebra $\mathfrak{H}_C$ whose matrix is the given one. The procedure of CHEVALLEY described in § 3 may now be applied to give a Lie algebra over $\mathfrak{F}$. A given fundamental system Σ_C relative to $\mathfrak{H}_C$ induces in $\mathfrak{L}/\mathfrak{Z}$ a fundamental system relative to $\mathfrak{H}/\mathfrak{Z}$, with the same Cartan integers as has Σ_C [94, 355, 381]. The center $\mathfrak{Z}$ of $\mathfrak{L}$ is zero except when one of the simple summands of $\mathfrak{L}_C$ is of type A_n, where $n \equiv -1 \pmod{p}$, p being the characteristic of $\mathfrak{F}$. In this case, each such summand has a one-dimensional center spanned by the element $h_1 + 2h_2 + \cdots + n h_n$, where the $h_i \in \mathfrak{L}_C$ are chosen from $[(\mathfrak{L}_C)_{\alpha_i} (\mathfrak{L}_C)_{-\alpha_i}]$ with $\alpha_i(h_i) = 2$ as in § 7, and are in fact in $\mathfrak{L}_Z$ [76, 94, 370].

Thus each admissible matrix (i.e., composed of diagonal blocks $A_n, \ldots, E_8$, with zeros elsewhere) is the matrix of Cartan integers of a classical Lie algebra (viz., $\mathfrak{L}/\mathfrak{Z}$) over $\mathfrak{F}$ determined by a fundamental system of roots. Taking into account the isomorphism Th. 8.1, we can complete the proof that there is a one-one correspondence between admissible matrices of rational integers and isomorphism classes of classical Lie algebras over $\mathfrak{F}$ (characteristic $\neq 2, 3$) by proving that if $\mathfrak{L}_1, \mathfrak{H}_1, \Sigma_1$ and $\mathfrak{L}_2, \mathfrak{H}_2, \Sigma_2$ are as above, and if $\mathfrak{L}_1$ and $\mathfrak{L}_2$ are isomorphic as Lie algebras over $\mathfrak{F}$, then a suitable reordering of the roots in Σ_2 yields the Cartan matrix of Σ_1. Since the notions of classical Cartan subalgebra, Cartan integers, and fundamental system of roots are preserved under isomorphism, we may assume $\mathfrak{L}_1 = \mathfrak{L}_2$. Thus the proof may be completed by showing the existence of an automorphism of $\mathfrak{L}_1$ mapping $\mathfrak{H}_1$ onto $\mathfrak{H}_2$ (hence $\mathfrak{H}_1 = \mathfrak{H}_2$ may be assumed), and then of an automorphism mapping $\mathfrak{H}_1$ onto itself and Σ_1 onto Σ_2 (in the mapping induced on $\mathfrak{H}_1^*$). This will be done in the next chapter (Th. III.4.1 and Lemma III.1.2). Anticipating that result, we have the following theorem, which may be called the "classification theorem" for classical Lie algebras:

Theorem II.10.1. Let $\mathfrak{F}$ be a field of characteristic $\neq 2, 3$. Then there is a one-one correspondence between isomorphism classes of classical Lie algebras over $\mathfrak{F}$ and admissible matrices, the latter being determined up to rearrangement of the diagonal blocks of types $A_n, \ldots, E_8$. The correspondence assigns to the classical Lie algebra $\mathfrak{L}$ the matrix of Cartan integers of a fundamental system Σ of roots relative to a classical Cartan subalgebra $\mathfrak{H}$, the elements of the separate components of Σ being suitably rearranged.

By the fact that there exists a classical complex Lie algebra for each admissible Cartan matrix, and by the correspondence of roots and Weyl groups over different fields, as established in § 9, we may deduce the following properties of roots and the Weyl group from the complex case:

Theorem II.10.2. Let $\mathfrak{L}$ be a classical Lie algebra, $\mathfrak{H}$ a classical Cartan subalgebra, Σ a fundamental system of roots. Then:

a) If $\alpha \neq 0$ is a root, and if $h \in [\mathfrak{L}_\alpha \mathfrak{L}_{-\alpha}]$ has $\alpha(h) = 2$, then the mapping $S_\alpha : \mathfrak{H}^* \to \mathfrak{H}^*$ defined by $\varphi S_\alpha = \varphi - \varphi(h)\,\alpha$ is in $\mathfrak{W} = \mathfrak{W}(\mathfrak{L}, \mathfrak{H}, \Sigma)$; hence $\mathfrak{W}$ is independent of the choice of fundamental system Σ of roots relative to $\mathfrak{H}$.

b) If $\alpha \neq 0$ is a root, then there is an $S \in \mathfrak{W}$ such that $\alpha S \in \Sigma$. If T is a second fundamental system of roots, there is a unique $S \in \mathfrak{W}$ such that $\Sigma S = T$.

c) If Σ is connected (i.e., if $\mathfrak{L}$ is simple), then all non-zero roots are conjugate under $\mathfrak{W}$ unless Σ is of type B, C, F or G, in which cases there are two conjugate classes of roots under $\mathfrak{W}$, with representatives α_1 and α_n, in the notation of § 7.

Except for a), it is clear that it suffices to know the assertions in the complex case. The mapping S_α of a) sends a root β into $\beta - \beta(h)\,\alpha = \beta - A_{\beta,\alpha}\,\alpha$, by Lemma 4.2. Since the Cartan integers are preserved under the mapping τ of § 9 and since S_α is determined by its effect on the roots, we may also reduce a) to the complex case. But for the complex field, all the results are well known: for a) and b) one may see [234, pp. 240—242]; also cf. [76, p. 21], where a proof of c) is given.

If $\mathfrak{L}, \mathfrak{H}, \Sigma$ are as above over $\mathfrak{F}$, and if $\mathfrak{K}$ is any extension field of $\mathfrak{F}$, it is immediate that $\mathfrak{H}_\mathfrak{K}$ is a classical Cartan subalgebra of $\mathfrak{L}_\mathfrak{K}$ and that the roots of $\mathfrak{L}_\mathfrak{K}$ relative to $\mathfrak{H}_\mathfrak{K}$ are the unique $\mathfrak{K}$-linear extensions of the roots of $\mathfrak{L}$ relative to $\mathfrak{H}$. Thus the roots of Σ may be regarded as the restrictions to $\mathfrak{H}$ of a fundamental system relative to $\mathfrak{H}_\mathfrak{K}$, having the same Cartan integers. In particular, $\mathfrak{L}_\mathfrak{K}$ is simple if $\mathfrak{L}$ is, since this amounts to saying that Σ is connected.

Finally, it may be of interest to comment as to what is gained in generality by considering classical Lie algebras rather than Lie algebras with non-singular Killing form or one of the generalizations of this condition which prefaced our introduction of the notion of classical Lie algebra. If $\mathfrak{L}_C$ is a simple Lie algebra over C, then it is known that the discriminant of the Killing form of $\mathfrak{L}_C$ relative to Chevalley's basis is divisible by all primes $p > 2, 3$ dividing: $n+1$, if $\mathfrak{L}_C$ is of type A_n or C_n; $2n-1$, if $\mathfrak{L}_C$ is of type B_n; $n-1$, if $\mathfrak{L}_C$ is of type D_n; 5, if $\mathfrak{L}_C$ is of type E_8; and by no other such primes (cf. [128, 355]). It follows that the algebra $\mathfrak{L}$ has non-singular Killing form if and only if the characteristic of $\mathfrak{F}$ does not divide the appropriate integer in the list above. When $\mathfrak{L}_C$ is of type A_n, and $\mathfrak{F}$ is of prime characteristic p, where $n \equiv -1 \pmod{p}$, $\mathfrak{L}$ is not classical, while $\mathfrak{L}/\mathfrak{Z}$ is classical and simple; in this case the Killing form of $\mathfrak{L}/\mathfrak{Z}$ is zero (as is, in all other singular cases, that of the simple algebra $\mathfrak{L}$). Thus all classical Lie algebras with non-singular Killing form (in case $\mathfrak{F}$ is perfect, all Lie algebras with non-singular Killing form and a Cartan subalgebra $\mathfrak{H}$ such that $\mathfrak{F}$ contains all characteristic roots of each ad h, $h \in \mathfrak{H}$) have been determined when the characteristic is not 2 or 3; they are direct sums of the simple algebras with this property.

If one requires only the existence of a representation with non-singular trace form, Block [35] has shown that such representations fail to exist for algebras of type A_n, $n \equiv -1 \pmod{p}$. On the other hand, one may easily show that the Lie algebra of $2n$ by $2n$ skew-symplectic matrices over $\mathfrak{F}$ is a classical simple Lie algebra of type C_n

if $n \geqq 3$, and that the ordinary trace form $\mathrm{Tr}(XY)$ is non-singular for $p \neq 2, 3$. Similarly, the Lie algebra of linear transformations of a vector space over $\mathfrak{F}$ of dimension $2n-1$ (resp. $2n+2$) which are skew with respect to a non-singular symmetric bilinear form of maximal Witt index is classical simple of type B_{n-1} (resp. D_{n+1}) if $n \geqq 3$, and the ordinary trace form is non-singular. When $p = 5$, the question of the existence of a non-singular trace form on the classical algebra E_8 appears to be open. The assertions on B_n, C_n, D_n above are verified in [216, 234, 353].

If the characteristic of $\mathfrak{F}$ is a prime $p > 2, 3$, there is a basis for a classical Lie algebra consisting of root-vectors e_α ($\alpha \neq 0$), and of elements $h_i \in [\mathfrak{L}_{\alpha_i}, \mathfrak{L}_{-\alpha_i}]$ with $\alpha_i(h_i) = 2$. It follows that $(\mathrm{ad}\, e_\alpha)^p = 0$ and, by Lemma 4.2 and the commutativity of the Cartan subalgebra $\mathfrak{H}$, that $\mathrm{ad}\, h_i$ is semisimple with all its characteristic roots in the prime field, hence that $(\mathrm{ad}\, h_i)^p = \mathrm{ad}\, h_i$. Hence, by Th. I.3.3, there is a unique structure of restricted Lie algebra on $\mathfrak{L}$ in which $e_\alpha^{[p]} = 0$, $h_i^{[p]} = h_i$. If $\mathfrak{L}$ is of type B, C, D, then $\mathfrak{L}$ is simple, and the trace form of the preceding paragraph is that of a restricted representation.

§ 11. Generalizations of the theory

The axioms of § 3 for a classical Lie algebra represent only one of the many sets of properties of complex semisimple Lie algebras which might be chosen as axiomatic for a class of Lie algebras over a general field. Another choice may very well lead to another class of algebras; an example of such a choice is that of Block [38], who replaced iii) b) and c) by the requirements that each $\mathfrak{L}_\alpha$, $\alpha \neq 0$, have dimension one, and that $\alpha([\mathfrak{L}_\alpha \mathfrak{L}_{-\alpha}]) \neq 0$. He then concluded that $\mathfrak{L}$ is a direct sum of simple algebras satisfying the same hypotheses (analogous to Th. 6.1), and that these simple algebras either satisfy iii) c) (hence are classical), or the condition that the Cartan subalgebra $\mathfrak{H}$ has dimension one. Over a perfect field of characteristic $p > 5$, he further showed that the non-classical simple algebras of this type are certain algebras of dimension p^n which we shall define in Chapter V, § 4, C, b. Some additional investigations on simple Lie algebras having Cartan subalgebras of dimension one or two have been carried out by Kaplansky [247], who obtained a determination of: 1) all simple Lie algebras over algebraically closed fields having a regular element u such that all characteristic roots of $\mathrm{ad}\, u$ are in the prime field, the root 0 having multiplicity one; 2) all simple restricted Lie algebras over an algebraically closed field, having a one-dimensional Cartan subalgebra defined by a regular element. In these cases, Kaplansky assumed characteristic $p > 3$; for $p = 2, 3$, he also obtained some sharp results. For non-classical algebras $\mathfrak{L}$ of dimension greater than three over algebraically closed fields with $p > 3$, having a Cartan subalgebra $\mathfrak{H}$ of dimension one such that the pairing of $\mathfrak{L}_\alpha \times \mathfrak{L}_{-\alpha}$ into $\mathfrak{H}$: $(x_\alpha, x_{-\alpha}) \rightarrow [x_\alpha\, x_{-\alpha}]$ is nonsingular for each root $\alpha \neq 0$, BLOCK [36] showed that only the algebras of dimension p^n of Chapter V, § 4, C, b result.

Rather than relaxing the axioms of § 3, one may choose to relax the conditions of § 1. For example, one has the result of BOREL and MOSTOW [43] in characteristic zero that a linear Lie algebra $\mathfrak{L}$ with nonsingular trace form is the direct sum of its center and a semisimple algebra. Over algebraically closed fields of characteristic $p > 3$, BLOCK and ZASSENHAUS [39] have shown that the same hypotheses lead to the conclusion that $\mathfrak{L}$ is the direct sum of ideals which are either abelian, classical simple, or isomorphic with a Lie algebra $\mathfrak{E}(\mathfrak{B})$, where the dimension of $\mathfrak{B}$ is divisible by p. One may also replace the requirement of a nonsingular trace form by requiring only a nonsingular symmetric associative bilinear form. In this degree of generality, few conclusive results have been obtained; however, ZASSENHAUS [424] has considered the case where the algebra $\mathfrak{L}$ is of the form $\mathfrak{M}/\mathfrak{M}^\perp$, where $\mathfrak{M}$ is a linear Lie algebra, $\mathfrak{M}^\perp$ the radical of its trace form, and where the form on $\mathfrak{L}$ is that induced by the trace form of $\mathfrak{M}$. He has called such a form a *quotient trace form*. Over algebraically closed fields of characteristic $p > 3$, a full description of Lie algebras having a quotient trace form has been given by ZASSENHAUS [424] and BLOCK [35, 37]. In particular, the simple ones are classical, and all classical simple algebras (except perhaps for E_8 when $p = 5$) admit a quotient trace form.

Chapter III

Automorphisms of the Classical Algebras

The automorphism groups of classical Lie algebras, in the sense of the previous chapter, have been studied for the four "great classes" A—D by JACOBSON [216], considering the most natural realizations of these algebras. A unified approach has been made by the author [357], substituting certain combinations of algebraic operations for the exponential functions used in the fundamental work of GANTMACHER [145] in the complex case. Where the author's results are incomplete (in case the ground field is not algebraically closed), they have been completed by STEINBERG [381]. Indeed, STEINBERG is able to deal with characteristics 2 and 3 as well, since he obtains his Lie algebras by Chevalley's process (Chap. II, § 3) from a complex semisimple Lie algebra. We reproduce here the results of STEINBERG, restricted to the case of classical algebras in our sense, as well as giving essentially Chevalley's results on the general structure of the groups of CHEVALLEY, when regarded as subgroups of the automorphism groups of classical algebras. Finally, we give interpretations for these results in terms of the natural realizations for types A—D, as well as for the exceptional algebras.

§ 1. The Chevalley groups

Let $\mathfrak{L}$ be a classical Lie algebra over $\mathfrak{F}$ (characteristic $\neq 2, 3$), $\mathfrak{H}$ a classical Cartan subalgebra, $\mathfrak{L} = \mathfrak{H} + \sum_{\alpha} \mathfrak{L}_\alpha$ the Cartan decomposition relative to $\mathfrak{H}$. Let $\alpha_1, \ldots, \alpha_r$ be a fundamental system of roots relative to $\mathfrak{H}$, and let h_i be that unique element of $[\mathfrak{L}_{\alpha_i} \mathfrak{L}_{-\alpha_i}]$ such that $\alpha_i(h_i) = 2$. For each root $\alpha \neq 0$, let $0 \neq e_\alpha \in \mathfrak{L}_\alpha$, so that e_α is a basis for $\mathfrak{L}_\alpha$, and the $e_{\pm\alpha_i}$ generate $\mathfrak{L}$. We call a root $\alpha \neq 0$ *positive* if the level, $L(\alpha)$, defined as in Chapter II, § 5, satisfies $L(\alpha) > 0$, otherwise *negative* (then $L(\alpha) < 0$). We speak of an *admissible ordering* of the roots as one in which $L(\alpha) < L(\beta)$ for roots α, β implies $\alpha < \beta$, $\alpha < \beta$ implies $-\beta < -\alpha$. Such (linear) orderings clearly exist.

By Th. II.8.1, we have for every r-tuple $(\mu) = (\mu_1, \ldots, \mu_r)$ of elements from $\mathfrak{F}^*$ a unique automorphism $D = D(\mu)$ of $\mathfrak{L}$ mapping

e_{α_i} onto $\mu_i e_{\alpha_i}$, $e_{-\alpha_i}$ onto $\mu_i^{-1} e_{-\alpha_i}$ for each i; D leaves fixed the elements of $\mathfrak{H}$, and every automorphism of $\mathfrak{L}$ leaving $\mathfrak{H}$ fixed has the form $D(\mu)$ for some (μ). Evidently $D(\mu) D(\nu) = D(\lambda)$, where $(\lambda) = (\mu_1 \nu_1, \ldots, \mu_r \nu_r)$, so that the group H of automorphisms of $\mathfrak{L}$ leaving $\mathfrak{H}$ fixed is isomorphic to the direct product of r copies of $\mathfrak{F}^*$.

Now let $\alpha \neq 0$ be a root. By Lemma II.3.4, $(\operatorname{ad} x_\alpha)^4 = 0$ for $x_\alpha \in \mathfrak{L}_\alpha$, so that $E(x_\alpha) = \exp(\operatorname{ad} x_\alpha)$ is a non-singular (unipotent) linear transformation of $\mathfrak{L}$. We further have $E(0) = I$, $E(x_\alpha + y_\alpha) = E(x_\alpha) E(y_\alpha)$ by calculation with the exponentials (note that $[x_\alpha y_\alpha] \in \mathfrak{L}_{2\alpha} = 0$). Similar calculation for characteristics $\neq 5$, based on the fact that $\operatorname{ad} x_\alpha$ is a derivation, shows that $E(x_\alpha)$ is an automorphism of $\mathfrak{L}$. This assertion is valid for characteristic 5 as well, although more detailed information must be used in proving it in that case (cf. [357]). For fixed $\alpha \neq 0$, the mapping $\lambda \to E(\lambda e_\alpha)$ is an isomorphism of the additive group of $\mathfrak{F}$ into the automorphisms of $\mathfrak{L}$ (for $h \in \mathfrak{H}$, we have $hE(\lambda e_\alpha) = h - \alpha(h) \lambda e_\alpha$).

We denote by $\mathfrak{E}(\alpha)$ the group of all $E(x_\alpha)$, $x_\alpha \in \mathfrak{L}_\alpha$, and by G' the group of automorphisms of $\mathfrak{L}$ generated by all $\mathfrak{E}(\alpha)$, $\alpha \neq 0$ a root relative to $\mathfrak{H}$. Let G be the group of automorphisms of $\mathfrak{L}$ generated by G' and by H. If $D \in H$, $x_\alpha \in \mathfrak{L}_\alpha$, then $x_\alpha D \in \mathfrak{L}_\alpha$, and $D^{-1} E(x_\alpha) D = E(x_\alpha D)$; thus G' is an invariant subgroup of G, and $G = G'H = HG'$. Also, $G/G' = G'H/G' \cong H/G' \cap H$ is abelian, so that G' contains the commutator subgroup of G.

If $\{\alpha_1, \ldots, \alpha_r\}$ decomposes into components $\Pi_1, \ldots, \Pi_s$, so that $\mathfrak{L} = \mathfrak{L}_1 \oplus \cdots \oplus \mathfrak{L}_s$, the $\mathfrak{L}_i$ being simple ideals as in § II.6, we may regard H as the direct product $H_1 \times \cdots \times H_s$, where H_k is the subgroup of H leaving fixed all $e_{\alpha_i} \notin \mathfrak{L}_k$, and we may regard G' as the direct product $G_1' \times \cdots \times G_s'$, where G_k' is the subgroup generated by all $\mathfrak{E}(\alpha)$, $\mathfrak{L}_\alpha \subseteqq \mathfrak{L}_k$. Since H_j and G_k', as well as G_j' and G_k', commute elementwise for $j \neq k$, we have $G = G_1 \times \cdots \times G_s$, $G_k = G_k' H_k = H_k G_k'$, where G_k leaves fixed all $\mathfrak{L}_j$, $j \neq k$, and maps $\mathfrak{L}_k$ into itself. Then G_k may be identified with the group "G" constructed from the simple classical algebra $\mathfrak{L}_k$ relative to the Cartan subalgebra $\mathfrak{H}_k = \mathfrak{H} \cap \mathfrak{L}_k$.

Let $e_i \in \mathfrak{L}_{\alpha_i}$, $f_i \in \mathfrak{L}_{-\alpha_i}$, $[e_i f_i] = h_i$; let $\mathfrak{T}_i$ be the 3-dimensional subalgebra $\{e_i, f_i, h_i\}$. We study the adjoint representation of $\mathfrak{T}_i$ on $\mathfrak{L}$. First let $\alpha \neq 0$ be a root, $\alpha \neq \pm\alpha_i$; let $\alpha + \alpha_i, \ldots, \alpha + j\alpha_i = \beta$ be roots, $\alpha + (j+1)\alpha_i$ not a root, so that $x_\beta = x_\alpha (\operatorname{ad} e_i)^j \neq 0$ for $0 \neq x_\alpha \in \mathfrak{L}_\alpha$, while $[x_\beta e_i] = 0$. Then $j \leqq 3$, and $x_\beta (\operatorname{ad} f_i)^k \neq 0$, $x_\beta (\operatorname{ad} f_i)^{k+1} = 0$ holds for some k, $j \leqq k \leqq 3$ (cf. proof of Lemma II.2.2), and $x_\alpha = \lambda x_\beta (\operatorname{ad} f_i)^j$, for some λ. If $\alpha > 0$, all $\mathfrak{L}_\beta (\operatorname{ad} f_i)^r$ are of the form $\mathfrak{L}_\gamma$, $\gamma > 0$, and if $\alpha < 0$, these are root-spaces of negative roots. Also by the proof of Lemma II.2.2, the sum of these $\mathfrak{L}_\beta (\operatorname{ad} f_i)^r$ is the unique minimal $\mathfrak{T}_i$-invariant subspace of $\mathfrak{L}$ containing any root-space

$\mathfrak{L}_\gamma$ contained in it (it is in fact irreducible and may be described as the linear span of all $e_\gamma (\mathrm{ad}\, e_i)^q (\mathrm{ad}\, f_i)^r$ for each $\mathfrak{L}_\gamma$ it contains). Denoting by $\sum^{(i)} \mathfrak{L}_\alpha$ the sum of all $\mathfrak{L}_\alpha$ for $\alpha \neq 0, \pm\alpha_i$, we see that $\sum^{(i)} \mathfrak{L}_\alpha$ is the direct sum of $\mathfrak{T}_i$-invariant subspaces as described above, and that $\mathfrak{L}$ is the direct sum $\sum^{(i)} \mathfrak{L}_\alpha \oplus \mathfrak{T}_i \oplus \mathfrak{H}^{(i)}$, where $\mathfrak{H}^{(i)}$ is the subspace of $\mathfrak{H}$ annihilated by α_i, hence by all $\mathrm{ad}\, x$, $x \in \mathfrak{T}_i$. Those invariant summands (including $\mathfrak{T}_i$) not annihilated by all $\mathrm{ad}\, x$, $x \in \mathfrak{T}_i$, have bases $x_\beta, x_\beta (\mathrm{ad}\, f_i), \ldots, x_\beta (\mathrm{ad}\, f_i)^j$, $1 \leqq j \leqq 3$, where $[x_\beta e_i] = 0 = x_\beta (\mathrm{ad}\, f_i)^{j+1}$. As in Lemmas II.2.1, II.2.2, $x_\beta (\mathrm{ad}\, f_i)^k (\mathrm{ad}\, e_i) = -k(j-k+1)\, x_\beta (\mathrm{ad}\, f_i)^{k-1}$, $1 \leqq k \leqq j$, so that, relative to the above bases, the matrices of $\mathrm{ad}\, f_i$, $\mathrm{ad}\, e_i$ are as follows:

j	f_i	e_i
1	$\begin{pmatrix} 0 & 1 \\ 0 & 0 \end{pmatrix}$	$\begin{pmatrix} 0 & 0 \\ -1 & 0 \end{pmatrix}$
2	$\begin{pmatrix} 0 & 1 & 0 \\ 0 & 0 & 1 \\ 0 & 0 & 0 \end{pmatrix}$	$\begin{pmatrix} 0 & 0 & 0 \\ -2 & 0 & 0 \\ 0 & -2 & 0 \end{pmatrix}$
3	$\begin{pmatrix} 0 & 1 & 0 & 0 \\ 0 & 0 & 1 & 0 \\ 0 & 0 & 0 & 1 \\ 0 & 0 & 0 & 0 \end{pmatrix}$	$\begin{pmatrix} 0 & 0 & 0 & 0 \\ -3 & 0 & 0 & 0 \\ 0 & -4 & 0 & 0 \\ 0 & 0 & -3 & 0 \end{pmatrix}$

Now let $0 \neq \lambda \in \mathfrak{F}$, and let $A_i(\lambda) = E(\lambda f_i)\, E(\lambda^{-1} e_i)\, E(\lambda f_i)$. Relative to the bases as above, $A_i(\lambda)$ has matrix:

$$j = 1\colon \begin{pmatrix} 0 & \lambda \\ -\lambda^{-1} & 0 \end{pmatrix}; \quad j = 2\colon \begin{pmatrix} 0 & 0 & \frac{\lambda^2}{2} \\ 0 & -1 & 0 \\ 2\lambda^{-2} & 0 & 0 \end{pmatrix};$$

$$\text{(1)} \qquad j = 3\colon \begin{pmatrix} 0 & 0 & 0 & \frac{1}{3!}\lambda^3 \\ 0 & 0 & -\frac{\lambda}{2} & 0 \\ 0 & 2\lambda^{-1} & 0 & 0 \\ -6\lambda^{-3} & 0 & 0 & 0 \end{pmatrix}.$$

Moreover, $A_i(\lambda) = E(\lambda^{-1} e_i)\, E(\lambda f_i)\, E(\lambda^{-1} e_i)$.

Thus $A_i(\lambda)$ leaves fixed all $h \in \mathfrak{H}$ with $\alpha_i(h) = 0$ and maps h_i into $-h_i$ (case $j = 2$ above). The group of all automorphisms of $\mathfrak{H}$ has the contragredient representation $A \to A^{*-1}$ in the dual space $\mathfrak{H}^*$, where for $\varphi \in \mathfrak{H}^*$, φA^* is the linear function on $\mathfrak{H}$ sending h into $\varphi(hA)$. We write φ^A for φA^{*-1}, and note for $A = A_i(\lambda)$ that $\varphi^A(h) = \varphi(h)$ if $\alpha_i(h) = 0$, $\varphi^A(h_i) = \varphi(-h_i)$; thus $\varphi^A = \varphi - \varphi(h_i)\,\alpha_i = \varphi\, S_i$, where S_i is the Weyl reflection determined by α_i as in § II.9. Since the Weyl group $\mathfrak{W}$ is generated by the S_i, it follows that for each $S \in \mathfrak{W}$ we may choose an element $\omega(S) \in G'$ such that $\mathfrak{H}\,\omega(S) = \mathfrak{H}$, $\varphi^{\omega(S)} = \varphi\, S$ for all $\varphi \in \mathfrak{H}^*$. We assume one such choice, with $\omega(I) = I$, $\omega(S_i) = A_i(\lambda)$, to be fixed for the rest of the discussion. For $S, T \in \mathfrak{W}$, $\omega(ST)^{-1}\,\omega(S)\,\omega(T)$ induces the identity on $\mathfrak{H}^*$, hence on $\mathfrak{H}$: $\omega(S)\,\omega(T) = \omega(ST)\,D$, where $D \in G' \cap H$. For $\alpha \neq 0$, $S \in \mathfrak{W}$, one verifies easily that $\mathfrak{L}_\alpha\,\omega(S) = \mathfrak{L}_{\alpha S}$, hence that $\omega(S)^{-1}\,\mathfrak{E}(\alpha)\,\omega(S) = \mathfrak{E}(\alpha S)$. Also, $\omega(S)\,H = H\,\omega(S)$.

Finally, we note from (1) and the above that $D_i(\lambda) = A_i(\lambda)\,A_i(1) \in H$, and that for $j \neq i$, $e_{\alpha_j} D_i(\lambda) = (-A_{ji})!\,(-\lambda)^{A_{ji}}\, x_{\alpha_j S_i} A_i(1) = (-\lambda)^{A_{ji}}\, e_{\alpha_j}$. (For e_{α_j} is either fixed under $A_i(\lambda)$ or corresponds to the last row of one of the matrices (1), where $x_{\alpha_j S_i}$ corresponds to the first row.) When $j = i$, we have the 3 by 3 matrix of (1), with e_{α_i} corresponding to the first row: $e_{\alpha_i} D_i(\lambda) = \frac{\lambda^2}{2}\, x_{-\alpha_i} A_i(1) = \lambda^2\, e_{\alpha_i}$, where $x_{-\alpha_i}$ corresponds to the last row. Thus $D_i(\lambda) = D(\mu_1, \ldots, \mu_r)$, $\mu_j = (-\lambda)^{A_{ji}}$. Letting H' be the subgroup of H generated by the $D_i(\lambda)$, $1 \leq i \leq r$, $0 \neq \lambda \in \mathfrak{F}$, we have $H' \subseteqq G' \cap H$. We shall see later that $H' = G' \cap H$.

Lemma III.1.1. Let $\alpha \neq \beta$ be non-zero roots, and let $\mathfrak{F}$ have more than 5 elements; then there is $D \in H'$ such that e_α and e_β belong to distinct characteristic roots of D.

For let $\alpha = \sum m_i(\alpha)\,\alpha_i$, $\beta = \sum m_i(\beta)\,\alpha_i$, where $m_i(\gamma)$ are the uniquely determined rational integers such that $A_{\gamma, \alpha_j} = \sum_i m_i(\gamma)\, A_{ij}$ for all j; by Chapter II, § 5, all the m_i are either non-positive or all are non-negative, and e_α is a multiple of a product $[[\ldots[x\,y]\ldots]\,z]$, involving $|m_i|$ factors equal to f_i in the former case, m_i factors equal to e_i in the latter. Thus $e_\alpha D_j(\lambda_j) = \prod_i (-\lambda_j)^{A_{ij} m_i(\alpha)}\, e_\alpha = (-\lambda_j)^{A_{\alpha, \alpha_j}}\, e_\alpha$. Since $\alpha \neq \beta$, we have $A_{\alpha, \alpha_j} \neq A_{\beta, \alpha_j}$ for some j; since both of these are between -3 and 3, and since $\mathfrak{F}$ has more than 5 elements, the conclusion follows unless $|A_{\alpha, \alpha_j} - A_{\beta, \alpha_j}| = 6$ and $\mathfrak{F}$ has 7 elements; but then α_j is part of a system G_2, and if α_k is the other root in this system, we may assume $\alpha = \alpha_k + 3\alpha_j$ or $\alpha = -\alpha_k$, while $\beta = \alpha_k - 3\alpha_j$ or $\beta = \alpha_k$, respectively. The corresponding values for A_{α, α_k} are $-1, -2$; those for A_{β, α_k} are $1, 2$; the conclusion follows by using α_k in place of α_j.

In any case we have for each root $\alpha \neq 0$, $0 < |A_{\alpha, \alpha_j}| \leq 3$ for some j, from which $(-\lambda_j)^{A_{\alpha, \alpha_j}}$ takes on a value other than 1 as λ_j ranges over the non-zero elements of $\mathfrak{F}$. That is, we have an element $D \in H' \subseteqq G' \cap H$ such that $e_\alpha D = \varrho\, e_\alpha \neq e_\alpha$. Thus $D^{-1} E(\lambda\, e_\alpha)\, D$

$= E(\lambda\varrho e_\alpha)$, $D^{-1}E(\lambda e_\alpha)DE(\lambda e_\alpha)^{-1} = E(\lambda(\varrho - 1)e_\alpha)$. Letting λ run over $\mathfrak{F}$, we see that $\mathfrak{E}(\alpha)$ is contained in the commutator subgroup of G', hence that G' is its own commutator subgroup as well as that of G. The following lemma summarizes a number of our conclusions:

Lemma III.1.2. Let H be the group of automorphisms of $\mathfrak{L}$ leaving $\mathfrak{H}$ fixed, G' the group generated by the $\mathfrak{E}(\alpha)$, $\mathfrak{W}$ the Weyl group, $G = G'H$ the group generated by H and G'. Let H' be the subgroup of H generated by the $D_j(\lambda)$, $1 \leqq j \leqq r$, $0 \neq \lambda \in \mathfrak{F}$. Then $H' \subseteqq G' \cap H$, and G' is the commutator subgroup of G, and that of G' itself. If $\mathfrak{L}$ decomposes into a direct sum of simple ideals $\mathfrak{L}_i$, then G, H, G' decompose into direct products of the groups G_i, H_i, G_i' associated with these $\mathfrak{L}_i$ according to the corresponding decomposition of $\mathfrak{H}$ and of the fundamental system $\{\alpha_j\}$. For each $S \in \mathfrak{W}$, we have a canonically chosen $\omega(S) \in G'$ with $\mathfrak{H}\omega(S) = \mathfrak{H}$, $\mathfrak{L}_\alpha\omega(S) = \mathfrak{L}_{\alpha S}$ for all $\alpha \neq 0$, $\omega(I) = I$. Thus $\omega(S)^{-1}\mathfrak{E}(\alpha)\omega(S) = \mathfrak{E}(\alpha S)$, $\omega(S)^{-1}H\omega(S) = H$ for all roots $\alpha \neq 0$, all $S \in \mathfrak{W}$. If $A \in G$ satisfies $\mathfrak{H}A = \mathfrak{H}$, and if $\mathfrak{L}_\alpha A = \mathfrak{L}_{\alpha S}$ for all roots $\alpha \neq 0$, where $S \in \mathfrak{W}$, then $A = \omega(S)D$ for some $D \in H$; in particular, $\omega(S)\omega(T) = \omega(ST)D$ for $D \in H$, where $S, T \in \mathfrak{W}$.

Lemma III.1.3. If $\alpha = \pm\alpha_i$, $1 \leqq i \leqq r$, then $\mathfrak{E}(\alpha)\mathfrak{E}(-\alpha)\mathfrak{E}(\alpha) \subseteqq \subseteqq \{(\mathfrak{E}(\alpha)\omega(S_i)H'\mathfrak{E}(\alpha) \cup H'\mathfrak{E}(\alpha)) \cap \mathfrak{E}(-\alpha)H'\mathfrak{E}(\alpha)\mathfrak{E}(-\alpha)\}$.

Namely, we first let $\alpha = -\alpha_i$, $A = E(\lambda f_i)E(\mu e_i)E(\nu f_i)$. If $\mu = 0$, this element is in $\mathfrak{E}(\alpha)$, and we are done. If $\mu \neq 0$, then $A = E((\lambda - \mu^{-1})f_i)A_i(\mu^{-1})E((\nu - \mu^{-1})f_i)$, and if $\omega(S_i) = A_i(\varkappa)$, then one easily sees from (1) that $\omega(S_i)^{-1}A_i(\mu^{-1})$ fixes $\mathfrak{H}$ and sends e_j into $(-\mu\varkappa)^{A_{ji}}e_j$. That is, $A_i(\mu^{-1}) = \omega(S_i)D_i(\mu\varkappa) \in \omega(S_i)H'$, and $A \in \mathfrak{E}(\alpha)\omega(S_i)H'\mathfrak{E}(\alpha)$. If $\lambda\nu = 0$, then $A \in \mathfrak{E}(-\alpha)H'\mathfrak{E}(\alpha)\mathfrak{E}(-\alpha)$, and we are done; hence we may assume $\lambda\mu\nu \neq 0$. If $\lambda\mu = 1$ or if $\mu\nu = 1$, then $A = A_i(\lambda)E(\varrho f_i)$ or $A = E(\varrho f_i)A_i(\nu)$ for some ϱ, from which $A = E(\sigma e_i)A_i(\lambda)$ or $A = A_i(\nu)E(\sigma e_i)$, respectively, for some σ. But now $A_i(\varkappa) = E(\varkappa^{-1}e_i)E(\varkappa f_i)E(\varkappa^{-1}e_i)$ shows that $A \in \mathfrak{E}(-\alpha)\mathfrak{E}(\alpha)\mathfrak{E}(-\alpha) \subseteqq \mathfrak{E}(-\alpha)H'\mathfrak{E}(\alpha)\mathfrak{E}(-\alpha)$. Finally, if $\lambda\mu \neq 1 \neq \mu\nu$, then $A = E(\varrho f_i)A_i(\mu^{-1})E(\sigma f_i)$ for some $\varrho, \sigma \neq 0$, $A = E(\varrho f_i)E(\tau e_i)A_i(\lambda^{-1})$ for $\varrho, \tau \neq 0$, and $A = E(-\varrho^{-1}e_i) \times \times A_i(\varrho)E((\tau - \varrho^{-1})e_i)A_i(\lambda^{-1}) = E(-\varrho^{-1}e_i)A_i(\varrho)A_i(\lambda^{-1})E(\xi f_i) \in \in \mathfrak{E}(-\alpha)H'\mathfrak{E}(\alpha) \subseteqq \mathfrak{E}(-\alpha)H'\mathfrak{E}(\alpha)\mathfrak{E}(-\alpha)$. For $\alpha = \alpha_i$, the argument is symmetric.

Lemma III.1.4. G' is generated by the $\mathfrak{E}(\alpha)$, $\alpha = \pm\alpha_i$, $1 \leqq i \leqq r$.

For if G_0 is the group generated by these $\mathfrak{E}(\alpha)$, then $\omega(S_i) \in G_0$ for all i, so that if $S = S_{i_1} \ldots S_{i_k} \in \mathfrak{W}$, then $\sigma(S) = \omega(S_{i_1}) \ldots \omega(S_{i_k}) \in G_0$ has the same effect on $\mathfrak{H}$ as does $\omega(S)$, thus maps $\mathfrak{L}_{\alpha_j}$ onto $\mathfrak{L}_{\alpha_j S}$. Therefore $\mathfrak{E}(\alpha_j)$ and $\mathfrak{E}(\alpha_j S)$ are conjugate under G_0 for $1 \leqq j \leqq r$, $S \in \mathfrak{W}$. The assertion now follows by Th. II.10.2.b.

§ 2. The fundamental decomposition of G. Consequences

Let α, β be non-zero roots relative to $\mathfrak{H}$, $\beta \neq -\alpha$. If $\alpha + \beta$ is not a root, then $\mathfrak{E}(\alpha)$ and $\mathfrak{E}(\beta)$ commute elementwise. If $\alpha + \beta$ is a root, we have the following

Lemma III.2.1. Let $x_\alpha \in \mathfrak{L}_\alpha$, $x_\beta \in \mathfrak{L}_\beta$, $\beta \neq -\alpha$; then the commutator:

$$E(x_\alpha)^{-1} E(x_\beta)^{-1} E(x_\alpha) E(x_\beta) \tag{2}$$

is a product of elements $E(x_\gamma)$, where $\gamma \neq \alpha, \beta$ and $\mathfrak{L}_\gamma$ is contained in the subalgebra of $\mathfrak{L}$ generated by $\mathfrak{L}_\alpha$ and $\mathfrak{L}_\beta$.

Namely, the commutator in question is equal to

$$\begin{aligned} & E([x_\alpha\, x_\beta])\, E\left(\tfrac{1}{2}[[x_\alpha\, x_\beta]\, x_\beta]\right) E\left(\tfrac{1}{2}[[x_\alpha\, x_\beta]\, x_\alpha]\right) \times \\ & \times E\left(\tfrac{1}{3!}[[[x_\alpha\, x_\beta]\, x_\beta]\, x_\beta]\right) E\left(\tfrac{1}{3!}[[[x_\alpha\, x_\beta]\, x_\alpha]\, x_\alpha]\right) \times \\ & \times E\left(\tfrac{1}{3}\left[[[x_\alpha\, x_\beta]\, x_\alpha]\, x_\alpha]\, x_\beta\right]\right) E\left(-\tfrac{1}{3!}\left[[[x_\alpha\, x_\beta]\, x_\beta]\, x_\beta]\, x_\alpha\right]\right). \end{aligned} \tag{3}$$

For it is not hard to show that (2) and (3) have the same effect on each $h \in \mathfrak{H}$. Next we note that, relative to a suitable basis for $\mathfrak{L}$, both $\operatorname{ad} x_\alpha$ and $\operatorname{ad} x_\beta$ have matrices with zeros on and below the diagonal, hence that (2) and (3) have matrices with zeros below the diagonal and ones on the diagonal. Then the product of (3) by the inverse of (2) is in H and is unipotent, hence is the identity.

To see the existence of such a basis, it suffices to show that there is a fundamental system relative to $\mathfrak{H}$ such that both α and β are positive roots; for then one takes an ordering of the roots which is admissible relative to this fundamental system, and chooses a basis for $\mathfrak{L}$ consisting of bases for the root-spaces, proceeding from lowest to highest. Since $\operatorname{ad} x_\alpha$ and $\operatorname{ad} x_\beta$ raise levels of root-vectors, we have the desired form. Replacing β by $\beta - j\alpha$, where $\beta - (j+1)\alpha$ is not a root, we may assume $\beta - \alpha$ is not a root. If both α, β are positive, or if both are negative, relative to our fundamental system $\alpha_1, \ldots, \alpha_r$, then we are done. Thus we may assume $\alpha > 0$, $\beta < 0$, and it suffices to show there is an $S \in \mathfrak{W}$ such that both $\alpha S > 0$ and $\beta S > 0$. The following cases are exhaustive (cf. proof of Lemma II.7.1), and one checks the solutions S by proceeding in the order indicated:

Case 1: $A_{\alpha\beta} = 0$. Take $S = S_\beta$.

Case 2: $A_{\alpha\beta} = -3$. Take $S = S_\beta,\ S_\beta S_{\alpha+3\beta},\ S_\beta S_{\alpha+3\beta} S_{\alpha+2\beta},$

$S_\beta S_{\alpha+3\beta} S_{\alpha+2\beta} S_{2\alpha+3\beta}$ or $S_\beta S_{\alpha+3\beta} S_{\alpha+2\beta} S_{2\alpha+3\beta} S_{\alpha+\beta}$.

Case 3: $A_{\beta\alpha} = -3$. Take $S = S_\beta,\ S_\beta S_{\alpha+\beta},\ S_\beta S_{\alpha+\beta} S_{3\alpha+2\beta},$

$S_\beta S_{\alpha+\beta} S_{3\alpha+2\beta} S_{2\alpha+\beta}$ or $S_\beta S_{\alpha+\beta} S_{3\alpha+2\beta} S_{2\alpha+\beta} S_{3\alpha+\beta}$.

Case 4: $A_{\alpha\beta} = -2$. Take $S = S_\beta$, $S_\beta S_{\alpha+2\beta}$, or $S_\beta S_{\alpha+2\beta} S_{\alpha+\beta}$.

Case 5: $A_{\beta\alpha} = -2$. Take $S = S_\beta$, $S_\beta S_{\alpha+\beta}$, or $S_\beta S_{\alpha+\beta} S_{2\alpha+\beta}$.

Case 6: $A_{\alpha\beta} = -1 = A_{\beta\alpha}$. Take $S = S_\beta$ or $S_\beta S_{\alpha+\beta}$.

With respect to our fixed admissible ordering of the roots, let $\mathfrak{U}$ be the subgroup of G' generated by all $\mathfrak{E}(\alpha)$, $\alpha > 0$, $\mathfrak{V}$ the subgroup generated by all $\mathfrak{E}(\alpha)$, $\alpha < 0$. For each integer $k > 0$, let $\mathfrak{U}^{(k)}$ be the subgroup generated by all $\mathfrak{E}(\alpha)$, $L(\alpha) \geqq k$. For each $S \in \mathfrak{W}$, let $\mathfrak{U}_S$ be the subgroup of $\mathfrak{U}$ generated by all $\mathfrak{E}(\alpha)$ with $\alpha > 0$, $\alpha S < 0$; let $\mathfrak{U}'_S$ be the subgroup generated by all $\mathfrak{E}(\alpha)$ with $\alpha > 0$, $\alpha S > 0$. If $S = S_i$ is a fundamental Weyl reflection, we write $\mathfrak{U}_i$ resp. $\mathfrak{U}'_i$ for $\mathfrak{U}_S$ resp. $\mathfrak{U}'_S$, and have $\mathfrak{U}_i = \mathfrak{E}(\alpha_i)$.

Lemma III.2.2. If $k \geqq j$, $\mathfrak{U}^{(k)}$ is a normal subgroup of $\mathfrak{U}^{(j)}$, and the commutator group $(\mathfrak{U}^{(j)}, \mathfrak{U}^{(k)})$ is contained in $\mathfrak{U}^{(j+k)}$. If Γ is a subset of the positive roots such that α and β in Γ, $\alpha + \beta$ a root, implies $\alpha + \beta$ in Γ, let $\mathfrak{U}_\Gamma$ be the subgroup of $\mathfrak{U}$ generated by all $\mathfrak{E}(\alpha)$, $\alpha \in \Gamma$. Then every element of $\mathfrak{U}_\Gamma$ can be written in one and only one way as a product $E(x_{\beta_1}) \ldots E(x_{\beta_s})$, and in one and only one way as a product $E(x_{\beta_s}) \ldots E(x_{\beta_1})$, where $\Gamma = \{\beta_1, \ldots, \beta_s\}$ and where $\beta_1 < \cdots < \beta_s$. The analogous statement holds with $\mathfrak{U}$ replaced by $\mathfrak{V}$, "positive" by "negative".

For Lemma 1 shows that if $L(\alpha) \geqq j$, $L(\beta) \geqq k$, then $(E(x_\alpha), E(x_\beta)) \in \mathfrak{U}^{(j+k)}$; in particular, $E(x_\alpha)^{-1} \mathfrak{E}(\beta) E(x_\alpha) \subseteqq \mathfrak{U}^{(k)}$. Thus conjugation by generators for $\mathfrak{U}^{(j)}$ maps generators for $\mathfrak{U}^{(k)}$ into $\mathfrak{U}^{(k)}$, from which the normality of $\mathfrak{U}^{(k)}$ in $\mathfrak{U}^{(j)}$ follows. Now let $V \in \mathfrak{U}^{(k)}$, and let $x_\alpha \in \mathfrak{L}_\alpha$, $L(\alpha) \geqq j$; let $V = E(y_1) \ldots E(y_t)$, $y_i \in \mathfrak{L}_{\beta_i}$, $L(\beta_i) \geqq k$. Then $(E(x_\alpha), V) = E(x_\alpha)^{-1} E(y_t)^{-1} \ldots E(y_2)^{-1} (E(y_1), E(-x_\alpha)) E(x_\alpha) E(y_2) \ldots E(y_t)$; now $(E(y_1), E(-x_\alpha)) \in \mathfrak{U}^{(j+k)}$, which is normal in $\mathfrak{U}^{(j)}$, so that $(E(x_\alpha), V) \equiv (E(x_\alpha), E(y_2) \ldots E(y_t)) \bmod \mathfrak{U}^{(j+k)}$. It follows by induction on t that $(E(x_\alpha), V) \in \mathfrak{U}^{(j+k)}$ for all $V \in \mathfrak{U}^{(k)}$, $L(\alpha) \geqq j$. Repeating the argument with an induction on u shows that $(E(x_1) \ldots E(x_u), V) \in \mathfrak{U}^{(j+k)}$ for all $V \in \mathfrak{U}^{(k)}$, $x_i \in \mathfrak{L}_{\gamma_i}$, $L(\gamma_i) \geqq j$, hence $(\mathfrak{U}^{(j)}, \mathfrak{U}^{(k)}) \subseteqq \mathfrak{U}^{(j+k)}$. In particular, $\mathfrak{U}^{(k)}/\mathfrak{U}^{(k+1)}$ is commutative.

With Γ as in the statement, let $\mathfrak{U}_\Gamma^{(k)}$ be the subgroup of $\mathfrak{U}_\Gamma$ generated by all $\mathfrak{E}(\alpha)$, $\alpha \in \Gamma$, $L(\alpha) \geqq k$. For k maximal among levels of roots in Γ, it follows that $\mathfrak{U}_\Gamma^{(k)}$ is commutative, from which every element of $\mathfrak{U}_\Gamma^{(k)}$ has a representation of the first form with $L(\beta_i) = k$ for all i. We proceed downward by induction on k to show each element of $\mathfrak{U}_\Gamma^{(k)}$ has such a representation with $L(\beta_i) \geqq k$ for all β_i involved. Thus we may assume the result for $\mathfrak{U}_\Gamma^{(k+1)}$, and let $U \in \mathfrak{U}_\Gamma^{(k)}$. By the fact that $\mathfrak{U}_\Gamma^{(k)}/\mathfrak{U}_\Gamma^{(k+1)}$ is commutative, which follows as for $\mathfrak{U}^{(k)}/\mathfrak{U}^{(k+1)}$, a representation for U can be reduced to the form $E(x_{\gamma_1}) \ldots E(x_{\gamma_r}) V$,

$\gamma_1 < \cdots < \gamma_r$, $L(\gamma_i) = k$, $\gamma_i \in \Gamma$, $V \in \mathfrak{U}_\Gamma^{(k+1)}$. By induction, $V = E(x_{\delta_1}) \ldots E(x_{\delta_t})$, where $\delta_1 < \cdots < \delta_t$ are roots in Γ of level greater than k. Thus $\gamma_r < \delta_1$, and our induction is complete; for $k = 1$, the existence of the first representation follows. As for uniqueness, let $E(x_{\beta_1}) \ldots E(x_{\beta_s}) = E(y_{\beta_1}) \ldots E(y_{\beta_s})$; apply to $h \in \mathfrak{H}$ with $\beta_1(h) \neq 0$ to get $h - \beta_1(h)\, x_{\beta_1} + w = h - \beta_1(h)\, y_{\beta_1} + z$, where $w, z \in \sum_{\alpha \neq \beta_1} \mathfrak{L}_\alpha$. Thus $x_{\beta_1} = y_{\beta_1}$, and repetition yields uniqueness. For the reversed representation, take inverses. The negative case is symmetric.

In particular, we may take $\Gamma = \{\alpha \mid \alpha > 0, \alpha S < 0\}$, or $\{\alpha \mid \alpha > 0, \alpha S > 0\}$, or $\{\alpha \mid \alpha > 0\}$, or $\{\alpha \mid L(\alpha) \geqq k\}$. The uniqueness of the representation for elements of $\mathfrak{U}$ yields that $\mathfrak{U}_S$ and $\mathfrak{U}'_S$ intersect only in 1.

Lemma III.2.3. For each $S \in \mathfrak{W}$, $\mathfrak{U} = \mathfrak{U}_S \mathfrak{U}'_S = \mathfrak{U}'_S \mathfrak{U}_S$, and $\mathfrak{U}_S \cap \mathfrak{U}'_S = 1$.

We prove only the first representation, the remaining assertions being either symmetric or shown above. Again we induce downward on k to show $\mathfrak{U}^{(k)} = \mathfrak{U}_S^{(k)} \mathfrak{U}_S'^{(k)}$, the assertion being evident if k is the maximal level. Now assume $\mathfrak{U}^{(j)} = \mathfrak{U}_S^{(j)} \mathfrak{U}_S'^{(j)}$ shown for $j > k$, and let $U \in \mathfrak{U}^{(k)}$. Collecting factors $E(x_\alpha)$, $L(\alpha) = k$, and using commutativity of $\mathfrak{U}^{(k)}/\mathfrak{U}^{(k+1)}$ shows that $U = U_1 U_2 U_3$, $U_1 \in \mathfrak{U}_S^{(k)}$, $U_2 \in \mathfrak{U}_S'^{(k)}$, $U_3 \in \mathfrak{U}^{(k+1)}$. Then $U_3 = V_1 V_2$, $V_1 \in \mathfrak{U}_S^{(k+1)}$, $V_2 \in \mathfrak{U}_S'^{(k+1)}$, and $U = U_1 V_1 (V_1, U_2^{-1}) U_2 V_2$, with $(V_1, U_2^{-1}) \in \mathfrak{U}^{(2k+1)}$. Thus $U = U'_1 U'_2 U'_3$, $U'_1 \in \mathfrak{U}_S^{(k)}$, $U'_2 \in \mathfrak{U}_S'^{(k)}$, $U'_3 \in \mathfrak{U}^{(2k+1)}$; repetition gives $U = U_1^{(m)} U_2^{(m)} U_3^{(m)}$, $U_1^{(m)} \in \mathfrak{U}_S^{(k)}$, $U_2^{(m)} \in \mathfrak{U}_S'^{(k)}$, $U_3^{(m)} \in \mathfrak{U}^{(mk+k+1)}$, which is 1 for sufficiently large m. This completes the proof.

Lemma III.2.4. G is the union of the sets $\mathfrak{U}\, \omega(S)\, H\, \mathfrak{U}$, where S runs over the Weyl group $\mathfrak{W}$.

For with $S = I$ we see that the union contains $\mathfrak{U}$ and H, and it is clearly closed under left and right multiplication by elements of $\mathfrak{U}$ and H. Since G is generated by H, $\mathfrak{U}$ and the $\mathfrak{E}(-\alpha_i)$ (Lemma 1.4), it suffices to show the union is closed under left and right multiplication by the $\mathfrak{E}(-\alpha_i)$. Lemma 1 and Lemma 3 with $S = S_i$ give

$$\mathfrak{E}(-\alpha_i)\, \mathfrak{U}\, \omega(S)\, H\, \mathfrak{U} \subseteqq \mathfrak{U}'_i\, \mathfrak{E}(-\alpha_i)\, \mathfrak{E}(\alpha_i)\, \omega(S)\, H\, \mathfrak{U},$$

which by Lemma 1.3 is contained in

$$\mathfrak{U}\, \omega(S_i)\, H\, \mathfrak{E}(\alpha_i)\, \omega(S)\, H\, \mathfrak{U} \cup \mathfrak{U}'_i\, H\, \mathfrak{E}(\alpha_i)\, \omega(S)\, H\, \mathfrak{U}.$$

The second member of this union is $\mathfrak{U}\, \omega(S)\, H\, \mathfrak{U}$ (Lemma 1.2). For the first, we have $\mathfrak{E}(\alpha_i)\, \omega(S) = \omega(S)\, \mathfrak{E}(\alpha_i S)$. If $\alpha_i S > 0$, the first member is $\mathfrak{U}\, \omega(S_i S)\, H\, \mathfrak{U}$, and we are done. If $\alpha_i S < 0$, then let $S = S_i T$, so that $\alpha_i T = -\alpha_i S > 0$, and $\omega(S_i)\, H\, \mathfrak{E}(\alpha_i)\, \omega(S)\, H = \omega(S_i) \times H\, \mathfrak{E}(\alpha_i)\, \omega(S_i)\, \omega(T)\, H = H\, \mathfrak{E}(-\alpha_i)\, \omega(T)\, H = \mathfrak{E}(-\alpha_i)\, \omega(T)\, H$. Now $\mathfrak{U}\, \mathfrak{E}(-\alpha_i)\, \omega(T)\, H\, \mathfrak{U} = \mathfrak{U}'_i\, \mathfrak{E}(\alpha_i)\, \mathfrak{E}(-\alpha_i)\, \omega(T)\, H\, \mathfrak{U} \subseteqq \mathfrak{U}'_i\, \mathfrak{E}(\alpha_i)\, \omega(S_i) \times$

$\times H\, \mathfrak{E}(\alpha_i)\, \omega(T)\, H\, \mathfrak{U} \cup \mathfrak{U}_i'\, H\, \mathfrak{E}(\alpha_i)\, \omega(T)\, H\, \mathfrak{U}$, by Lemma 1.3, and $\mathfrak{E}(\alpha_i)\, \omega(T)\, H\, \mathfrak{U} \subseteqq \omega(T)\, H\, \mathfrak{U}$ since $\alpha_i T > 0$; thus the above is contained in $\mathfrak{U}\, \omega(S)\, H\, \mathfrak{U} \cup \mathfrak{U}\, \omega(T)\, H\, \mathfrak{U}$. Closure of the union under right multiplication by the $\mathfrak{E}(-\alpha_i)$ follows by taking inverses. This completes the proof.

Noting that $\mathfrak{U}\, \omega(S)\, H\, \mathfrak{U} = \mathfrak{U}_S\, \mathfrak{U}_S'\, \omega(S)\, H\, \mathfrak{U} \subseteqq \mathfrak{U}_S\, \omega(S)\, H\, \mathfrak{U}$, we have only to prove uniqueness to obtain the important *Bruhat–Chevalley decomposition* [76]:

Theorem III.2.1. *G* is the disjoint union of the sets $\mathfrak{U}_S\, \omega(S)\, H\, \mathfrak{U}$, as S ranges over $\mathfrak{W}$. Each element of $\mathfrak{U}_S\, \omega(S)\, H\mathfrak{U}$ has a unique representation in the form $U'\omega(S)\, D\, U$, where $U' \in \mathfrak{U}_S$, $D \in H$, $U \in \mathfrak{U}$.

To prove the assertions, we show that if $S, T \in \mathfrak{W}$; $U' \in \mathfrak{U}_S$, $W' \in \mathfrak{U}_T$; $D, E \in H$; $U, W \in \mathfrak{U}$, then $A = U'\omega(S)\, D\, U = W'\omega(T)\, E\, W$ only if $S = T$, $U' = W'$, $D = E$, $U = W$. If A is as above, let $\mathfrak{L}^+ = \sum_{\alpha > 0} \mathfrak{L}_\alpha$; then $\mathfrak{L}^+ U' = \mathfrak{L}^+ = \mathfrak{L}^+ W'$, so that $\mathfrak{L}^+ A = \mathfrak{L}^+\, \omega(S)\, D\, U = \mathfrak{L}^+ \omega(T) E\, W$. Since $\mathfrak{L}^+\, \omega(S)$ is a sum of root spaces, $\mathfrak{L}^+\, \omega(S)\, D = \mathfrak{L}^+\, \omega(S)$, and $\mathfrak{L}^+\, \omega(S)\, U = \mathfrak{L}^+\, \omega(T)\, W$, $\mathfrak{L}^+\, \omega(S)\, U W^{-1}\, \omega(T)^{-1} = \mathfrak{L}^+$. Now if $\alpha > 0$ is a root, $0 \neq e_\alpha \in \mathfrak{L}_\alpha$, then $e_\alpha\, \omega(S) = x_{\alpha S} \neq 0$ in $\mathfrak{L}_{\alpha S}$, $e_\alpha\, \omega(S)\, U W^{-1} = x_{\alpha S} + y$, $y \in \sum_{\beta \neq \alpha S} \mathfrak{L}_\beta$, $e_\alpha\, \omega(S)\, U W^{-1}\, \omega(T)^{-1} = w_{\alpha S T^{-1}} + z$, $z \in \sum_{\gamma \neq \alpha S T^{-1}} \mathfrak{L}_\gamma$, where $w_{\alpha S T^{-1}} = x_{\alpha S}\, \omega(T)^{-1} \neq 0$. For this to lie in $\mathfrak{L}^+$, we must have $\alpha S T^{-1} > 0$, so that $S T^{-1} \in \mathfrak{W}$ maps positive roots onto positive roots. It follows that $S = T$ (see note below), or $A = U'\omega(S)\, D\, U = W'\omega(S)\, E\, W$, $U', W' \in \mathfrak{U}_S$. But then $\omega(S)^{-1} \times W'^{-1} U' \omega(S) \in \mathfrak{B}$, and this is equal to $E\, W\, U^{-1} D^{-1} \in H\, \mathfrak{U}$. Relative to a basis for $\mathfrak{L}$ consistent with our ordering of the roots, the elements of $H\, \mathfrak{U}$ have upper triangular matrices, while those of $\mathfrak{B}$ have unipotent lower triangular matrices; hence $\mathfrak{B} \cap H\, \mathfrak{U} = 1$, and $U' = W'$, $W U^{-1} = E^{-1} D \in H \cap \mathfrak{U}$ is both diagonal and unipotent, therefore 1. This completes the proof.

(Note. The fact that if $S \in \mathfrak{W}$ is such that $\alpha > 0$ implies $\alpha\, S > 0$, then $S = I$, is well known in the customary setting in characteristic zero. Rather than trying to make the methods of Chapter II, § 10 carry over to the case at hand, we sketch an argument. Let $\gamma_i = \alpha_i\, S$, $\alpha_i \in \Sigma$, so $\gamma_i > 0$ and $A_{\gamma_i, \alpha_j} = \sum_k m_{ik} A_{kj}$, where $\gamma_i = \sum_k m_{ik} \alpha_k$, m_{ik} non-negative integers as in Chapter II, § 5. Similarly $A_{\alpha_j, \gamma_i} = \sum_k n_{jk} A_{\gamma_k, \gamma_i} = \sum_k n_{jk} A_{ki}$, where $\alpha_j = \sum_k n_{jk}\, \gamma_k$, all n_{jk} having the same sign, and the construction which yields the n_{jk}, together with Lemma II.4.4, shows that $A_{ji} = \sum_k n_{jk} A_{\gamma_k, \alpha_i} = \sum_{k,s} n_{jk}\, m_{ks}\, A_{si}$, or $(A_{ij}) = (n_{ij})\, (m_{ij})\, (A_{ij})$. Thus $n_{ij}) = (m_{ij})^{-1}$, from which it follows that (n_{ij}) cannot have a non-positive row, and hence that all $n_{ij} \geqq 0$. Now one sees easily that (m_{ij}) is a permutation matrix, or that $\Sigma\, S = \Sigma$; thus $S = I$ by Th. II.10.2.)

As in the proof above, one sees that $A = U'\omega(S)\,DU$ stabilizes $\mathfrak{B} = \mathfrak{H} + \sum_{\alpha>0} \mathfrak{L}_\alpha$ only if $S = I$, hence only if $A \in H\,\mathfrak{U}$. The converse is clear; thus $H\,\mathfrak{U}$ is the stabilizer of $\mathfrak{B}$ in G.

Lemma III.2.5. If $\mathfrak{F}$ has more than 5 elements, the centralizer of H' in G is H.

For let $T = U'\omega(S)\,DU$ centralize H', where $S \in \mathfrak{W}$, $U' \in \mathfrak{U}_S$, $D \in H$, $U \in \mathfrak{U}$. Let $F \in H'$; $FT = TF$ means $(FU'F^{-1})\,\omega(S) \times \times (\omega(S)^{-1}F\,\omega(S)\,D)\,U = U'\omega(S)\,DF(F^{-1}UF)$. By the uniqueness, U and U' centralize H', and so if $U = E(x_{\beta_1}) \ldots E(x_{\beta_s})$, $F^{-1}UF = E(x_{\beta_1}F) \ldots E(x_{\beta_s}F)$, or $x_{\beta_i}F = x_{\beta_i}$ for all i, all $F \in H'$. But for each $\alpha \neq 0$, there is $D_j(\lambda) \in H'$ such that $\mathfrak{L}_\alpha$ belongs to the characteristic root $(-\lambda)^{A_{\alpha,\alpha_j}} \neq 1$ of $D_j(\lambda)$. Thus $x_{\beta_i} = 0$ for all i in the above, $U = 1$, and likewise $U' = 1$, $T = \omega(S)\,D$, so that $\omega(S)$ centralizes H'. If $S \neq I$, and if α is a root, $\alpha \neq \alpha S$, then by Lemma 1.1 there is $F \in H'$ with $e_\alpha F = \lambda\, e_\alpha$, $e_\alpha\,\omega(S)\,F = \mu\, e_\alpha\,\omega(S)$, $\lambda \neq \mu$. It follows that $F\,\omega(S) \neq \omega(S)\,F$, and the lemma is proved.

Lemma III.2.6. The centralizer of $\mathfrak{U}$ in G is the group generated by all $\mathfrak{E}(\alpha)$, where $\alpha > 0$ and where $\alpha + \alpha_i$ is not a root for any $\alpha_i \in \Sigma$.

For if α is such a root and if $\beta > 0$ is a root, then $\alpha + \beta$ is not a root, since a basis vector for $\mathfrak{L}_\beta$ is in the subalgebra generated by the $\mathfrak{L}_{\alpha_i}$, from which $[\mathfrak{L}_\alpha\, \mathfrak{L}_\beta] = 0$. It follows that the elements in question centralize $\mathfrak{U}$.

Now let $T = U'\omega(S)\,DU$ as above centralize $\mathfrak{U}$. Let $W \in \mathfrak{U}_S$. Then $WT = TW$ implies, by uniqueness in Th. 1, that $W = 1$, $\mathfrak{U}_S = 1$, and $S = I$ by the note above. Hence $T = DU \in H\,\mathfrak{U}$. If $\alpha > 0$, $x_\alpha \in \mathfrak{L}_\alpha$, $E(x_\alpha) = T^{-1}E(x_\alpha)T = U^{-1}E(x_\alpha D)\,U$; if $D \neq 1$, we have $x_\alpha D = \lambda\, x_\alpha \neq x_\alpha$ for some $\alpha > 0$, $E((1-\lambda)\,x_\alpha) = (E(\lambda\, x_\alpha), U) \equiv 1 \pmod{\mathfrak{U}^{(k+1)}}$ by Lemma 2, where $L(\alpha) = k$. Again by Lemma 2, this is impossible, so that $D = 1$ and $T = U \in \mathfrak{U}$.

We now prove by induction downward on k that the centralizer in $\mathfrak{U}^{(k)}$ of all $\mathfrak{E}(\alpha_i)$ is the subgroup generated by all $\mathfrak{E}(\alpha)$, $L(\alpha) \geqq k$, for which no $\alpha + \alpha_i$ is a root. The lemma is an immediate consequence. For k the maximal level this is trivial. Now let $\beta_1, \ldots, \beta_t$ be of level $k > 0$, and let $U = E(x_{\beta_1}) \ldots E(x_{\beta_t})\,W$ centralize all $\mathfrak{E}(\alpha_i)$, where $W \in \mathfrak{U}^{(k+1)}$. By Lemmas 1 and 2 we have for $0 \neq e_i \in \mathfrak{L}_{\alpha_i}$, $1 = (E(e_i), U) \equiv E([e_i\, x_{\beta_1}]) \ldots E([e_i\, x_{\beta_t}]) \pmod{\mathfrak{U}^{(k+2)}}$, and if $\alpha_i + \beta_j$ is a root and $x_{\beta_j} \neq 0$, then $[e_i\, x_{\beta_j}] \neq 0$ and $L(\alpha_i + \beta_j) = k + 1$. By Lemma 2 it follows that $x_{\beta_j} = 0$ whenever $\beta_j + \alpha_i$ is a root for some i, and hence that W centralizes all $\mathfrak{E}(\alpha_i)$. The assertion follows by the induction hypothesis.

Corollary. If $\mathfrak{L}$ is simple, the centralizer of $\mathfrak{U}$ in G is $\mathfrak{E}(\alpha)$, where α is the unique root of maximal level.

For if α is a root such that $\alpha + \alpha_i$ is not a root for any i, one easily checks that the sum of all $\mathfrak{L}_\alpha (\operatorname{ad} e_{-\alpha_{i_1}}) \ldots (\operatorname{ad} e_{-\alpha_{i_k}})$ is closed under $\operatorname{ad} e_{\pm\alpha_i}$, hence is an ideal in $\mathfrak{L}$, hence is $\mathfrak{L}$ by simplicity. Now each of the summands is contained in a root-space $\mathfrak{L}_\beta$, and the only such β with $L(\beta) = L(\alpha)$, or with no $\beta + \alpha_i$ a root, is $\beta = \alpha$. Clearly $L(\beta) \leqq L(\alpha)$ for all β.

If $\mathfrak{K}$ is an extension of the field $\mathfrak{F}$, then $\mathfrak{L}_\mathfrak{K}$ is again classical relative to the decomposition $\mathfrak{L}_\mathfrak{K} = \mathfrak{H}_\mathfrak{K} + \sum_\alpha (\mathfrak{L}_\alpha)_\mathfrak{K}$, and we denote by $G_\mathfrak{K}$, $G'_\mathfrak{K}$, $H_\mathfrak{K}$, $\mathfrak{U}_\mathfrak{K}$, etc., the groups associated as above with $\mathfrak{L}_\mathfrak{K}$ relative to this decomposition and the same ordering of the roots. We may identify G, G', etc., with subgroups of $G_\mathfrak{K}$, $G'_\mathfrak{K}$, etc., since each automorphism in H extends to a unique automorphism of $\mathfrak{L}_\mathfrak{K}$ leaving $\mathfrak{H}_\mathfrak{K}$ fixed, and since $E(x_\alpha)$, $x_\alpha \in \mathfrak{L}_\alpha$, may be equally well applied to $\mathfrak{L}_\mathfrak{K}$. Now ONO has shown [311] that those elements of several of the groups associated with $\mathfrak{L}_\mathfrak{K}$, which map $\mathfrak{L}$ into itself, are the extensions to $\mathfrak{L}_\mathfrak{K}$ of elements of the groups denoted by the corresponding letters, acting in $\mathfrak{L}$. For example:

Lemma III.2.7. If $T \in G_\mathfrak{K}$ and if $\mathfrak{L}T \subseteqq \mathfrak{L}$, then $T \in G$.

First we note that $\omega(S) \in G'$ by our construction. Now let $T = U'\omega(S)DU$, $U' \in (\mathfrak{U}_S)_\mathfrak{K}$, $D \in H_\mathfrak{K}$, $U \in \mathfrak{U}_\mathfrak{K}$. Then $\omega(S)^{-1}T$ also maps $\mathfrak{L}$ into $\mathfrak{L}$, and $\omega(S)^{-1}T = VDU$, $V \in \mathfrak{V}_\mathfrak{K}$. Let $\{e_\alpha, h_i\}$ be a basis for $\mathfrak{L}$ compatible with our choice of an admissibly ordered set of roots; then the matrix of D is diagonal, that of U has 1 in all diagonal positions and 0 in all subdiagonal positions, and that of V has 1 in diagonal positions and 0 in superdiagonal positions. The condition that VDU map $\mathfrak{L}$ into $\mathfrak{L}$ yields that the product of these three matrices has entries in $\mathfrak{F}$, from which one easily checks that all three have entries in $\mathfrak{F}$. Thus it suffices to consider the three cases $T = V, D, U$ and, by symmetry, only the cases $T = D$, $T = U$. In the former case, $e_\alpha D = \lambda_\alpha e_\alpha$ for all roots $\alpha \neq 0$, where $0 \neq e_\alpha \in \mathfrak{L}_\alpha$, and $\lambda_\alpha e_\alpha \in \mathfrak{L}$ only if $\lambda_\alpha \in \mathfrak{F}$; hence $\mathfrak{L}D \subseteqq \mathfrak{L}$ implies $D \in H$. If $U = E(\lambda_1 e_{\beta_1}) \ldots E(\lambda_s e_{\beta_s})$ maps $\mathfrak{L}$ into $\mathfrak{L}$, where $0 \neq \lambda_i \in \mathfrak{K}$, $0 < \beta_1 < \cdots < \beta_s$, let $h \in \mathfrak{H}$, $\beta_1(h) \neq 0$; then $\beta_1(h) \in \mathfrak{F}$, and $hU = h - \lambda_1 \beta_1(h) e_{\beta_1} + z \in \mathfrak{L}$, $z \in \sum_{\alpha > \beta_1} (\mathfrak{L}_\alpha)_\mathfrak{K}$. It follows that $\lambda_1 \in \mathfrak{F}$ and, by induction, that $U \in \mathfrak{U}$.

§ 3. Structure of the Chevalley group

We assume that $\mathfrak{L}$ is as in §§ 1, 2, and furthermore that $\mathfrak{L}$ is *simple*.

Lemma III.3.1. Let $\mathfrak{J}$ be a subgroup of G which is self-conjugate under G'. Suppose that $\mathfrak{J} \cap \mathfrak{E}(\alpha) \neq \{1\}$ for some root α. Then $G' \subseteqq \mathfrak{J}$.

For by Th. II.10.2.b and the fact that $\omega(S) \in G'$ for all $S \in \mathfrak{W}$, we may assume $\alpha = \alpha_i$ is a member of the fundamental system of roots. Let α_j be another fundamental root, with $A_{ij} < 0$; we claim that $\mathfrak{J} \cap \mathfrak{E}(\alpha_j) \neq \{1\}$. If $A_{ij} = -1$, then $\alpha_i S_j = \alpha_i + \alpha_j, \alpha_i S_i = -\alpha_i$, so that $\mathfrak{J} \cap \mathfrak{E}(\alpha_i + \alpha_j) \neq \{1\} \neq \mathfrak{J} \cap \mathfrak{E}(-\alpha_i)$; let $0 \neq x \in \mathfrak{L}_{-\alpha_i}$, $0 \neq y \in \mathfrak{L}_{\alpha_i + \alpha_j}$, $E(x), E(y) \in \mathfrak{J}$. By (3) (proof of Lemma 2.1), the commutator of $E(x)$ and $E(y)$ is equal to $E([x\,y]) \neq 1$, $E([x\,y]) \in \mathfrak{E}(\alpha_j)$. If $A_{ij} = -2$ or -3, then $A_{ji} = -1$, and in these cases the commutator of $E(x)$, $x \in \mathfrak{L}_{\alpha_i}$, and $E(\lambda e_{\alpha_j})$, $0 \neq \lambda \in \mathfrak{F}$, is in $\mathfrak{J}$ if $E(x)$ is, and is equal to:

$$A_{ij} = -2: \; E(\lambda[x\,e_{\alpha_j}])\,E\left(\tfrac{1}{2}\lambda^2[[x\,e_{\alpha_j}]\,e_{\alpha_j}]\right);$$

$$A_{ij} = -3: \; E(\lambda[x\,e_{\alpha_j}])\,E\left(\tfrac{1}{2}\lambda^2[[x\,e_{\alpha_j}]\,e_{\alpha_j}]\right).$$

$$E\left(\frac{1}{3!}\lambda^3[[[x\,e_{\alpha_j}]\,e_{\alpha_j}]\,e_{\alpha_j}]\right)E\left(-\frac{1}{3!}\lambda^3\big[[[[x\,e_{\alpha_j}]\,e_{\alpha_j}]\,e_{\alpha_j}]\,x\big]\right).$$

In case $A_{ij} = -3$, we have $\alpha_i S_j = \alpha_i + 3\alpha_j$, $\alpha_i S_j S_i = 2\alpha_i + 3\alpha_j$, so that $\mathfrak{J} \cap \mathfrak{E}(\alpha_i + 3\alpha_j) \neq \{1\} \neq \mathfrak{J} \cap \mathfrak{E}(2\alpha_i + 3\alpha_j)$. Now we shall see below that $\mathfrak{J} \cap \mathfrak{E}(\alpha) \neq \{1\}$ is equivalent with $\mathfrak{E}(\alpha) \subseteqq \mathfrak{J}$. It follows that for all $\lambda \in \mathfrak{F}$, $E(\lambda[x\,e_{\alpha_j}])\,E\left(\tfrac{1}{2}\lambda^2[[x\,e_{\alpha_j}]\,e_{\alpha_j}]\right) \in \mathfrak{J}$. Taking $\lambda = \pm 1$ we find that $1 \neq E(2[x\,e_{\alpha_j}]) \in \mathfrak{J} \cap \mathfrak{E}(\alpha_i + \alpha_j)$: now $(\alpha_i + \alpha_j)\,S_i = \alpha_j$, and since the fundamental system is connected (by the simplicity of $\mathfrak{L}$), it only remains to show that $\mathfrak{J} \cap \mathfrak{E}(\alpha) \neq \{1\}$ implies $\mathfrak{E}(\alpha) \subseteqq \mathfrak{J}$; again we may assume $\alpha = \alpha_i$. For $\varrho \neq 0$, $D_i(\varrho)^{-1} E(x_{\alpha_i})\,D_i(\varrho) = E(\varrho^2 x_{\alpha_i})$ by Lemma 1.1, where $D_i(\varrho) \in H' \subseteqq G'$. It follows that if $\mathfrak{J}$ contains $E(x_{\alpha_i}) \neq 1$, then $\mathfrak{J}$ contains all $E(\lambda x_{\alpha_i})$, where λ is in the additive subgroup of $\mathfrak{F}$ generated by squares; but this subgroup is $\mathfrak{F}$, and the lemma follows.

Lemma III.3.2. Let $\mathfrak{J}$ be a subgroup of G, self-conjugate under G', such that $\mathfrak{J} \cap H\,\mathfrak{U} \neq \{1\}$. Then $G' \subseteqq \mathfrak{J}$.

It suffices to show $\mathfrak{J} \cap \mathfrak{E}(\alpha) \neq \{1\}$, where α is the unique highest root. If $\mathfrak{J} \cap H\,\mathfrak{U}$ centralizes $\mathfrak{U}$, this is clear by the corollary to Lemma 2.6. Otherwise, since $\mathfrak{U}$ is normal in $H\,\mathfrak{U}$, forming commutators with $\mathfrak{U}$ yields $\mathfrak{J} \cap \mathfrak{U} \neq \{1\}$. Now $(\mathfrak{U}^{(k)}, \mathfrak{U})$ is contained in $\mathfrak{U}^{(k+1)}$ by Lemma 2.2. It follows that for each $k > 0$, either $\mathfrak{J} \cap \mathfrak{U}^{(k+1)} \neq \{1\}$ or $\mathfrak{J} \cap H\,\mathfrak{U}$ contains an element $\neq 1$ which centralizes $\mathfrak{U}$. Thus we may assume $\mathfrak{J} \cap \mathfrak{U}^{(k+1)} \neq \{1\}$ for all k; but $\mathfrak{U}^{(k+1)} = \{1\}$ if k is the maximal level of roots, and the lemma is proved.

Theorem III.3.1 (Chevalley [76]). Assume that either: a) $\mathfrak{F}$ has more than five elements, or: b) $\mathfrak{L}$ is not of type A_1. If $\mathfrak{J}$ is a subgroup of G, self-conjugate under G', and if $\mathfrak{J} \neq \{1\}$, then $G' \subseteqq \mathfrak{J}$; in particular, G' is a simple group.

a) Let $T = U'\omega(S)\,DU \neq 1$ be in $\mathfrak{J}$, where $S \in \mathfrak{W}$, $U' \in \mathfrak{U}_S$, $D \in H$, $U \in \mathfrak{U}$. Then $U'^{-1}TU' \neq 1$ is in $\mathfrak{J}$, from which we may assume $T = \omega(S)\,DU$. If $\omega(S) \neq 1$, then $\omega(S) \notin H$, so by Lemma 2.5 there is $D' \in H' \subseteqq G' \cap H$ such that $D'^{-1}\omega(S)\,D' \neq \omega(S)$. Since $\omega(S)\,H = H\,\omega(S)$, $D'^{-1}\,\omega(S)\,D' = \omega(S)\,E$, $1 \neq E \in H$; thus $\mathfrak{J}$ contains $T_1 = D'^{-1}TD' = \omega(S)\,E\,D\,D'^{-1}UD' = \omega(S)\,(ED)\,(D'^{-1}UD') \neq \omega(S)\,DU$, since $D'^{-1}UD' \in \mathfrak{U}$, $E \neq 1$. Now $1 \neq T^{-1}T_1 = U^{-1}EU''$, $U'' \in \mathfrak{U}$, $T^{-1}T_1 \in \mathfrak{J} \cap \mathfrak{U}H\mathfrak{U} = \mathfrak{J} \cap H\mathfrak{U}$. The theorem follows by Lemma 2.

b) (STEINBERG [379]) Let $1 \neq T = \omega(S)\,DU \in \mathfrak{J}$ as above; we may assume $S \neq I$ by Lemma 2. If $S = S_i$ for some i, then since $\mathfrak{L}$ is not of type A_1 there is a fundamental root α_j with $\alpha_j \neq \alpha_j S_i > 0$, hence $L(\alpha_j S_i) > 1$. Let $1 \neq Y \in \mathfrak{E}(\alpha_j)$; then from $\omega(S)^{-1}Y\omega(S) \in \mathfrak{U}$ one sees that $(Y, T) = Y^{-1}T^{-1}YT \in \mathfrak{J} \cap \mathfrak{U}$, in fact that $(Y, T) \equiv Y^{-1} \pmod{\mathfrak{U}^{(2)}}$ since $\omega(S)^{-1}Y\omega(S) \in \mathfrak{U}^{(2)}$, an invariant subgroup of $H\mathfrak{U}$. Since $Y \notin \mathfrak{U}^{(2)}$, we have $1 \neq (Y, T) \in \mathfrak{J} \cap \mathfrak{U}$.

Finally, suppose $S \neq I$, S_i for all i. Since $S \neq I$, $\alpha_i S < 0$ for some i; let $Y \neq 1$ in $\mathfrak{E}(-\alpha_i)$, and form (Y, T) as above. Now $\omega(S)^{-1}Y\omega(S) \in \mathfrak{U}$ gives $(Y, T) = Y^{-1}D'U'$, $D' \in H$, $U' \in \mathfrak{U}$, so that $(Y, T) \neq 1$ since $\mathfrak{V} \cap H\mathfrak{U} = \{1\}$. Moreover, (Y, T) is in $\mathfrak{E}(-\alpha_i)\,\mathfrak{E}(\alpha_i)\,H\mathfrak{U}$, and by Lemma 1.3 $\mathfrak{E}(-\alpha_i)\,\mathfrak{E}(\alpha_i) \subseteq \mathfrak{E}(\alpha_i)\,\omega(S_i)\,H'\mathfrak{E}(\alpha_i) \cup H'\mathfrak{E}(\alpha_i)$. The considerations for the earlier cases $S = I$, $S = S_i$ may now be applied to $(Y, T) \in \mathfrak{J}$. This completes the proof.

(In the remaining case, that of A_1 over a field of five elements, the conclusion also holds; this will be seen in § 6 from the identification of G with the projective linear group $\mathrm{PGL}(2, \mathfrak{F})$ when $\mathfrak{L}$ is of type A_1 over $\mathfrak{F}$, and of G' with $\mathrm{PSL}(2, \mathfrak{F})$. That the case $|\mathfrak{F}| = 5$ is exceptional for A_1 may be regarded as a reflection of some technical difficulties involved in proving simplicity for $\mathrm{PSL}(2,5)$—e.g., see [18, p. 167].)

Lemma III.3.3. $G' = \mathfrak{V}H'\mathfrak{U}\mathfrak{V}$.

For the set $\mathfrak{V}H'\mathfrak{U}\mathfrak{V}$ contains generators for G' and is contained in G', so that it suffices to show it closed under left and right multiplication by the generators $\mathfrak{E}(\pm\alpha_i)$ of G'. Since $H'\mathfrak{U} = \mathfrak{U}H'$ and since all factors are subgroups, taking inverses reduces the problem to showing $\mathfrak{V}H'\mathfrak{U}\mathfrak{V}$ is closed under right multiplication by the $\mathfrak{E}(\pm\alpha_i)$, and indeed by the $\mathfrak{E}(\alpha_i)$. By Lemma 2.3 we have $\mathfrak{U}_i'\,\mathfrak{E}(\alpha_i) = \mathfrak{U} = \mathfrak{E}(\alpha_i)\,\mathfrak{U}_i'$, $\mathfrak{V}_i'\,\mathfrak{E}(-\alpha_i) = \mathfrak{V} = \mathfrak{E}(-\alpha_i)\,\mathfrak{V}_i'$, and conjugation by $\omega(S_i)$ gives $\mathfrak{U}_i'\,\mathfrak{E}(-\alpha_i) = \mathfrak{E}(-\alpha_i)\,\mathfrak{U}_i'$, $\mathfrak{V}_i'\,\mathfrak{E}(\alpha_i) = \mathfrak{E}(\alpha_i)\,\mathfrak{V}_i'$. Thus $\mathfrak{V}H'\mathfrak{U}\mathfrak{V}\,\mathfrak{E}(\alpha_i) = H'\mathfrak{V}\,\mathfrak{U}_i'\,\mathfrak{E}(\alpha_i)\,\mathfrak{E}(-\alpha_i)\,\mathfrak{V}_i'\,\mathfrak{E}(\alpha_i) = H'\mathfrak{V}\,\mathfrak{U}_i'\,\mathfrak{E}(\alpha_i)\,\mathfrak{E}(-\alpha_i)\,\mathfrak{E}(\alpha_i)\,\mathfrak{V}_i' \subseteq$ $\subseteq H'\mathfrak{V}\,\mathfrak{U}_i'\,\mathfrak{E}(-\alpha_i)\,H'\mathfrak{E}(\alpha_i)\,\mathfrak{E}(-\alpha_i)\,\mathfrak{V}_i'$ by Lemma 1.3, and this is contained in $H'\mathfrak{V}\mathfrak{U}H'\mathfrak{V} = \mathfrak{V}H'\mathfrak{U}\mathfrak{V}$ by the remarks above.

Lemma III.3.4. $H' = G' \cap H$.

For let $D \in G' \cap H$. By Lemma 3, $D = VD'UW$, where $V, W \in \mathfrak{V}$, $D' \in H'$, $U \in \mathfrak{U}$. Thus $1 = D^{-1}VD'UW = (D^{-1}VD)\,(D^{-1}D')\,UW$, or $(D^{-1}VD)^{-1}W^{-1} = (D^{-1}D')\,U \in \mathfrak{V} \cap H\mathfrak{U} = \{1\}$. Hence $U = DD'^{-1} \in$ $\in \mathfrak{U} \cap H = \{1\}$, or $D = D' \in H'$. Since $H' \subseteq G' \cap H$ is known, the lemma is proved.

One can now describe the structure of the commutator quotient group G/G'. We have seen that $G/G' \cong H/(G' \cap H)$, i.e., $G/G' \cong H/H'$ (Lemma 4). Now H may be identified with the direct product of r

copies of $\mathfrak{F}^*$, and H' with the subgroup generated by the r-tuples $(\lambda^{A_{1j}}, \ldots, \lambda^{A_{rj}})$, $\lambda \in \mathfrak{F}^*$, $1 \leqq j \leqq r$. If $d_1, \ldots, d_r$ are the invariant factors of the integral matrix (A_{ij}), with $d_1 | d_2 | \ldots | d_r$, then $d_r \neq 0$, and H' is isomorphic with $\mathfrak{F}^{*d_1} \times \cdots \times \mathfrak{F}^{*d_r}$. This group may be regarded as embedded in $H \cong \mathfrak{F}^* \times \cdots \times \mathfrak{F}^*$ by the identity mapping of $\mathfrak{F}^{*d_i}$ into $\mathfrak{F}^*$ (cf. [381]). Thus $G/G' \cong H/H' \cong \mathfrak{F}^*/\mathfrak{F}^{*d_1} \times \cdots \times \mathfrak{F}^*/\mathfrak{F}^{*d_r}$. The invariant factors of the matrices (A_{ij}) are readily computed: for type A_r, $d_r = r+1$, $d_i = 1$ for $i < r$; for types B_r, C_r, E_7, $d_r = 2$, $d_i = 1$ for $i < r$; for type D_{2k+1}, $d_{2k+1} = 4$, $d_i = 1$, $i < 2k+1$; for type D_{2k}, $d_{2k} = 2 = d_{2k-1}$, $d_i = 1$, $i < 2k-1$; for type E_6, $d_r = 3$, $d_i = 1$, $i < r$; for types G_2, F_4, E_8, $d_i = 1$, all i.

§ 4. Conjugacy of Cartan subalgebras

The following theorem was proved in the algebraically closed case by the author [355], with the group G' replaced by the group generated by all $G' = G'(\mathfrak{H})$ relative to *all* classical Cartan subalgebras $\mathfrak{H}$ of $\mathfrak{L}$. That this latter group coincides with G' is a consequence of the improved version given here, which is also valid for non-algebraically-closed ground fields, and is due to STEINBERG [381].

Theorem III.4.1. Let $\mathfrak{L}$ be a classical Lie algebra, and let $\mathfrak{H}_1, \mathfrak{H}_2$ be classical Cartan subalgebras. Let $G' = G'(\mathfrak{H}_1)$ be the Chevalley group G' constructed relative to $\mathfrak{H}_1$. Then there exists $\sigma \in G'$ with $\mathfrak{H}_1^\sigma = \mathfrak{H}_2$.

First suppose the ground field $\mathfrak{F}$ is algebraically closed, and let $\mathfrak{H}$ be any classical Cartan subalgebra. Let $\alpha_1, \ldots, \alpha_s$ be the non-zero roots relative to $\mathfrak{H}$, and let $h_1, \ldots, h_r$ be a basis for $\mathfrak{H}$. Let $0 \neq e_\alpha \in \mathfrak{L}_\alpha$ for each $\alpha \neq 0$. Regarding $\mathfrak{L}$ as the n-dimensional affine space $\mathfrak{F}^n$ with coordinate system given by the basis $\{h_i, e_\alpha\}$, consider the mapping φ: $\mathfrak{F}^n \to \mathfrak{F}^n$ given by

$$\varphi(\lambda_i; \lambda_\alpha) = \varphi\Big(\sum_i \lambda_i h_i + \sum_\alpha \lambda_\alpha e_\alpha\Big) = \Big(\sum \lambda_i h_i\Big) \prod_\alpha E(\lambda_\alpha e_\alpha), \tag{4}$$

where the product is taken in some fixed order. The mapping is a polynomial mapping, i.e., there are fixed polynomials $p_i(X), p_\alpha(X)$ in n variables such that $\varphi(\lambda_i; \lambda_\alpha) = (p_i(\lambda); p_\alpha(\lambda))$ in our coordinate system. Now let $h \in \mathfrak{H}$ have $\alpha(h) \neq 0$ for all $\alpha \neq 0$, $h = \sum \mu_i h_i$, and consider the Jacobian matrix of φ at h,

$$\left.\frac{\partial(p_i(X); p_\alpha(X))}{\partial(X_i; X_\alpha)}\right|_{(\mu_i; 0)}; \quad \text{this is equal to} \quad \left(\begin{array}{c|c} I_r & 0 \\ \hline 0 & \begin{matrix} \alpha_1(h) & & 0 \\ & \ddots & \\ 0 & & \alpha_s(h) \end{matrix} \end{array}\right),$$

hence is non-singular. It follows that $\mathfrak{F}(X_i; X_\alpha)$ admits no non-zero derivations vanishing on $\mathfrak{F}(p_i(X); p_\alpha(X))$, hence that the former field is a separable algebraic extension of the latter (cf. [223, Chap. IV, or 407, Chap. I]). In particular, $(p_i(X); p_\alpha(X))$ is a generic point for the closure of $\varphi(\mathfrak{F}^n)$ in the Zariski topology of $\mathfrak{F}^n$, and has dimension n. Therefore $\varphi(\mathfrak{L}) = \varphi(\mathfrak{F}^n)$ is dense in the Zariski topology of $\mathfrak{F}^n$. Since $\mathfrak{F}$ is algebraically closed, $\varphi(\mathfrak{F}^n)$ is *épais* (cf. [72], p. 186), i.e., contains a dense open subset of its closure. Hence $\varphi(\mathfrak{F}^n)$ contains an open dense subset of $\mathfrak{F}^n$. Now the regular elements of $\mathfrak{L}$ constitute a non-empty open (dense) subset of $\mathfrak{F}^n$, and any two non-empty open subsets of $\mathfrak{F}^n$ intersect.

Now let φ_1, φ_2 be constructed relative to $\mathfrak{H}_1, \mathfrak{H}_2$ as was φ above. Then it follows that $\varphi_1(\mathfrak{L}) \cap \varphi_2(\mathfrak{L})$ contains a regular element u of $\mathfrak{L}$; by the definition of φ, we have $u = g_1^{\sigma_1} = g_2^{\sigma_2}$, where $g_i \in \mathfrak{H}_i$, $\sigma_i \in G'(\mathfrak{H}_i)$. It follows further that g_i is regular, $g_2 = g_1^{\sigma_1 \sigma_2^{-1}}$, so that $\sigma_1 \sigma_2^{-1}$ maps the centralizer of g_1 in $\mathfrak{L}$ onto that of g_2, and these centralizers contain $\mathfrak{H}_1$ resp. $\mathfrak{H}_2$. By the fact that g_i is regular, and because $\mathfrak{H}_i$ contains an element (cf. h above) whose centralizer is $\mathfrak{H}_i$, the centralizers of the g_i are the $\mathfrak{H}_i$, and $\mathfrak{H}_2 = \mathfrak{H}_1^{\sigma_1 \sigma_2^{-1}}$, $\sigma_i \in G'(\mathfrak{H}_i)$. Hence if α is a non-zero root relative to $\mathfrak{H}_2$, $0 \neq e_\alpha \in \mathfrak{L}_\alpha$, we have $e_\alpha = x_\beta \sigma_1 \sigma_2^{-1}$, where x_β is a root-vector relative to $\mathfrak{H}_1$, or $e_\alpha \sigma_2 = x_\beta \sigma_1$. Thus, as sets, $\{\sigma_2^{-1} E(\lambda e_\alpha) \sigma_2\} = \{\sigma_1^{-1} E(\lambda x_\beta) \sigma_1\}$; but the former set generates $G'(\mathfrak{H}_2)$, and the latter is contained in $G'(\mathfrak{H}_1)$. Hence $G'(\mathfrak{H}_2) = G'(\mathfrak{H}_1)$, and the theorem is proved in the algebraically closed case.

Next suppose $\mathfrak{K}$ is the algebraic closure of $\mathfrak{F}$, and consider $\mathfrak{L}_\mathfrak{K}$, which is classical relative to $(\mathfrak{H}_1)_\mathfrak{K}$, $(\mathfrak{H}_2)_\mathfrak{K}$. Hence there exists $\sigma \in G'((\mathfrak{H}_1)_\mathfrak{K})$ mapping $(\mathfrak{H}_1)_\mathfrak{K}$ onto $(\mathfrak{H}_2)_\mathfrak{K}$. If $\alpha_1, \ldots, \alpha_r$ is a fundamental system of roots for $\mathfrak{L}$ relative to $\mathfrak{H}_1$, and if $0 \neq e_i \in \mathfrak{L}_{\alpha_i}$, then $e_i \sigma = \lambda_i x_i$, $0 \neq x_i \in \mathfrak{L}_{\beta_i}$, where $\beta_1, \ldots, \beta_r$ is a fundamental system of roots relative to $\mathfrak{H}_2$, and where $0 \neq \lambda_i \in \mathfrak{K}$. Let $D \in H((\mathfrak{H}_1)_\mathfrak{K})$ map e_i onto $\lambda_i^{-1} e_i$, $1 \leqq i \leqq r$. Then $(\mathfrak{H}_1)_\mathfrak{K}^{D\sigma} = (\mathfrak{H}_2)_\mathfrak{K}$, $e_i D\sigma = x_i$, and $D\sigma \in G((\mathfrak{H}_1)_\mathfrak{K}) = G'((\mathfrak{H}_1)_\mathfrak{K})$ since $\mathfrak{K}$ is algebraically closed (§ 3). Thus we may replace σ by $D\sigma$ to assume $e_i \sigma \in \mathfrak{L}$, $1 \leqq i \leqq r$. If $0 \neq f_i \in \mathfrak{L}_{-\alpha_i}$, $0 \neq y_i \in \mathfrak{L}_{-\beta_i}$, then $f_i \sigma = \mu y_i$, $\mu \in \mathfrak{K}$, and $[e_i f_i] \sigma = h_i \sigma = \mu [x_i y_i]$, $h_i \in \mathfrak{H}_1$, and $0 \neq [e_i h_i] \sigma = \alpha_i(h_i) e_i \sigma = \mu [x_i [x_i y_i]] \in \mathfrak{L}$, from which $\mu \in \mathfrak{F}$. Since the e_i, f_i generate $\mathfrak{L}$, $\mathfrak{L}^\sigma = \mathfrak{L}$, and $\sigma \in G((\mathfrak{H}_1)_\mathfrak{K})$. By Ono's theorem (Lemma 2.7), $\sigma \in G(\mathfrak{H}_1)$. Now $\sigma = D\sigma_1$, $D \in H(\mathfrak{H}_1)$, $\sigma_1 \in G'(\mathfrak{H}_1)$, by § 1, and $\mathfrak{H}_2 = (\mathfrak{H}_2)_\mathfrak{K} \cap \mathfrak{L} = (\mathfrak{H}_1)_\mathfrak{K}^\sigma \cap \mathfrak{L}^\sigma = ((\mathfrak{H}_1)_\mathfrak{K} \cap \mathfrak{L})^\sigma = \mathfrak{H}_1^\sigma = \mathfrak{H}_1^{\sigma_1}$. This completes the proof.

Corollary. Let $\mathfrak{H}_1, \mathfrak{H}_2$ be classical Cartan subalgebras of $\mathfrak{L}$. Then the groups $G'(\mathfrak{H}_1)$ and $G'(\mathfrak{H}_2)$ coincide, as do $G(\mathfrak{H}_1)$ and $G(\mathfrak{H}_2)$.

For if $\mathfrak{H}_2 = \mathfrak{H}_1^\sigma$, $\sigma \in G'(\mathfrak{H}_1)$, we have $G(\mathfrak{H}_2) = \sigma^{-1} G(\mathfrak{H}_1) \sigma$, $G'(\mathfrak{H}_2) = \sigma^{-1} G'(\mathfrak{H}_1) \sigma$.

§ 5. Structure of the automorphism group

Let $\mathfrak{L}$ be a classical Lie algebra over $\mathfrak{F}$, with classical Cartan subalgebra $\mathfrak{H}$. Let $\mathfrak{A}(\mathfrak{L})$ be the group of automorphisms of $\mathfrak{L}$. Let $\{e_\alpha\}$ be a basis as before for $\sum_{\alpha \neq 0} \mathfrak{L}_\alpha$, so chosen that $\alpha([e_\alpha e_{-\alpha}]) = 2$. Let $\alpha_1, \ldots, \alpha_r$ be a fundamental system of roots relative to $\mathfrak{H}$. If π is any permutation of $\{1, 2, \ldots, r\}$ such that $A_{i\pi, j\pi} = A_{ij}$ for all i, j, we refer to π as an *automorphism of the Cartan matrix* (A_{ij}); in the customary representation of (A_{ij}) by a mixed graph (Schläfli or Dynkin diagram) of r vertices α_i, with the edges $\{\alpha_i, \alpha_j\}$ of multiplicity $A_{ij} A_{ji}$, this edge being directed from α_i to α_j if and only if $A_{ij} < A_{ji}$, these are the *automorphisms of the diagram*, or *graph automorphisms*. By Th. II.8.1, for each such π there is a unique automorphism of $\mathfrak{L}$ mapping e_{α_i} onto $e_{\alpha_{i\pi}}$ and $e_{-\alpha_i}$ onto $e_{-\alpha_{i\pi}}$ for all i. We denote this automorphism also by π; such automorphisms form a finite group Γ isomorphic with the group of graph automorphisms, and we refer to Γ as the *group of graph automorphisms of* $\mathfrak{L}$.

Now let $\sigma \in \mathfrak{A}(\mathfrak{L})$; then $\mathfrak{H}^\sigma$ is a classical Cartan subalgebra, so that $\mathfrak{H}^\sigma = \mathfrak{H}^\tau$ for some $\tau \in G'(\mathfrak{H}) = G'$. Thus $\mathfrak{H}^\varrho = \mathfrak{H}$, where $\varrho = \sigma\tau^{-1}$. Now $\mathfrak{L}^\varrho_{\alpha_i} = \mathfrak{L}_{\beta_i}$, $1 \leqq i \leqq r$, where $\{\beta_i\}$ is a fundamental system relative to $\mathfrak{H}$, so that for some $S \in \mathfrak{W}$, $\theta = \varrho\,\omega(S)^{-1}$ has $\mathfrak{H}^\theta = \mathfrak{H}$, $\mathfrak{L}^\theta_{\alpha_i} = \mathfrak{L}_{\alpha_{i\pi}}$, π being a permutation of $\{1, \ldots, r\}$. Since the A_{ij} are determined by the vanishing or non-vanishing of products of the e_{α_i}, it follows that π is an automorphism of the Cartan matrix. Let $D \in H = H(\mathfrak{H})$ be the automorphism mapping $e_{\alpha_{i\pi}}$ onto $\lambda_i e_{\alpha_{i\pi}}$, where $e_{\alpha_i}\theta = \lambda_i e_{\alpha_{i\pi}}$, $1 \leqq i \leqq r$. Then πD, where π is a graph automorphism of $\mathfrak{L}$, maps e_{α_i} onto $\lambda_i e_{\alpha_{i\pi}}$, $e_{-\alpha_i}$ onto $\lambda_i^{-1} e_{-\alpha_{i\pi}}$, h_i onto $h_{i\pi}$, where $h_i = [e_{\alpha_i} e_{-\alpha_i}]$, and θ maps h_i onto that $h \in [\mathfrak{L}_{\alpha_{i\pi}}, \mathfrak{L}_{-\alpha_{i\pi}}]$ such that $\lambda_i [e_{\alpha_{i\pi}} h] = [e^\theta_{\alpha_i} h^\theta_i] = 2e^\theta_{\alpha_i} = 2\lambda_i e_{\alpha_{i\pi}}$, viz., $h = h_{i\pi}$. It follows that $e^\theta_{-\alpha_i} = \lambda_i^{-1} e_{-\alpha_{i\pi}}$, or that $\theta = \pi D$. That is, $\sigma\tau^{-1}\omega(S)^{-1} = \pi D$, or $\sigma = \pi D \omega(S)\tau \in \pi G(\mathfrak{H}) = \pi G$. We have thus proved part of the

Theorem III.5.1 (Steinberg). Let σ be an automorphism of $\mathfrak{L}$. Then σ can be written uniquely as $\sigma = \pi\theta$, where $\theta \in G = G(\mathfrak{H})$, and where π is a graph automorphism as above.

It remains only to prove uniqueness, or that the only graph automorphism $\pi \in G$ is the identity. If π is such an automorphism, then since π stabilizes $\mathfrak{B} = \mathfrak{H} + \mathfrak{L}^+$, we have $\pi = DU$, $D \in H$, $U \in \mathfrak{U}$, by remarks preceding Lemma 2.5. Similarly, $\pi = EV$, $E \in H$, $V \in \mathfrak{V}$, so that $V = (E^{-1}D)\,U \in \mathfrak{V} \cap H\,\mathfrak{U} = \{1\}$, $V = 1 = U$, and $\pi = D \in H$. Thus $e_{\alpha_{i\pi}} = e_{\alpha_i}\pi = \lambda_i e_{\alpha_i}$ for all i, which implies $\pi = 1$, and completes the proof.

It is clear that a graph automorphism π permutes root-spaces relative to $\mathfrak{H}$, hence that $\pi^{-1} G' \pi = G'$. Similarly, $\mathfrak{H}^\pi = \mathfrak{H}$ implies

$\pi^{-1} H \pi = H$, so that G is a normal subgroup of $\mathfrak{A}(\mathfrak{L})$. Thus $\mathfrak{A}(\mathfrak{L})$ has the chain of normal subgroups $\mathfrak{A}(\mathfrak{L}) \supseteqq G \supseteqq G' \supseteqq 1$, where $\mathfrak{A}(\mathfrak{L})/G$ is isomorphic with the group of graph automorphisms, $G/G' \cong H/H' \cong \mathfrak{F}^*/\mathfrak{F}^{*d_1} \times \cdots \times \mathfrak{F}^*/\mathfrak{F}^{*d_r}$ as in § 3, and where G' is a simple group, or a product of simple groups. The group $\mathfrak{A}(\mathfrak{L})$ is the semi-direct product of G and Γ. Also, Γ is easily determined: If the graph of $\mathfrak{L}$ has distinct isomorphic components, i.e., if $\mathfrak{L}$ has some isomorphic simple summands, then the group Γ is generated by a complete set of permutations of roots, preserving the Cartan integers and inducing all permutations of components mapping isomorphic components onto one another, together with the direct product of the groups of graph automorphisms of the separate components. In case no two components are isomorphic, the latter direct product is the full group Γ. In the notation (7) of Chapter II, § 7, the groups Γ for connected diagrams of the various types are:

Types $A_1, B, C, E_7, E_8, F, G$: $\Gamma = 1$.

Type A_r $(r > 1)$: $\Gamma = \{1, \pi\}$, where $i\pi = r + 1 - i$.

Type D_r $(r > 4)$: $\Gamma = \{1, \pi\}$, where $i\pi = i$, $i \leqq r - 2$;

$$(r-1)\pi = r;\ r\pi = r - 1.$$

Type D_4: $\pi \in \Gamma$ may effect every permutation of $\{1, 3, 4\}$, while

$$2\pi = 2;\ \Gamma \text{ is the symmetric group } \mathfrak{S}_3.$$

Type E_6: $\Gamma = \{1, \pi\}$, where $6\pi = 6$, $i\pi = 6 - i$ if $i < 6$.

Thus $\mathfrak{A}(\mathfrak{L}) = G$ if and only if $\mathfrak{L}$ is a direct sum of simple algebras of distinct types among $A_1, B, C, E_7, E_8, F_4, G_2$. For *simple* algebras, these and the remarks of § 3 show that $\mathfrak{A}(\mathfrak{L}) = G'$ if $\mathfrak{L}$ is of type E_8, F_4, G_2; $\mathfrak{A}(\mathfrak{L}) = G'$ if $\mathfrak{L}$ is of type A_1, B, C, E_7, and if every element of $\mathfrak{F}$ is a square; $\mathfrak{A}(\mathfrak{L}) \neq G'$ otherwise.

§ 6. Realizations

A_n: Let $\mathfrak{F}$ be a field of characteristic $\neq 2, 3$, and let $\mathfrak{V}$ be an $(n+1)$-dimensional vector space over $\mathfrak{F}$. Let $\mathfrak{L}$ be the quotient Lie algebra by its center of the Lie algebra $\mathfrak{M}$ of endomorphisms of $\mathfrak{V}$ having trace zero. Then $\mathfrak{L} = \mathfrak{M}$ unless $\mathfrak{F}$ has prime characteristic p dividing $n + 1$, in which case the center of $\mathfrak{M}$ consists of the scalars. Relative to a fixed basis $v_1, \ldots, v_{n+1}$ for $\mathfrak{V}$, the cosets, by the center, of elements of $\mathfrak{M}$ having diagonal matrices form a Cartan subalgebra $\mathfrak{H}$, the cosets of matrix units E_{ij}, $i \neq j$, are root-vectors belonging to distinct non-zero roots relative to $\mathfrak{H}$, and $\mathfrak{L}$ is classical relative to the classical Cartan subalgebra $\mathfrak{H}$. The roots α_i to which the $E_{i,i+1}$ belong constitute a fundamental system, which is connected of type A_n. Thus $\mathfrak{L}$ is a simple Lie algebra of type A_n.

For each automorphism U of $\mathfrak{B}$, $U^{-1}\mathfrak{M}U = \mathfrak{M}$, and it follows that conjugation by U induces an automorphism σ_U of $\mathfrak{L}$. The homomorphism $U \to \sigma_U$ maps the full linear group $GL(\mathfrak{B})$ into $\mathfrak{A}(\mathfrak{L})$, with kernel consisting of scalars. If $U = I + \lambda E_{ij}$, $i \neq j$, $\lambda \in \mathfrak{F}$, then $\sigma_U = E(\lambda \bar{E}_{ij})$, where $\bar{E}_{ij}$ is the coset in $\mathfrak{L}$ of E_{ij}, so that the group $G' = G'(\mathfrak{H})$ is the image of the group generated by the $I + \lambda E_{ij}$, under the mapping $U \to \sigma_U$. Since the latter group is the special linear group $SL(\mathfrak{B})$, we have $G' \cong PSL(\mathfrak{B})$, the projective special linear group. Letting $D = \operatorname{diag}\{1, \mu_1, \mu_1\mu_2, \ldots, \mu_1 \ldots \mu_n\}$, $\mu_i \neq 0$, one sees that σ_D leaves $\mathfrak{H}$ fixed and maps $\bar{E}_{i,i+1}$ onto $\mu_i \bar{E}_{i,i+1}$. Thus the group H is contained in the image of $GL(\mathfrak{B})$ under $U \to \sigma_U$, so that $G = G(\mathfrak{H})$ is also contained in this image, which is isomorphic with $PGL(\mathfrak{B})$. Finally, there is an automorphism τ of $\mathfrak{L}$ sending $\bar{E}_{ij}$ onto $-\bar{E}_{ji}$ for all $i \neq j$, and $\tau \neq \sigma_U$ for all $U \in GL(\mathfrak{B})$ unless $n = 1$. Since G is of index 1 (if $n = 1$) or 2 (if $n > 1$) in $\mathfrak{A}(\mathfrak{L})$, we have $G \cong PGL(\mathfrak{B})$, and $\mathfrak{A}(\mathfrak{L})$ has $PGL(\mathfrak{B})$ as a subgroup of index 1 or 2, there being another coset (with τ as representative) if and only if $n > 1$ (cf. [216, 358]).

C_r: Let $\mathfrak{B}$ be a vector space of dimension $2r$ over $\mathfrak{F}$, and denote by (x, y) an alternate non-singular bilinear form on $\mathfrak{B}$; i.e., $(x, x) = 0 = (x, y) + (y, x)$ for all x, y. Let $\mathfrak{L}$ be the Lie algebra of endomorphisms T of $\mathfrak{B}$ which are skew with respect to this form: $(xT, y) = -(x, yT)$ for all x, y. Then $\mathfrak{B}$ has a "symplectic basis" $v_1, \ldots, v_{2r}$ with $(v_i, v_j) = 0$ unless $|i - j| = r$, while $(v_i, v_{i+r}) = 1$, $1 \leqq i \leqq r$. Relative to this basis for $\mathfrak{B}$, $\mathfrak{L}$ has a basis consisting of the matrix-unit combinations

$$\begin{aligned} &E_{ii} - E_{i+r,i+r};\ E_{i,i+r};\ E_{i+r,i};\\ (5)\qquad &E_{ij} - E_{j+r,i+r},\ i \neq j;\\ &E_{i,j+r} + E_{j,i+r},\ E_{i+r,j} + E_{j+r,i},\ i < j;\\ &1 \leqq i, j \leqq r. \end{aligned}$$

The $E_{ii} - E_{i+r,i+r}$ span a Cartan subalgebra $\mathfrak{H}$, relative to which the remaining basis elements are root vectors belonging to distinct roots. One verifies at once that $\mathfrak{L}$ is classical relative to $\mathfrak{H}$. If α_i, $1 \leqq i \leqq r - 1$, is the root to which $E_{i,i+1} - E_{i+r+1,i+r}$ belongs, and α_r that to which $E_{r,2r}$ belongs, one finds that $\alpha_1, \ldots, \alpha_r$ is a fundamental system relative to $\mathfrak{H}$. If $r = 1$, this system is of type A_1; if $r = 2$, it is of type B_2; if $r \geqq 3$, it is of type C_r; in each case the results of § 5 yield that the group $\mathfrak{A}(\mathfrak{L})$ of automorphisms of $\mathfrak{L}$ is the group G of Chevalley. In particular, $\mathfrak{A}(\mathfrak{L})$ contains all mappings σ_U: $X \to U^{-1}XU$, where U is in the group $\mathfrak{S}$ of *similitudes* of $\mathfrak{B}$: $(xU, yU) = \beta(x, y)$ for all $x, y \in \mathfrak{B}$, where $0 \neq \beta = \beta(U) \in \mathfrak{F}$. Let T be a *symplectic transvection* of $\mathfrak{B}$: $xT = x + \lambda(a, x)\,a$, where $0 \neq a$ is fixed in $\mathfrak{B}$. We may take $a = u_1$ as the first vector in a symplectic basis

$u_1, \ldots, u_{2r}$ for $\mathfrak{V}$, in which case T has the matrix $I + \lambda E_{r+1,1}$. As with our original basis $\{v_i\}$, $E_{r+1,1}$, relative to the $\{u_i\}$, is a root vector relative to a classical Cartan subalgebra, so that $\exp(\lambda \operatorname{ad} E_{r+1,1}) = \sigma_T$ is in the group G' by § 5. Now the symplectic transvections generate the symplectic group $\mathrm{Sp}(\mathfrak{V})$ [18, 102, 112], so that the image of $\mathrm{Sp}(\mathfrak{V})$ under the homomorphism $U \to \sigma_U$ is contained in G'; on the other hand, each of the root vectors e_α of (5) is an endomorphism of square zero belonging to $\mathfrak{L}$, from which $I + \lambda e_\alpha \in \mathrm{Sp}(\mathfrak{V})$ for all $\lambda \in \mathfrak{F}$, and $E(\lambda e_\alpha) = \sigma_{I+\lambda e_\alpha}$ shows that every element of G' has the form σ_U for some $U \in \mathrm{Sp}(\mathfrak{V})$.

Next let $D \in H$, $(E_{i,i+1} - E_{i+r+1,i+r}) D = \beta_i (E_{i,i+1} - E_{i+r+1,i+1})$, $1 \leqq i \leqq r-1$; $E_{r,2r} D = \beta_r E_{r,2r}$, $0 \neq \beta_i \in \mathfrak{F}$. Then D is determined by the β_i. Let U be that endomorphism of $\mathfrak{V}$ with $v_i U = \lambda_i v_i$, $1 \leqq i \leqq 2r$, where $\lambda_1 = 1$, $\lambda_i = \beta_1 \ldots \beta_{i-1}$, $2 \leqq i \leqq r$; $\lambda_{2r} = \beta_1 \ldots \beta_r$, $\lambda_{2r-i} = \beta_1 \ldots \beta_r \beta_{r-1} \ldots \beta_{r-i}$, $1 \leqq i < r$. Then U is a similitude, $\beta(U) = \beta_1^2 \ldots \beta_{r-1}^2 \beta_r$, and $\sigma_U = D$. It follows that the image of $\mathfrak{S}$ under $U \to \sigma_U$ contains $G = \mathfrak{A}(\mathfrak{L})$, i.e., that every automorphism of $\mathfrak{L}$ is of the form $X \to U^{-1} X U$, $U \in \mathfrak{S}$. From these observations one sees at once that $G' \cong P\,\mathrm{Sp}(\mathfrak{V})$, $G = \mathfrak{A}(\mathfrak{L}) \cong P\mathfrak{S}$, where $P\,\mathrm{Sp}$, $P\mathfrak{S}$ denote the corresponding projective groups. The mapping assigning to each $U \in \mathfrak{S}$ the coset $\beta(U)\,\mathfrak{F}^{*2}$ in $\mathfrak{F}^*/\mathfrak{F}^{*2}$ induces on G/G' an isomorphism onto $\mathfrak{F}^*/\mathfrak{F}^{*2}$; this may be regarded as another way of obtaining the result $G/G' \cong \mathfrak{F}^*/\mathfrak{F}^{*2}$ of § 5 (cf. [216, 326, 358]).

B_r: Let $\mathfrak{V}$ be a vector space of dimension $2r+1$ $(r \geqq 1)$ over $\mathfrak{F}$, and assume given on $\mathfrak{V}$ a non-singular symmetric bilinear form (x, y) of maximal Witt index, r. We assume that the discriminant of the form is $(-1)^r$, modulo squares (this assumption is inessential, since any two spaces of dimension $2r+1$ with forms of maximal Witt index are related by a similitude), so that $\mathfrak{V}$ has a basis $v_1, \ldots, v_{2r+1}$ with $(v_i, v_j) = 0$ if $|i-j| \neq r$, $(v_i, v_{i+r}) = 1$ for $1 \leqq i, j \leqq 2r$, and with $(v_i, v_{2r+1}) = \delta_{i,2r+1}$, $1 \leqq i \leqq 2r+1$. Let $\mathfrak{L}$ be the Lie algebra of endomorphisms of $\mathfrak{V}$ which are skew relative to (x, y). Then $\mathfrak{L}$ has a basis

$$
\begin{aligned}
&E_{ii} - E_{i+r,i+r};\ E_{ij} - E_{j+r,i+r},\ i \neq j;\\
(6)\qquad &E_{i,2r+1} - E_{2r+1,i+r};\ E_{i+r,2r+1} - E_{2r+1,i};\\
&E_{i,j+r} - E_{j,i+r},\ i < j;\ E_{i+r,j} - E_{j+r,i},\ i < j;\\
&1 \leqq i, j \leqq r.
\end{aligned}
$$

The $E_{ii} - E_{i+r,i+r}$ span a classical Cartan subalgebra $\mathfrak{H}$ relative to which the remaining basis elements are root vectors belonging to distinct roots. A fundamental system of roots consists of the roots α_i, $1 \leqq i \leqq r-1$, to which the $E_{i,i+1} - E_{i+r+1,i+r}$ belong, and the root α_r to which $E_{r,2r-1} - E_{2r+1,2r}$ belongs. If $r = 1$, this system is of type

A_1; if $r \geqq 2$, it is of type B_r. Thus $\mathfrak{L}$ is a classical simple Lie algebra of type B_r if $r \geqq 2$ and we have $\mathfrak{A}(\mathfrak{L}) = G$ for all values of r by § 5.

Again $\mathfrak{A}(\mathfrak{L})$ contains all mappings σ_U, where U is a similitude of $\mathfrak{V}$, and the mapping $U \to \sigma_U$ is a homomorphism of the group $\mathfrak{S}$ of similitudes onto a subgroup of $\mathfrak{A}(\mathfrak{L})$, the kernel being the scalar automorphisms of $\mathfrak{V}$. The displayed (in (6)) root vectors e_α relative to $\mathfrak{H}$ satisfy $e_\alpha^3 = 0$ as endomorphisms of $\mathfrak{V}$, with $e_\alpha^2 = 0$ except for those e_α involving the index $2r+1$. It follows that $E(\lambda e_\alpha) = \sigma_U$, where $U = I + \lambda e_\alpha + \frac{1}{2}\lambda^2 e_\alpha^2 \in \mathfrak{O}^+(\mathfrak{V})$, the proper orthogonal group of $\mathfrak{V}$, hence that G' is contained in the image of $\mathfrak{O}^+(\mathfrak{V})$ under the mapping $U \to \sigma_U$. Next let $D \in H$, $e_{\alpha_i} D = \beta_i e_{\alpha_i} \neq 0$, $1 \leqq i \leqq r$; then $D = \sigma_U$, where $v_j U = \lambda_j v_j$ for all j, with $\lambda_{2r+1} = 1$, $\lambda_{2r-j} = \beta_r \beta_{r-1} \ldots \beta_{r-j}$, $0 \leqq j \leqq r-1$, $\lambda_i = \lambda_{i+r}^{-1}$, $1 \leqq i \leqq r$. Moreover, $U \in \mathfrak{O}^+(\mathfrak{V})$. It follows that $\mathfrak{A}(\mathfrak{L}) = G$ consists of automorphisms σ_U, $U \in \mathfrak{O}^+(\mathfrak{V})$, and hence that if $W \in \mathfrak{S}$, then $W = \mu U$ for some $U \in \mathfrak{O}^+(\mathfrak{V})$, $\mu \in \mathfrak{F}^*$, so that the factor of similitude $\beta(W)$ is $\mu^2 \in \mathfrak{F}^{*2}$; *every factor of similitude in $\mathfrak{V}$ is a square.* Since $\mathfrak{O}^+(\mathfrak{V})$ has trivial center, we have $G = \mathfrak{A}(\mathfrak{L}) \cong \mathfrak{O}^+(\mathfrak{V})$; since G' is the commutator subgroup of G and $\mathfrak{O}'(\mathfrak{V})$ that of $\mathfrak{O}^+(\mathfrak{V})$, we have $G' \cong \mathfrak{O}'(\mathfrak{V})$. An isomorphism of G/G' with $\mathfrak{F}^*/\mathfrak{F}^{*2}$ may be obtained via the spinorial norm isomorphism of $\mathfrak{O}^+(\mathfrak{V})/\mathfrak{O}'(\mathfrak{V})$ with $\mathfrak{F}^*/\mathfrak{F}^{*2}$ [18, 112, 129, 358].

D_r: Let $\mathfrak{V}$ be a vector space of dimension $2r$ over $\mathfrak{F}$, and let (x, y) be a non-singular symmetric bilinear form on $\mathfrak{V}$ of maximal Witt index. Let $\mathfrak{L}$ be the Lie algebra of endomorphisms of $\mathfrak{V}$ which are skew with respect to (x, y). Then $\mathfrak{V}$ has a basis $v_1, \ldots, v_{2r}$ with $(v_i, v_j) = 0$ if $|i-j| \neq r$, and with $(v_i, v_{i+r}) = 1$. $\mathfrak{L}$ has as basis the

$$\begin{aligned}
&E_{ii} - E_{i+r,\,i+r};\ E_{i,j} - E_{j+r,\,i+r},\ i \neq j;\\
(7)\qquad &E_{i,\,j+r} - E_{j,\,i+r},\ i<j;\ E_{i+r,\,j} - E_{j+r,\,i},\ i<j;\\
&1 \leqq i, j \leqq r.
\end{aligned}$$

If $r \geqq 2$, the $E_{ii} - E_{i+r,\,i+r}$ span a classical Cartan subalgebra $\mathfrak{H}$, relative to which the other basis elements are root-vectors; letting $E_{i,\,i+1} - E_{i+r+1,\,i+r}$ belong to the root α_i, $1 \leqq i \leqq r-1$, and $E_{r-1,\,2r} - E_{r,\,2r-1}$ to α_r, we see that $\alpha_1, \ldots, \alpha_r$ is a fundamental system of roots (for $r = 1$, $\mathfrak{L}$ has dimension one). When $r = 2$, the system α_1, α_2 decomposes, and $\mathfrak{L}$ is the direct sum of two ideals of type A_1; when $r = 3$, the system $\alpha_1, \alpha_2, \alpha_3$ is of type A_3; when $r \geqq 4$, the fundamental system is of type D_r. Now the group of graph automorphisms has order 2 unless $r = 4$, in which case it is the symmetric group on 3 symbols. Again $\mathfrak{A}(\mathfrak{L})$ contains all σ_U, U a similitude, and the group G' is contained in the group of all σ_U for $U \in \mathfrak{O}^+(\mathfrak{V})$. Since $G' = (G', G')$, G' is contained in the group of all σ_U for $U \in \mathfrak{O}'(\mathfrak{V})$. If $D \in H$, $e_{\alpha_i} D = \beta_i e_{\alpha_i}$, let U be the automorphism of $\mathfrak{V}$ with $v_1 U = v_1$, $v_i U = \beta_1 \ldots \beta_{i-1} v_i$,

$1 < i \leqq r$; $v_{r+1}U = \gamma\beta_r v_{r+1}$, $v_{r+i}U = \gamma\beta_r(\beta_1 \ldots \beta_{i-1})^{-1} v_{r+i}$, $1 < i \leqq r$, where $\gamma = \beta_{r-1}(\beta_1 \ldots \beta_{r-2})^2$. Then U is a similitude with factor $\gamma\beta_r$ such that $D = \sigma_U$. Moreover, $\det U = (\gamma\beta_r)^r$; now if W is a similitude with factor β, $\det W$ is equal to $\pm\beta^r$. W is called *proper* if $\det W = \beta^r$, otherwise *improper*. Since all elements of $\mathfrak{O}^+(\mathfrak{V})$, as well as U defined above, are proper, the group G is contained in the group of automorphisms of $\mathfrak{L}$ of the form σ_U, where U is a proper similitude.

Now let $T \in \mathfrak{O}(\mathfrak{V})$ leave v_i fixed for $i \neq r, 2r$, with $v_r T = v_{2r}$, $v_{2r}T = v_r$; then $\det T = -1$, so that T is improper, and σ_T has the form $D\pi$, where $D \in H \subseteqq G$, and where π is the graph automorphism interchanging α_{r-1} and α_r as in § 5. If an element of the coset $G\pi$ is of the form σ_U for U a proper similitude, then so is σ_T, so that $T = \lambda U$, $\lambda \in \mathfrak{F}$, $\lambda^2\beta(U) = 1$, and $\det T = \lambda^{2r}\beta(U)^r = 1$, which is absurd. Hence $G\pi = G\sigma_T$, and we have proved *for $r \neq 4$ that every automorphism of $\mathfrak{L}$ is of the form σ_U for some similitude U of $\mathfrak{V}$. The group G consists of those σ_U for $U \in \mathfrak{S}^+$, the group of proper similitudes; the group G' consists of those σ_U for $U \in \mathfrak{O}'(\mathfrak{V})$, the commutator subgroup of the orthogonal group.* The last remark follows from the fact that $U \to \sigma_U$ maps the commutator subgroup $(\mathfrak{S}^+, \mathfrak{S}^+)$ onto the commutator subgroup G'. Since $G' \subseteqq \sigma_{\mathfrak{O}'}$, it suffices to note that $\mathfrak{O}' \subseteqq (\mathfrak{S}^+, \mathfrak{S}^+)$, which is immediate from the fact that $\mathfrak{O}' = (\mathfrak{O}^+, \mathfrak{O}^+)$ [18, 102].

For $r = 4$, the bilinear form (x, y) admits composition, i.e., a bilinear product xy can be introduced in $\mathfrak{V}$ relative to which $(xy, xy) = (x, x)(y, y)$ for all x and y [73, 229]. There is in $\mathfrak{L}$ a *principle of triality*: if $A \in \mathfrak{L}$, then there exist uniquely determined elements $A^\varphi, A^\psi \in \mathfrak{L}$, such that $(xy)A = (xA^\varphi)y + x(yA^\psi)$. for all $x, y \in \mathfrak{V}$; furthermore, there is no automorphism B of $\mathfrak{V}$ such that $A^\varphi = B^{-1}AB$ for all $A \in \mathfrak{L}$, nor such that $A^\psi = B^{-1}AB$ for all A, nor such that $A^\varphi = B^{-1}A^\psi B$ for all A [234, 358]. Thus the automorphisms I, φ, ψ belong to distinct cosets in $\mathfrak{A}(\mathfrak{L})$ of the group G^* of automorphisms of the form σ_U, $U \in \mathfrak{S}$. Now we have seen that $(\mathfrak{A}(\mathfrak{L}) : G) = 6$, $(G^* : G) = 2$. Thus $G^*, \varphi G^*, \psi G^*$ are the left cosets of G^* in $\mathfrak{A}(\mathfrak{L})$, and *every automorphism of $\mathfrak{L}$ has one and only one of the forms, σ_U, $\varphi\sigma_U$, $\psi\sigma_U$, where $U \in \mathfrak{S}$.*

G_2: Letting $\mathfrak{C}$ be the algebra (the "split Cayley algebra") on the vector space $\mathfrak{V}$ utilized above in describing the automorphisms of D_4, let $\mathfrak{L}$ be the Lie algebra of derivations of $\mathfrak{C}$. If A is an automorphism of $\mathfrak{C}$, then σ_A is an automorphism of $\mathfrak{L}$. Both the center of the automorphism group $\mathfrak{A}(\mathfrak{C})$ and the centralizer of $\mathfrak{L}$ in $\mathfrak{A}(\mathfrak{C})$ reduce to the identity. Further, a classical Cartan decomposition of type G_2 of $\mathfrak{L}$ may be chosen in which the root-vectors e_α satisfy $e_\alpha^3 = 0$, acting in $\mathfrak{C}$. Since the characteristic is not 2 or 3, each $\exp(\lambda e_\alpha)$ is in $\mathfrak{A}(\mathfrak{C})$, and $E(\lambda e_\alpha) = \sigma_{\exp(\lambda e_\alpha)}$. Hence, by § 5, $G' \subseteqq \sigma_{\mathfrak{A}(\mathfrak{C})} \subseteqq \mathfrak{A}(\mathfrak{L}) = G = G'$. It follows that *the group* $\mathfrak{A}(\mathfrak{L})$, *which coincides with the simple group* G', *is isomorphic with the group* $\mathfrak{A}(\mathfrak{C})$ *of automorphisms of* $\mathfrak{C}$, *via the mapping* $A \to \sigma_A$ *from* $\mathfrak{A}(\mathfrak{C})$ *onto* $\mathfrak{A}(\mathfrak{L})$ [214, 234, 358].

F_4: We speak of "the split exceptional Jordan algebra" $\mathfrak{J}$ over $\mathfrak{F}$ as the space of 3 by 3 hermitian $\mathfrak{C}$-matrices, relative to the involution $x \to 2(x, 1)\,1 - x$ in $\mathfrak{C}$, and with the symmetrized product $u\,v + v\,u$ (written simply $u\,v$ in the sequel). $\mathfrak{J}$ is a 27-dimensional simple Jordan algebra over $\mathfrak{F}$, and the bilinear form $(u, v) = \mathrm{Trace}(u\,v) \in \mathfrak{F}$ is symmetric and non-singular on $\mathfrak{J}$. Let $\mathfrak{L}$ be the Lie algebra of derivations of $\mathfrak{J}$, $\mathfrak{A}(\mathfrak{J})$ the group of automorphisms of $\mathfrak{J}$. Then $\mathfrak{L}$ is a classical Lie algebra of type F_4 and has a classical Cartan decomposition with root-vectors e_α satisfying $e_\alpha^3 = 0$. From the fact that $\mathfrak{A}(\mathfrak{L}) = G = G'$ in this case, and because the centralizer of $\mathfrak{L}$ in $\mathfrak{A}(\mathfrak{J})$ is the identity, one sees as for type G_2 that $\mathfrak{A}(\mathfrak{L}) = G'$ *is isomorphic with* $\mathfrak{A}(\mathfrak{J})$ *by the mapping* $A \to \sigma_A$, $A \in \mathfrak{A}(\mathfrak{J})$ [81, 234, 359, 400].

E_6: With $\mathfrak{J}$ as above, every element $x \in \mathfrak{J}$ satisfies a relation $x^3 - t(x)\,x^2 + s(x)\,x - n(x)\,1 = 0$, where $t(x), s(x), n(x) \in \mathfrak{F}$, and where n is a cubic form on $\mathfrak{J}$. Polarization of n yields a symmetric trilinear form (x, y, z) on $\mathfrak{J}$. Let $\mathfrak{L}$ be the Lie algebra of linear transformations B of $\mathfrak{J}$ which are skew with respect to this form, i.e., which satisfy $(xB, y, z) + (x, yB, z) + (x, y, zB) = 0$ for all $x, y, z \in \mathfrak{J}$. Then $\mathfrak{L}$ is a classical Lie algebra over $\mathfrak{F}$, and has a fundamental system of roots of type E_6. $\mathfrak{L}$ has a classical Cartan decomposition in which the root vectors e_α satisfy $e_\alpha^2 = 0$, from which it follows as above that the group G' is contained in the group of all σ_U, where U is in the group $\mathfrak{T}$ of non-singular linear transformations of $\mathfrak{J}$ preserving the form (x, y, z). Such σ_U are clearly automorphisms of $\mathfrak{L}$, as are all σ_S, for S a similitude of $\mathfrak{J}$ with respect to (x, y, z). By considerations similar to those of earlier cases one sees that G is contained in the group of all σ_S, S a similitude. Furthermore, the mapping $\varphi: B \to -B^*$, where $(xB^*, y) = (x, yB)$ for all $x, y \in \mathfrak{J}$ (the bilinear form being the trace form of the preceding paragraph), is an automorphism of $\mathfrak{L}$ and is not of the form σ_U for any linear transformation U of $\mathfrak{J}$. Since $(\mathfrak{A}(\mathfrak{L}) : G) = 2$, it follows that $G = \{\sigma_S \mid S \text{ a similitude}\}$, and that $\mathfrak{A}(\mathfrak{L}) = G \cup \varphi\,G$ is the coset decomposition of $\mathfrak{A}(\mathfrak{L})$ relative to G. One can show that the $\exp(\lambda\,e_\alpha)$ as above generate $\mathfrak{T}$, so that $G' = \{\sigma_U \mid U \in \mathfrak{T}\}$, and that the mapping which assigns to the similitude S the coset modulo $\mathfrak{F}^{*3}$ of the factor of similitude of S induces on G a homomorphism onto $\mathfrak{F}^*/\mathfrak{F}^{*3}$ with kernel G'. The conclusions: *Let* $\mathfrak{S}$ *be the group of similitudes,* $\mathfrak{T}$ *the group preserving the trilinear ("norm") form on* $\mathfrak{J}$; *then the group* $\mathfrak{A}(\mathfrak{L})$ *has the coset decomposition* $G \cup \varphi\,G$, *where* $\varphi: B \to -B^*$ *is as above. The group* G *is the set of automorphisms* σ_S, $S \in \mathfrak{S}$, *and* G' *is the set of* σ_T, $T \in \mathfrak{T}$ [137, 227, 359, 376].

E_7: Upon multiplying the trilinear form of the last paragraph by a suitable scalar, we may assume $3n(x) = (x, x, x)$ for all $x \in \mathfrak{J}$. Combining with the trace form (u, v) on $\mathfrak{J}$ yields a commutative bilinear multiplication $x \times y$ in $\mathfrak{J}$ determined by $(x \times y, z) = (x, y, z)$ for all $z \in \mathfrak{J}$. Let $\mathfrak{B}$ be the 56-dimensional space $\mathfrak{J} \oplus \mathfrak{J} \oplus \mathfrak{F} \oplus \mathfrak{F}$, and let q be the quartic form on $\mathfrak{B}$ defined by

(8) $q(x, y, \xi, \eta) = (x \times x, y \times y) - 2\xi\, n(x) - 2\eta\, n(y) - 2(\frac{1}{2}(x, y) - \xi\eta)^2$.

Denote by (t, u, v, w) a polarization of q, and let $\mathfrak{L}$ be the Lie algebra of endomorphisms of $\mathfrak{B}$ which are skew with respect to this symmetric 4-linear form. Then $\mathfrak{L}$ has a classical Cartan decomposition with a fundamental system of roots of type E_7 and root-vectors e_α all satisfying $e_\alpha^2 = 0$ as mappings of $\mathfrak{B}$. It follows as with E_6 that the group G' is contained in the group of σ_U, $U \in \mathfrak{O}$, the group of q-preserving automorphisms of $\mathfrak{B}$, while $\mathfrak{A}(\mathfrak{L}) = G$ contains all σ_S, $S \in \mathfrak{S}$, the group of q-similitudes. In fact, one can show:

a) *If* $S \in \mathfrak{S}$, *the factor of similitude of* S *is in* $\mathfrak{F}^{*2}$, *and* $G = \{\sigma_S \mid S \in \mathfrak{S}\}$;

b) $G' = \{\sigma_U \mid U \in \mathfrak{O}\}$ *if and only if* $-1 \in \mathfrak{F}^{*2}$;

c) $G' = \{\sigma_U \mid U \in (\mathfrak{O}, \mathfrak{O})\}$.

These results have been established in [363].

E_8: In view of the representation theory in characteristic zero, it appears that the faithful representation of minimal degree of an algebra of type E_8 is the adjoint representation, in 248 dimensions. From the realization given by CARTAN [53], it seems that (if $p \neq 2, 3, 5$) a Lie algebra of type E_8 may be realized as linear transformations in 248 dimensions which are skew both with respect to a certain quintic form q and with respect to a symmetric bilinear form b (the Killing form, if one regards this as the adjoint representation). The fact that $\mathfrak{A}(\mathfrak{L}) = G = G'$ in this case will be reflected in coincidence of the groups of all σ_S, S a q- and b-similitude, of all σ_U, where U preserves q and b, and the group of all σ_T, where T is in the commutator subgroup of either the group of joint similitudes or the group of joint isometries. Thus the forms q and b should have the property that if S is a (q, b)-similitude with factors (β, γ), then $\beta \in \mathfrak{F}^{*5}$, $\gamma \in \mathfrak{F}^{*2}$.

In each of the cases listed above, the group G' is identified with the quotient, by its scalar members, of a certain linear group. In no case has the simplicity of this quotient been used in establishing the identification, although the quotient is known to be simple except in the cases of E_7 and E_8 (for types A—D see, e.g., [18 or 112]; for G_2, see [229, 394]; for F_4 and E_6, [232, 233]). Thus these identifications, together with the simplicity of G', establish the simplicity (over fields of characteristics $\neq 2, 3$) of a number of classes of projective linear groups.

Chapter IV

Forms of the Classical Lie Algebras

The discussion in this chapter may be regarded as motivated by the problem of determining all Lie algebras $\mathfrak{L}$ with non-singular Killing form over an arbitrary field $\mathfrak{F}$ (characteristic $\neq 2, 3$). By Chapter I, § 7, the problem reduces to the case where $\mathfrak{L}$ is simple. If $\mathfrak{Z}$ is the centroid of $\mathfrak{L}$, then $\mathfrak{L}$ is normal simple when regarded as a Lie algebra over $\mathfrak{Z}$; moreover, if S is a $\mathfrak{Z}$-linear transformation of $\mathfrak{L}$, we have $\mathrm{Tr}_{\mathfrak{F}}(S) = T_{\mathfrak{Z}/\mathfrak{F}}(\mathrm{Tr}_{\mathfrak{Z}}(S))$, where $T_{\mathfrak{Z}/\mathfrak{F}}$ is the trace in the field extension $\mathfrak{Z}/\mathfrak{F}$ (cf. [223, p. 66]). It follows that $\mathfrak{L}$ has non-singular Killing form over $\mathfrak{Z}$, and that $T_{\mathfrak{Z}/\mathfrak{F}}$ is not zero, hence that $\mathfrak{Z}/\mathfrak{F}$ is a separable extension. Thus the problem is reduced to the study of finite separable extensions of $\mathfrak{F}$, and to the case where $\mathfrak{L}$ is normal simple. Assuming $\mathfrak{L}$ normal simple over $\mathfrak{F}$, let Ω be an algebraically closed extension of $\mathfrak{F}$. Then $\mathfrak{L}_\Omega$ has non-singular Killing form, and so is a classical simple Lie algebra over Ω; therefore $\mathfrak{L}_\Omega$ belongs to one of an already determined set of isomorphism classes. The "problem of forms" is that of describing $\mathfrak{L}$ when $\mathfrak{L}_\Omega$ is known, i.e., of determining the $\mathfrak{F}$-isomorphism classes of algebras $\mathfrak{L}$ such that $\mathfrak{L}_\Omega$ belongs to a given Ω-isomorphism class.

§ 1. Forms and splitting fields

Let $\mathfrak{K}$ be a field, $\mathfrak{F}$ a subfield of $\mathfrak{K}$. Let $(\mathfrak{V}, \{f_i\})$ be a vector space $\mathfrak{V}$ over $\mathfrak{K}$, together with certain $\mathfrak{K}$-multilinear operations f_i on $\mathfrak{V}$, taking on values either in $\mathfrak{K}$ or in $\mathfrak{V}$. Then an $\mathfrak{F}$-*form* of $(\mathfrak{V}, \{f_i\})$ will be a vector space with multilinear operations $(\mathfrak{U}, \{g_i\})$ over $\mathfrak{F}$ such that, denoting again by g_i the unique extension of g_i to a $\mathfrak{K}$-multilinear mapping on $\mathfrak{U}_\mathfrak{K}$, there is a $\mathfrak{K}$-isomorphism of $(\mathfrak{U}_\mathfrak{K}, \{g_i\})$ onto $(\mathfrak{V}, \{f_i\})$, i.e., a $\mathfrak{K}$-isomorphism φ of $\mathfrak{U}_\mathfrak{K}$ onto $\mathfrak{V}$ such that for all $x_1, \ldots, x_n \in \mathfrak{U}_\mathfrak{K}$, $f_i(x_1\varphi, \ldots, x_n\varphi) = g_i(x_1, \ldots, x_n)\varphi$ whenever f_i is n-linear with values in $\mathfrak{V}$ and $f_i(x_1\varphi, \ldots, x_n\varphi) = g_i(x_1, \ldots, x_n)$ whenever f_i has values in $\mathfrak{K}$. (No attempt is made here to define the notion of "form" in a comprehensive manner; rather, a simple definition has been chosen which is more than adequate for the purposes of the theory to be presented. For other formulations, cf. [11, 62, 131, 237, 255].)

As an example, let $\mathfrak{L}$ be a Lie algebra over $\mathfrak{K}$, from this point of view a vector space $\mathfrak{L}$ over $\mathfrak{K}$ with a $\mathfrak{K}$-bilinear mapping of $\mathfrak{L}$ into $\mathfrak{L}$ satisfying the Lie identities. Then an $\mathfrak{F}$-form of $\mathfrak{L}$ is an algebra $\mathfrak{M}$ over $\mathfrak{F}$ such that the algebra $\mathfrak{M}_{\mathfrak{K}}$ is $\mathfrak{K}$-isomorphic with $\mathfrak{L}$; thus $\mathfrak{M}_{\mathfrak{K}}$ is a Lie algebra over $\mathfrak{K}$, and $\mathfrak{M}$ is a Lie algebra over $\mathfrak{F}$. More generally, if $\mathfrak{A}$ is a linear algebra over $\mathfrak{K}$, an $\mathfrak{F}$-form of $\mathfrak{A}$ is a linear algebra $\mathfrak{B}$ over $\mathfrak{F}$ such that $\mathfrak{B}_{\mathfrak{K}}$ is $\mathfrak{K}$-isomorphic with $\mathfrak{A}$. If $\mathfrak{A}$ is a linear algebra with involution θ over $\mathfrak{K}$, an $\mathfrak{F}$-form of $(\mathfrak{A}, \theta)$ is a linear algebra $\mathfrak{B}$ over $\mathfrak{F}$ with mapping (necessarily an involution) η of $\mathfrak{B}$ into $\mathfrak{B}$ such that there is an isomorphism φ of $\mathfrak{K}$-algebras of $\mathfrak{B}_{\mathfrak{K}}$ onto $\mathfrak{A}$ with $(b^{\eta})\,\varphi = (b\,\varphi)^{\theta}$ for all $b \in \mathfrak{B}_{\mathfrak{K}}$, where η is again used to denote its $\mathfrak{K}$-linear extension to $\mathfrak{B}_{\mathfrak{K}}$. We say in this case that φ is an isomorphism of the *involutorial* $\mathfrak{K}$*-algebra* $(\mathfrak{B}_{\mathfrak{K}}, \eta)$ onto $(\mathfrak{A}, \theta)$.

The classical Lie algebras of Chapter II are seen at once to enjoy the following properties:

1) For every type (A_r, B_r, etc.) there is a classical Lie algebra of this type defined over every field $\mathfrak{F}$ (characteristic $\neq 2, 3$).

2) If $\mathfrak{F}$ is a subfield of $\mathfrak{K}$ and if $\mathfrak{M}$ is an $\mathfrak{F}$-form of $\mathfrak{L}$, then if $\mathfrak{M}$ is classical simple of type X over $\mathfrak{F}$, so is $\mathfrak{L}$ over $\mathfrak{K}$. Hence if $\mathfrak{M}$ and $\mathfrak{N}$ are classical simple and are $\mathfrak{F}$-forms of the same algebra $\mathfrak{L}$ over $\mathfrak{K}$, then $\mathfrak{M}$ and $\mathfrak{N}$ are $\mathfrak{F}$-isomorphic.

3) If $\mathfrak{L}$ is classical over $\mathfrak{K}$, and if $\mathfrak{M}$ is an $\mathfrak{F}$-form of $\mathfrak{L}$, then there is an intermediate field $\mathfrak{E}$, finitely generated over $\mathfrak{F}$, such that $\mathfrak{M}_{\mathfrak{E}}$ is classical over $\mathfrak{E}$.

Now let $\mathfrak{K}$ be a field, and let $\mathfrak{F}$ run over all subfields of $\mathfrak{K}$. For each $\mathfrak{F}$, let $C_{\mathfrak{F}}$ be a class of $\mathfrak{F}$-vector spaces with multilinear operations, and let $S_{\mathfrak{F}}$ be a subclass of $C_{\mathfrak{F}}$ such that:

a) Each member of $C_{\mathfrak{F}}$ is an $\mathfrak{F}$-form of a member of $S_{\mathfrak{K}}$; in particular, each member of $C_{\mathfrak{K}}$ is isomorphic with a member of $S_{\mathfrak{K}}$.

b) For each $\mathfrak{F}$ and each $(\mathfrak{V}, \{f_i\})$ in $S_{\mathfrak{K}}$, there is a $(\mathfrak{U}, \{g_i\})$ in $S_{\mathfrak{F}}$ which is an $\mathfrak{F}$-form of $(\mathfrak{V}, \{f_i\})$, and any two such are $\mathfrak{F}$-isomorphic.

c) For each member $(\mathfrak{U}, \{g_i\})$ of $C_{\mathfrak{F}}$, there is a finite separable extension $\mathfrak{E}$ of $\mathfrak{F}$, $\mathfrak{E} \subseteq \mathfrak{K}$, such that $(\mathfrak{U}, \{g_i\})$ is an $\mathfrak{F}$-form of a member of $S_{\mathfrak{E}}$.

Then it is clear that the members of $S_{\mathfrak{F}}$ are determined to within isomorphism by the members of $S_{\mathfrak{P}}$, $\mathfrak{P}$ being the prime field of $\mathfrak{K}$. Taking $S_{\mathfrak{F}}$ to be the class of classical simple Lie algebras over $\mathfrak{F}$ and $C_{\mathfrak{F}}$ to be the class of $\mathfrak{F}$-forms of elements of $S_{\mathfrak{K}}$, we see from 1) and 2) that a) and b) are satisfied, and from 3) that c) is satisfied whenever $\mathfrak{K}$ is algebraic over $\mathfrak{F}$, except perhaps for the requirement of separability. In the setting above, we say that a field $\mathfrak{E}$, $\mathfrak{F} \subseteq \mathfrak{E} \subseteq \mathfrak{K}$, is a *splitting field* for $(\mathfrak{U}, \{g_i\})$, a member of $C_{\mathfrak{F}}$, if $(\mathfrak{U}, \{g_i\})$ is an $\mathfrak{F}$-form of a member of $S_{\mathfrak{E}}$; we say that $(\mathfrak{U}, \{g_i\})$ is *split* if $\mathfrak{F}$ is a splitting field for $(\mathfrak{U}, \{g_i\})$, i.e., if $(\mathfrak{U}, \{g_i\})$ is ($\mathfrak{F}$-) isomorphic with a member of $S_{\mathfrak{F}}$.

Thus we use the term "*split* Lie algebra" to refer to a classical simple Lie algebra, and, if $\mathfrak{L}$ is a Lie algebra over $\mathfrak{F}$ such that $\mathfrak{L}_{\mathfrak{E}}$ is classical simple for some extension $\mathfrak{E}$ of $\mathfrak{F}$, each field $\mathfrak{E}$ with this property is called a *splitting field* for $\mathfrak{L}$. The study of forms of certain classical simple Lie algebras, as well as that of the forms of certain other simple Lie algebras (the *Witt–Jacobson algebras*) has been carried out for modular fields by JACOBSON [216, 217] and BARNES [21], using both the classical Galois theory and the theory for purely inseparable extensions of exponent one; for the non-modular case, see Chapter 10 of [234], and its bibliography. In [21a], BARNES noted that for the study of algebras with non-singular Killing form, the classical Galois theory suffices; i.e., these algebras have separable splitting fields. In fact, this result extends to all forms of classical simple Lie algebras, with at most one exception:

Theorem IV.1.1. Let $\mathfrak{L}$ be a normal simple Lie algebra over a field $\mathfrak{F} \subseteq \mathfrak{K}$, and suppose that $\mathfrak{L}_{\mathfrak{K}}$ is classical, but not of type E_8 if $\mathfrak{F}$ has characteristic 5. Then $\mathfrak{L}_{\mathfrak{E}}$ is classical for a finite separable extension $\mathfrak{E}$ of $\mathfrak{F}$.

We need only consider the case where $\mathfrak{F}$ is infinite; for if $\mathfrak{F}$ is finite, we have an algebraic closure $\overline{\mathfrak{F}}$ of $\mathfrak{F}$ contained in Ω, the algebraic closure of $\mathfrak{K}$. The theorem applies to $\overline{\mathfrak{F}}$ and shows that $\mathfrak{L}_{\overline{\mathfrak{F}}}$ is classical, therefore that $\mathfrak{L}_{\mathfrak{E}}$ is classical for a finitely generated algebraic extension $\mathfrak{E}$ of $\mathfrak{F}$. Since $\mathfrak{F}$ is perfect, $\mathfrak{E}$ is the desired extension. Therefore we may assume $\mathfrak{L}$ has a Cartan subalgebra $\mathfrak{H}$. Now $\mathfrak{L}_{\Omega}$ is classical, and we prove the

Lemma IV.1.1. All Cartan subalgebras of $\mathfrak{L}_{\Omega}$ are classical.

By remarks in Chapter II, §§ 3, 10, it suffices to prove the assertion when $\mathfrak{L} = \mathfrak{L}_{\Omega}$ is of type A_n, with $n \equiv -1 \pmod{p}$. Now let $\mathfrak{M}$ be the Lie algebra of $(n+1)$ by $(n+1)$ matrices of trace zero over Ω, and let $(x, y) = \mathrm{Tr}(x\,y)$ on $\mathfrak{M}$. Let $\mathfrak{H}$ be a Cartan subalgebra of $\mathfrak{M}$, and let $z \in [\mathfrak{H}\,\mathfrak{H}]$, $[z\,\mathfrak{H}] = 0$. Let ϱ be an irreducible constituent, of degree d, of the representation of $\mathfrak{H}$ on $(n+1)$-space obtained from the given representation of $\mathfrak{M}$. Then $\varrho(z) = \varkappa I$ is a scalar, and $0 = \mathrm{Tr}(\varrho(z)) = d\varkappa$. If $y \in \mathfrak{H}$, then by Th. I.5.2, $\varrho(y)$ has a single characteristic root μ, so that $\mathrm{Tr}(\varrho(y)\,\varrho(z)) = d\varkappa\,\mu = 0$; it follows that $(z, \mathfrak{H}) = 0$, and by Lemma II.1.3 that $(z, \mathfrak{M}) = 0$. One easily sees that such z must be scalar: $z = \lambda I$. If $\mathfrak{H}$ consists of scalars, then $z \in [\mathfrak{H}\,\mathfrak{H}] = 0$; otherwise, since the characteristic p is not 2, there is an $x \in \mathfrak{H}$ with $(x, x) \neq 0$. We may assume ϱ is as above, $\mathrm{Tr}(\varrho(x)^2) = d\xi^2 \neq 0$, where ξ is the characteristic root of $\varrho(x)$; but $\mathrm{Tr}(\varrho(z)) = d\lambda = 0$, so $\lambda = 0$, and $z = 0$. As in Lemma II.1.4, $\mathfrak{H}$ is abelian (cf. [424] for a generalization of this argument).

Now let $\mathfrak{H}^*$ be a Cartan subalgebra of $\mathfrak{M}/\Omega I \cong \mathfrak{L}$; we identify $\mathfrak{M}/\Omega I$ with $\mathfrak{L}$. Then $\mathfrak{H}^* = \mathfrak{H}/\Omega I$, where $\mathfrak{H}$ is a subalgebra of $\mathfrak{M}$ containing ΩI. One verifies at once that $\mathfrak{H}$ is a Cartan subalgebra of $\mathfrak{M}$, hence is abelian; therefore $\mathfrak{H}^*$ is abelian. The p-power operation in $\mathfrak{L}$ is induced by the ordinary p-th power in $\mathfrak{M}$. Thus if $h \in \mathfrak{H}^*$, $h^p = 0$, we have $h = z + \Omega I$, $z \in \mathfrak{H}$, $z^p = \lambda I$; replacing z by $z - \lambda^{\frac{1}{p}} I$, we may assume $z^p = 0$. But then $y\,z$ is nilpotent for all $y \in \mathfrak{H}$, $(z, \mathfrak{H}) = 0$, and z is scalar; that is, $h = 0$. It follows that $\mathfrak{H}^*$ is classical; for (*) of Chapter II, § 3, follows as in the proofs of Ths. II.1.1 and II.1.2, and the form on $\mathfrak{L}$ induced by the trace form on $\mathfrak{M}$ is nonsingular.

To return to the proof of the theorem, let $e_1, \ldots, e_r$ be a basis for $\mathfrak{H}$, and consider the Killing polynomial $k(\xi, X) = \operatorname{Det}(\operatorname{ad}(\sum \xi_i e_i) - XI) = X^r(p_r(\xi) + p_{r+1}(\xi)X + \cdots \pm X^{n-r})$, $n = \dim \mathfrak{L}$, where the $p_i(\xi)$ are polynomial functions, $p_r(\xi)$ not identically zero. Denote by $q(\xi, X)$ the polynomial $X^{-r}k(\xi, X)$, and by $D(\xi)$ the discriminant of $g(\xi, X)$, a polynomial function in $\xi_1, \ldots, \xi_r$. Then $D(\xi)$ is not identically zero, since $\mathfrak{L}_\Omega$ is classical relative to $\mathfrak{H}_\Omega$ by the lemma; for $\mathfrak{L}_\Omega = \mathfrak{H}_\Omega + \sum_\alpha (\mathfrak{L}_\Omega)_\alpha$, the $(\mathfrak{L}_\Omega)_\alpha$ being one-dimensional and the α's distinct non-zero linear functions; hence there is $w = \sum \omega_i e_i \in \mathfrak{H}_\Omega$ such that $\alpha(w)$ are all distinct and non-zero. It follows that $p_r(\omega) D(\omega) \neq 0$, hence that $p_r(\lambda) D(\lambda) \neq 0$ for some $(\lambda_1, \ldots, \lambda_r)$ from the infinite field $\mathfrak{F}$. Let $h = \sum \lambda_i e_i \in \mathfrak{H}$. Then h is regular in $\mathfrak{L}$ and the non-zero characteristic roots of ad h have multiplicity one. These are the roots of the polynomial $q(\lambda, X)$, which is therefore separable; let $\mathfrak{E}$ be the splitting field in Ω of this polynomial.

We claim that $\mathfrak{L}_\mathfrak{E}$ is classical relative to $\mathfrak{H}_\mathfrak{E}$. Namely, if $y \in \mathfrak{H}_\mathfrak{E}$, then $[y\,h] = 0$ since $\mathfrak{H}_\Omega$ is abelian, and hence if $0 \neq x \in \mathfrak{L}_\mathfrak{E}$ belongs to the characteristic root $\mu \neq 0$ of ad h, we have $[x\,y] = \beta x$ for some $\beta \in \mathfrak{E}$. Thus all ad y, $y \in \mathfrak{H}_\mathfrak{E}$, act diagonally. Now the roots of $\mathfrak{L}_\Omega$ relative to $\mathfrak{H}_\Omega$ are the Ω-linear extensions of the roots of $\mathfrak{L}_\mathfrak{E}$ relative to $\mathfrak{H}_\mathfrak{E}$, and if $\alpha \neq 0$ is a root, $[(\mathfrak{L}_\mathfrak{E})_\alpha, (\mathfrak{L}_\mathfrak{E})_{-\alpha}]_\Omega = [(\mathfrak{L}_\Omega)_\alpha, (\mathfrak{L}_\Omega)_{-\alpha}] \neq 0$. Properties b) and c) of iii) in the definition of a classical Lie algebra now follow at once for $\mathfrak{L}_\mathfrak{E}$, and the proof is complete. The lemma and Th. III.4.1 together yield the

Theorem IV.1.2. Over an algebraically closed field of characteristic $\neq 5$, all Cartan subalgebras of a classical Lie algebra are conjugate.

It will be noted that simplicity is not assumed; however, if $\mathfrak{L} = \mathfrak{L}_1 \oplus \cdots \oplus \mathfrak{L}_s$, $\mathfrak{L}_i$ simple, and if $\mathfrak{H}$ is a Cartan subalgebra of $\mathfrak{L}$, then it follows from $[\mathfrak{L}_i\, \mathfrak{H}] \subseteq \mathfrak{L}_i$ that $\mathfrak{L}_i = \mathfrak{L}_i^{(0)} \oplus \mathfrak{L}_i^{(1)}$, where $\mathfrak{L}_i^{(0)}$ is the intersection of the kernels of all $(\operatorname{ad} h)^n|_{\mathfrak{L}_i}$ $(h \in \mathfrak{H})$, $n = \dim \mathfrak{L}$, and where $\mathfrak{L}_i^{(1)}$ is the sum of the images of $\mathfrak{L}_i$ under all these mappings (the *Fitting decomposition* of $\mathfrak{L}_i$—cf. [234], p. 39). Now $\mathfrak{L}_i^{(0)} \subseteq \mathfrak{L}^{(0)} = \mathfrak{H}$, and

$\mathfrak{L}_i^{(1)} \subseteq \mathfrak{L}^{(1)}$, with $\mathfrak{L} = \mathfrak{L}^{(0)} \oplus \mathfrak{L}^{(1)} = (\sum \mathfrak{L}_i^{(0)}) \oplus (\sum \mathfrak{L}_i^{(1)})$. Thus in particular $\mathfrak{H} = \sum \mathfrak{L}_i^{(0)}$, $\mathfrak{L}_i^{(0)} = \mathfrak{H} \cap \mathfrak{L}_i$, which is now evidently a Cartan subalgebra of $\mathfrak{L}_i$. If every $\mathfrak{H} \cap \mathfrak{L}_i$ is classical, then it is clear that $\mathfrak{H}$ is classical.

As a consequence of Th. 1, we see that when $\mathfrak{K}$ is an algebraically closed field, a)—c) are satisfied with $S_{\mathfrak{F}}$ the class of split Lie algebras over $\mathfrak{F}$ (excluding E_8 if $p = 5$) and $C_{\mathfrak{F}}$ the class of $\mathfrak{F}$-forms of members of $S_{\mathfrak{K}}$.

Note. Th. 1 and 2 are also valid generally for characteristics at least 5, by results of HUMPHREYS on algebraic Lie algebras (§ VI.2).

§ 2. Galois semi-automorphisms and 1-cohomology

Let $\mathfrak{K}/\mathfrak{F}$ be a finite Galois extension, with Galois group $\mathfrak{G}$. Let $(\mathfrak{V}, \{f_i\})$ be an $\mathfrak{F}$-vector space with multilinear operations, and let $\mathfrak{U}$ be the $\mathfrak{K}$-vector space $\mathfrak{V}_{\mathfrak{K}}$. Denoting again by f_i the $\mathfrak{K}$-multilinear extension of f_i to $\mathfrak{U}$, we see that $(\mathfrak{U}, \{f_i\})$ is a $\mathfrak{K}$-vector space with multilinear operations, and that $(\mathfrak{V}, \{f_i\})$ is an $\mathfrak{F}$-form of $(\mathfrak{U}, \{f_i\})$. If $\sigma \in \mathfrak{G}$, there is a unique σ-semilinear mapping V_σ of $\mathfrak{U}$ into $\mathfrak{U}$ leaving $\mathfrak{V}$ fixed; namely, if $x_1, \ldots, x_n$ is a basis for $\mathfrak{V}$, we have $(\sum \gamma_i x_i) V_\sigma = \sum \gamma_i^\sigma x_i$, $\gamma_i \in \mathfrak{K}$. From this formula it is clear that the fixed elements of $\mathfrak{U}$ under all V_σ, $\sigma \in \mathfrak{G}$, are just those of $\mathfrak{V}$, and that $V_1 = I$, $V_\sigma V_\tau = V_{\sigma\tau}$ for $1, \sigma, \tau \in \mathfrak{G}$. One also sees at once that for $y_1, \ldots, y_m \in \mathfrak{U}$, $f_i(y_1 V_\sigma, \ldots, y_m V_\sigma) = f_i(y_1, \ldots, y_m) V_\sigma$ if the original f_i is $\mathfrak{V}$-valued, $f_i(y_1 V_\sigma, \ldots, y_m V_\sigma) = f_i(y_1, \ldots, y_m)^\sigma$ if f_i is $\mathfrak{F}$-valued. It follows that if A is an automorphism of $(\mathfrak{U}, \{f_i\})$ in the sense of § 1, then so is $V_\sigma^{-1} A V_\sigma$, and that the mapping $A \to A^\sigma = V_\sigma^{-1} A V_\sigma$ defines an action of $\mathfrak{G}$ as a group of automorphisms of the automorphism group $\mathrm{Aut}(\mathfrak{U}, \{f_i\})$. Examination of the condition $x_j V_\sigma^{-1} A V_\sigma = x_j A$ for all j, $1 \leqq j \leqq n$, and for all $\sigma \in \mathfrak{G}$ shows that $A^\sigma = A$ for all $\sigma \in \mathfrak{G}$ if and only if $x_j A \in \mathfrak{V}$ for all j, i.e., if and only if A is the extension to $\mathfrak{U}$ of an automorphism of $(\mathfrak{V}, \{f_i\})$.

Now let $(\mathfrak{W}, \{g_i\})$ be an arbitrary $\mathfrak{F}$-form of $(\mathfrak{U}, \{f_i\})$, and let φ be a $\mathfrak{K}$-isomorphism of $(\mathfrak{W}_{\mathfrak{K}}, \{g_i\})$ onto $(\mathfrak{U}, \{f_i\})$. If $w_1, \ldots, w_n$ is a basis for $\mathfrak{W}$, and if $\sigma \in \mathfrak{G}$, the mapping U_σ: $(\sum \gamma_j w_j) \varphi \to (\sum \gamma_j^\sigma w_j) \varphi$ is a σ-semilinear mapping of $\mathfrak{U}$ into $\mathfrak{U}$, and $U_1 = I$, $U_\sigma U_\tau = U_{\sigma\tau}$, $f_i(y_1 U_\sigma, \ldots, y_m U_\sigma) = f_i(y_1, \ldots, y_m) U_\sigma$ or $f_i(y_1, \ldots, y_m)^\sigma$, according as f_i is $\mathfrak{U}$-valued or $\mathfrak{K}$-valued. The group of all $\{U_\sigma\}$ is called a group of *Galois semi-automorphisms* of $(\mathfrak{U}, \{f_i\})$. The fixed elements of $\mathfrak{U}$ under this group form an $\mathfrak{F}$-subspace $\mathfrak{X}$ whose inverse image under φ is $\mathfrak{W}$, and the $\mathfrak{F}$-space with multilinear operations $(\mathfrak{X}, \{f_i \mid \mathfrak{X}\})$ is isomorphic with $(\mathfrak{W}, \{g_i\})$ by means of φ.

With V_σ as above, let $Z_\sigma = U_\sigma^{-1} V_\sigma$; then $Z_\sigma \in \mathrm{Aut}(\mathfrak{U}, \{f_i\})$, $Z_1 = I$, and $Z_{\sigma\tau} = U_\tau^{-1} U_\sigma^{-1} V_\sigma V_\tau = Z_\tau (Z_\sigma)^\tau$. A mapping $\sigma \to Z_\sigma$ of $\mathfrak{G}$ into $\mathfrak{A} = \mathrm{Aut}(\mathfrak{U}, \{f_i\})$ satisfying

$$(1) \qquad Z_1 = I, \quad Z_{\sigma\tau} = Z_\tau (Z_\sigma)^\tau$$

will be called a 1-*cocycle* on $\mathfrak{G}$ with values in $\mathfrak{A}$.

Now let $(\mathfrak{Y}, \{h_i\})$ be a second $\mathfrak{F}$-form of $(\mathfrak{U}, \{f_i\})$, and let η be an $\mathfrak{F}$-isomorphism of $(\mathfrak{Y}, \{h_i\})$ onto $(\mathfrak{W}, \{g_i\})$, ψ a $\mathfrak{K}$-isomorphism of $(\mathfrak{Y}_\mathfrak{K}, \{h_i\})$ onto $(\mathfrak{U}, \{f_i\})$ and $\{T_\sigma\}$ a group of Galois semi-automorphisms of $(\mathfrak{U}, \{f_i\})$ associated with ψ as the $\{U_\sigma\}$ are associated with φ. Let $Y_\sigma = T_\sigma^{-1} V_\sigma$, the 1-cocycle associated with $\mathfrak{Y}$ (and with ψ). Denote by η the $\mathfrak{K}$-isomorphism of $(\mathfrak{Y}_\mathfrak{K}, \{h_i\})$ onto $(\mathfrak{W}_\mathfrak{K}, \{g_i\})$ extending the given η. Then $A = \psi^{-1} \eta \varphi \in \mathfrak{A}$, and for $\sigma \in \mathfrak{G}$, $A^{-1} Y_\sigma A^\sigma = A^{-1} T_{(\sigma^{-1})} A V_\sigma$. For $\tau \in \mathfrak{G}$, $T_\tau A$ is τ-semilinear, and sends $y\psi$, for $y \in \mathfrak{Y}$, into $y \psi A$; meanwhile $A U_\tau$ sends $y\psi$ into $y \eta \varphi U_\tau = y \eta \varphi = y \psi A$. Both being τ-semilinear and coinciding on $\mathfrak{Y}\psi$, which contains a basis for $\mathfrak{U}$, these mappings coincide. Thus $A^{-1} Y_\sigma A^\sigma = A^{-1} A U_{(\sigma^{-1})} V_\sigma = Z_\sigma$. In general two 1-cocycles $\sigma \to Y_\sigma$ and $\sigma \to Z_\sigma$ are called *cohomologous* if there is an $A \in \mathfrak{A}$ such that

$$(2) \qquad A^{-1} Y_\sigma A^\sigma = Z_\sigma \quad \text{for all} \quad \sigma \in \mathfrak{G}.$$

Cohomology is an equivalence relation on the set of 1-cocycles, and we have just seen that $\mathfrak{F}$-isomorphic forms of $(\mathfrak{U}, \{f_i\})$ give rise to cohomologous 1-cocycles. In particular, if we start with a given $\mathfrak{F}$-form of $(\mathfrak{U}, \{f_i\})$, then the cocycles obtained from any two $\mathfrak{K}$-isomorphisms of the extension with $(\mathfrak{U}, \{f_i\})$ are cohomologous. Thus we have a mapping of the $\mathfrak{F}$-isomorphism classes of $\mathfrak{F}$-forms of $(\mathfrak{U}, \{f_i\})$ into the *first cohomology set* $H^1(\mathfrak{G}, \mathfrak{A})$, i.e., the set of cohomology classes. (Note that the action of $\mathfrak{G}$ on $\mathfrak{A}$ is determined by the fixed choice of the $\{V_\sigma\}$.)

Theorem IV.2.1. The mapping described above establishes a 1—1 correspondence between the $\mathfrak{F}$-isomorphism classes of $\mathfrak{F}$-forms of $(\mathfrak{U}, \{f_i\})$ and the set $H^1(\mathfrak{G}, \mathfrak{A})$.

For if $\sigma \to Y_\sigma$ is a cocycle, define $T_\sigma = V_\sigma Y_\sigma^{-1}$; then T_σ is a σ-semi-automorphism of $(\mathfrak{U}, \{f_i\})$, $T_1 = I$, $T_{\sigma\tau} = T_\sigma T_\tau$. Let $\mathfrak{X}$ be the $\mathfrak{F}$-subspace of $\mathfrak{U}$ consisting of the fixed elements under $\{T_\sigma\}$. From the $\mathfrak{K}$-independence of the $\sigma \in \mathfrak{G}$ as $\mathfrak{K}$-valued functions on $\mathfrak{K}$, it follows that there is no non-zero $\mathfrak{K}$-valued $\mathfrak{K}$-linear function on $\mathfrak{U}$ vanishing on all elements $\sum_\sigma \gamma^\sigma (x T_\sigma)$ for $\gamma \in \mathfrak{K}$, $x \in \mathfrak{U}$. Since these elements are in $\mathfrak{X}$, we see that $\mathfrak{X}$ contains a basis for $\mathfrak{U}$, so that $\mathfrak{U} = \mathfrak{X}_\mathfrak{K}$, and the restrictions to $\mathfrak{X}$ of the f_i determine $(\mathfrak{X}, \{f_i \mid \mathfrak{X}\})$ as an $\mathfrak{F}$-form of $(\mathfrak{U}, \{f_i\})$, the identity mapping giving a $\mathfrak{K}$-isomorphism of $(\mathfrak{X}_\mathfrak{K}, \{f_i\})$ and $(\mathfrak{U}, \{f_i\})$. The Galois semi-automorphisms U_σ associated with this isomorphism are given by $(\sum y_j x_j) U_\sigma = \sum y_j^\sigma x_j$ for $x_j \in \mathfrak{X}$, $y_j \in \mathfrak{K}$. Since T_σ is σ-semilinear and fixes $\mathfrak{X}$, we have $U_\sigma = T_\sigma$. Thus $(\mathfrak{X}, \{f_i \mid \mathfrak{X}\})$ is an $\mathfrak{F}$-form of $(\mathfrak{U}, \{f_i\})$ whose cohomology class is that of $\sigma \to Y_\sigma$, so that every cohomology class corresponds to an $\mathfrak{F}$-form of $(\mathfrak{U}, \{f_i\})$.

Finally, let $(\mathfrak{W}, \{g_i\})$ and $(\mathfrak{Y}, \{h_i\})$ be $\mathfrak{F}$-forms of $(\mathfrak{U}, \{f_i\})$, with associated isomorphisms φ resp. ψ of $(\mathfrak{W}_\mathfrak{K}, \{g_i\})$ resp. $(\mathfrak{Y}_\mathfrak{K}, \{h_i\})$ onto $(\mathfrak{U}, \{f_i\})$. Let W_σ resp. Y_σ be the associated cocycles, and suppose these are cohomologous: there is an $A \in \mathfrak{A}$ with $A^{-1} Y_\sigma A^\sigma = W_\sigma$ for all $\sigma \in \mathfrak{G}$. Then $\eta = \varphi A^{-1} \psi^{-1}$ is a $\mathfrak{K}$-isomorphism of $(\mathfrak{W}_\mathfrak{K}, \{g_i\})$ onto $(\mathfrak{Y}_\mathfrak{K}, \{h_i\})$, so that it suffices to prove that $\mathfrak{W}\eta = \mathfrak{Y}$, that is, that if $x \in \mathfrak{W}_\mathfrak{K}$, $x\varphi$ fixed under all $U_\sigma = V_\sigma W_\sigma^{-1}$, then $(x\eta)\psi$ is fixed under all $T_\sigma = V_\sigma Y_\sigma^{-1}$. Now $(x\eta)\psi V_\sigma Y_\sigma^{-1} = x\varphi A^{-1} V_\sigma Y_\sigma^{-1} = x\varphi V_\sigma (A^\sigma)^{-1} \times$ $\times Y_\sigma^{-1} = x\varphi V_\sigma W_\sigma^{-1} A^{-1} = x\varphi A^{-1} = x\eta\psi$, and the proof is complete.

It will be noted that with $\mathfrak{K}/\mathfrak{F}$ as above, with $S_\mathfrak{K}$ the class of classical simple Lie algebras over $\mathfrak{K}$, $C_\mathfrak{E}$ the class of $\mathfrak{E}$-forms of members of $S_\mathfrak{K}$ for each subfield $\mathfrak{E}$ of $\mathfrak{K}$, and with $\mathfrak{L} = (\mathfrak{B}, \{f_i\})$ a fixed member of $S_\mathfrak{F}$, we obtain a canonical 1—1 correspondence between the $\mathfrak{F}$-isomorphism classes of $\mathfrak{F}$-forms of $\mathfrak{L}_\mathfrak{K}$ and the first cohomology set $H^1(\mathfrak{G}, \mathrm{Aut}\,\mathfrak{L}_\mathfrak{K})$. This correspondence depends in an essential way on the fact that $\mathfrak{L}$ is chosen from $S_\mathfrak{F}$, and this choice may be regarded as the choice of a distinguished "zero-element" in $H^1(\mathfrak{G}, \mathrm{Aut}\,\mathfrak{L}_\mathfrak{K})$.

§ 3. Simple involutorial algebras and the types $A-D$

We follow here the ideas of a paper by Weil [406]. Let $\mathfrak{K}$ be a field of characteristic not 2. By a *normal simple involutorial algebra* $(\mathfrak{M}, \eta)$ over $\mathfrak{K}$ we mean an associative $\mathfrak{K}$-algebra $\mathfrak{M}$ (finite-dimensional, with unit), together with a $\mathfrak{K}$-antiautomorphism η of period 2, such that $\mathfrak{M}_\mathfrak{E}$ has no proper η-invariant ideals for any field extension $\mathfrak{E}$ of $\mathfrak{K}$. We give a more concrete description of such algebras. First suppose $\mathfrak{K}$ is algebraically closed, and let $\mathfrak{B}$ be an ordinary proper ideal in $\mathfrak{M}$. From the η-invariance of $\mathfrak{B} \cap \mathfrak{B}^\eta$ and $\mathfrak{B} + \mathfrak{B}^\eta$ we have $\mathfrak{M} = \mathfrak{B} \oplus \mathfrak{B}^\eta$, and we see that $\mathfrak{B}$ must be a simple algebra over $\mathfrak{K}$. Thus $\mathfrak{B} \cong M_n(\mathfrak{K})$, the algebra of n by n $\mathfrak{K}$-matrices [5, 100, 218]; on the other hand, if $\mathfrak{M}$ contains no ordinary proper ideal, we have $\mathfrak{M} \cong M_n(\mathfrak{K})$. In the latter case, it is known [5, 226] that there is an identification of $\mathfrak{M}$ with the full algebra $\mathfrak{E}(\mathfrak{B})$ of endomorphisms of a finite-dimensional vector space $\mathfrak{B}$ over $\mathfrak{K}$ which identifies η with the adjoint operation with respect to either a symmetric or alternate scalar product on $\mathfrak{B}$. We say that $(\mathfrak{M}, \eta)$ is *of type B* if one such form on $\mathfrak{B}$ is symmetric and n is odd; *of type D* if such a form is symmetric and n is even; *of type C* if such a form is alternate (n is necessarily even). More generally, if $\mathfrak{F}$ is a subfield of $\mathfrak{K}$ and if $(\mathfrak{N}, \theta)$ is an $\mathfrak{F}$-form of $(\mathfrak{M}, \eta)$, we call $(\mathfrak{N}, \theta)$

of type X if $(\mathfrak{M}, \eta)$ is. (At this stage, it is conceivable that $(\mathfrak{M}, \eta)$ may belong to several types.) When $\mathfrak{B} \neq \mathfrak{M}$, so that $\mathfrak{M}$ is not a simple algebra, we say that $(\mathfrak{M}, \eta)$ is *of type A*. In this case, let φ be an isomorphism of $\mathfrak{B}$ onto $M_n(\mathfrak{K})$, and let $x \to {}^t x$ be the transpose mapping of $M_n(\mathfrak{K})$. Then $(M_n(\mathfrak{K}) \oplus M_n(\mathfrak{K}), \tau)$ is a simple involutorial $\mathfrak{K}$-algebra, where $(x \oplus y)^\tau = {}^t y \oplus {}^t x$, and the mapping $b \oplus c^\eta \to b\varphi \oplus {}^t(c\varphi)$ $(b, c \in \mathfrak{B})$ is an isomorphism of $(\mathfrak{M}, \eta)$ onto $(M_n(\mathfrak{K}) \oplus M_n(\mathfrak{K}), \tau)$.

When $\mathfrak{K}$ is not assumed algebraically closed, we say that a simple involutorial $\mathfrak{K}$-algebra is *split* if it is isomorphic to one of:

$$
(3)\qquad
\begin{array}{lll}
(M_n(\mathfrak{K}) \oplus M_n(\mathfrak{K}), \tau) & \ldots \text{ type } A_{n-1} & (n \geqq 2), \\
(M_{2n+1}(\mathfrak{K}), \theta_1) & \ldots \text{ type } B_n & (n \geqq 1), \\
(M_{2n}(\mathfrak{K}), \theta_2) & \ldots \text{ type } C_n & (n \geqq 1), \\
(M_{2n}(\mathfrak{K}), \theta_3) & \ldots \text{ type } D_n & (n \geqq 1),
\end{array}
$$

where θ_i is the involution $x \to s_i^{-1}({}^t x)\, s_i$ of $M_s(\mathfrak{K})$, and where

$$
(4)\quad s_1 = \left(\begin{array}{c|c|c} 0 & I_n & 0 \\ \hline I_n & 0 & 0 \\ \hline 0 & 0 & 1 \end{array}\right), \qquad s_2 = \left(\begin{array}{c|c} 0 & I_n \\ \hline -I_n & 0 \end{array}\right), \qquad s_3 = \left(\begin{array}{c|c} 0 & I_n \\ \hline I_n & 0 \end{array}\right).
$$

Then each of these involutorial algebras is normal simple, and each simple involutorial algebra over an algebraically closed field is split. Moreover, each normal simple involutorial algebra $(\mathfrak{M}, \eta)$ has a separable splitting field. For if $\mathfrak{M}$ is itself normal simple, then a finite separable extension $\mathfrak{E}$ gives $\mathfrak{M}_\mathfrak{E} \cong M_n(\mathfrak{E})$ for some n [5, p. 62]; if $\mathfrak{M}$ is simple, but not normal simple, then the center (= centroid) of $\mathfrak{M}$ is a quadratic extension $\mathfrak{L}$ of $\mathfrak{F}$, not fixed under η; if $\mathfrak{M} = \mathfrak{B} \oplus \mathfrak{B}^\eta$, then $\mathfrak{B}_\mathfrak{E} \cong M_n(\mathfrak{E})$ for some separable $\mathfrak{E}$. In the first case, we have an involution in $M_n(\mathfrak{E})$ over $\mathfrak{E}$, and the fixed elements in the center are those of $\mathfrak{E}$; in the second, $\mathfrak{M}_\mathfrak{L} = \mathfrak{B} \oplus \mathfrak{B}^\eta$, where $\mathfrak{B}$ is normal simple over $\mathfrak{L}$. The first case is that of an *involution of first kind*, and the involution may be identified with the adjoint relative to a symmetric or alternate form on $\mathfrak{E}^n$; in the alternate case, $\mathfrak{E}$ is a splitting field (type C); in the symmetric case, a finite sequence of quadratic extensions leads to a split algebra of type B or D. In the second case, where the involution is *of second kind*, then considering $\mathfrak{M}_\mathfrak{L}$ reduces the problem to the third case $\mathfrak{M}_\mathfrak{E} \cong \mathfrak{B} \oplus \mathfrak{B}^\eta$, $\mathfrak{B} \cong M_n(\mathfrak{E})$, in which case the algebra is split by $\mathfrak{E}$ and is of type A as before.

If $(\mathfrak{M}, \eta)$ is an involutorial algebra over $\mathfrak{K}$, we denote by $\mathfrak{S}(\mathfrak{M}, \eta)$ the set of all $x \in \mathfrak{M}$ with $x^\eta = -x$, a Lie subalgebra of $\mathfrak{M}$, by $\mathfrak{S}(\mathfrak{M}, \eta)'$ the derived algebra of $\mathfrak{S}(\mathfrak{M}, \eta)$, and by $\mathfrak{S}(\mathfrak{M}, \eta)^*$ the quotient of

$\mathfrak{S}(\mathfrak{M}, \eta)'$ by its center. When $(\mathfrak{M}, \eta)$ is split, the Lie algebra $\mathfrak{S}(\mathfrak{M}, \eta)$ is isomorphic to one of:

$$
\begin{aligned}
&\{x \oplus (-{}^t x),\ x \in M_n(\mathfrak{K})\} && (\text{type } A_{n-1}),\\
&\mathfrak{S}(M_{2n+1}(\mathfrak{K}), \theta_1) && (\text{type } B_n),\\
&\mathfrak{S}(M_{2n}(\mathfrak{K}), \theta_2) && (\text{type } C_n),\\
&\mathfrak{S}(M_{2n}(\mathfrak{K}), \theta_3) && (\text{type } D_n),
\end{aligned} \tag{5}
$$

the isomorphism being effected by means of restricting to $\mathfrak{S}(\mathfrak{M}, \eta)$ an isomorphism φ of $(\mathfrak{M}, \eta)$ onto the corresponding involutorial algebra from the list (3). Thus, for types B_n $(n \geqq 2)$, C_n $(n \geqq 3)$, D_n $(n \geqq 4)$, $\mathfrak{S}(\mathfrak{M}, \eta)$ is isomorphic with the classical simple Lie algebra given the same designation in Chapter III, § 6, provided the characteristic is not (2 or) 3. Thus $\mathfrak{S}(\mathfrak{M}, \eta)$ is a classical simple Lie algebra in these cases, and $\mathfrak{S}(\mathfrak{M}, \eta) = \mathfrak{S}(\mathfrak{M}, \eta)' = \mathfrak{S}(\mathfrak{M}, \eta)^*$. For type A, the map $x \oplus (-{}^t x) \to x$ is an isomorphism of $\mathfrak{S}(\mathfrak{M}, \eta)$ onto the Lie algebra $M_n(\mathfrak{K})$, hence induces an isomorphism of $\mathfrak{S}(\mathfrak{M}, \eta)'$ onto $M_n(\mathfrak{K})'$, the matrices of trace zero, and of $\mathfrak{S}(\mathfrak{M}, \eta)^*$ onto $M_n(\mathfrak{K})^*$, the matrices of trace zero modular scalars of trace zero. As seen in Chapter III, § 6, this is a classical simple Lie algebra of type A_{n-1} provided the characteristic is not 2 or 3, and $\mathfrak{S}(\mathfrak{M}, \eta)' = \mathfrak{S}(\mathfrak{M}, \eta)^*$ unless the characteristic divides n. Thus we have a correspondence $(\mathfrak{M}, \eta) \to \mathfrak{S}(\mathfrak{M}, \eta)^*$ between split simple involutorial algebras over $\mathfrak{K}$ (characteristic $\neq 2, 3$) and split simple Lie algebras of type $A-D$ over $\mathfrak{K}$. Clearly, isomorphic involutorial algebras give rise to isomorphic Lie algebras; the conclusions of Chapter II, § 10, and Chapter III, § 6, show that every classical simple Lie algebra of type $A-D$ over $\mathfrak{K}$ is isomorphic with an algebra $\mathfrak{S}(\mathfrak{M}, \eta)^*$ for a unique (up to isomorphism) split simple involutorial algebra $(\mathfrak{M}, \eta)$. (The restrictions $n \geqq 1$ for A_n, $n \geqq 2$ for B_n, $n \geqq 3$ for C_n, $n \geqq 4$ for D_n are always assumed.)

Lemma IV.3.1. Let A^* be an automorphism of $\mathfrak{S}(\mathfrak{M}, \eta)^*$, where $(\mathfrak{M}, \eta)$ is a split simple involutorial algebra over $\mathfrak{K}$, subject to restrictions on characteristic and dimension as above. Then, except in case D_4, A^* is induced by a unique automorphism A of $(\mathfrak{M}, \eta)$.

First suppose the type is B, C or D, where we may assume $(\mathfrak{M}, \eta) = (M_n(\mathfrak{K}), \theta_i)$ for $i = 1, 2, 3$ and for appropriate n. By Chapter III, § 6, A^* has the form $x \to s^{-1} x s$ $(x \in \mathfrak{S}(\mathfrak{M}, \eta)' = \mathfrak{S}(\mathfrak{M}, \eta)^*)$, where s is the matrix of a similitude of $\mathfrak{K}^n$ with respect to the bilinear form defined by θ_i, i.e., $s s^\eta = s s^{\theta_i} = \lambda I$, $0 \neq \lambda \in \mathfrak{K}$. Thus if $y \in \mathfrak{M}$, $(s^{-1} y s)^\eta = s^\eta y^\eta (s^\eta)^{-1} = s^{-1} y^\eta s$, so that $A: y \to s^{-1} y s$ is an automorphism of $(\mathfrak{M}, \eta)$ inducing A^*. The uniqueness follows by Chapter III, § 6.

For type A, we see by Chapter III, § 6, that A^* is induced by an automorphism A' of $\mathfrak{S}(\mathfrak{M}, \eta)'$, and that by combining A' with the isomorphism $x \oplus (-{}^t x) \to x$ of $\mathfrak{S}(\mathfrak{M}, \eta)'$ onto $M_n(\mathfrak{K})'$, either $(x \oplus (-{}^t x))A' = (b^{-1} x b) \oplus {}^t(-b^{-1} x b)$ for some $b \in GL(n, \mathfrak{K})$, or $(x \oplus (-{}^t x))A' = (-b^{-1}\,{}^t x\, b) \oplus {}^t(b^{-1}\,{}^t x\, b)$ for some $b \in GL(n, \mathfrak{K})$. One sees at once that the mappings $y \oplus z \to b^{-1} y b \oplus {}^t(b^{-1}\,{}^t z\, b)$ and $y \oplus z \to b^{-1} z b \oplus {}^t(b^{-1}\,{}^t y\, b)$ are automorphisms of $(\mathfrak{M}, \eta) = (M_n(\mathfrak{K}) \oplus \oplus M_n(\mathfrak{K}), \tau)$, and A' is the restriction to $\mathfrak{S}(\mathfrak{M}, \eta)'$ of one of these, which we take as our A inducing A^*. Uniqueness follows as in Chapter III, § 6.

Corollary. Under the assumptions as to type of the lemma, the mapping assigning to the automorphism A of the split involutorial simple algebra $(\mathfrak{M}, \eta)$ the induced automorphism A^* of $\mathfrak{S}(\mathfrak{M}, \eta)^*$ is an isomorphism of the group $\mathrm{Aut}(\mathfrak{M}, \eta)$ onto $\mathrm{Aut}(\mathfrak{S}(\mathfrak{M}, \eta)^*)$. (This follows from the Lemma, the considerations of Chapter III, § 6, and the known form of $\mathrm{Aut}(\mathfrak{M}, \eta)$ [213].)

Now let $\mathfrak{F}$ be a field of characteristic $\neq 2, 3$, and let $\mathfrak{L}$ be an $\mathfrak{F}$-form of a split simple Lie algebra (over some extension of $\mathfrak{F}$) of type A, B, C or D, where $n \geqq 5$ for type D_n. By Th. 1.1, there is a finite Galois extension $\mathfrak{K}$ of $\mathfrak{F}$, with group $\mathfrak{G}$, which is a splitting field for $\mathfrak{L}$. That is, we have a split simple involutorial algebra $(\mathfrak{N}, \theta)$ over $\mathfrak{F}$, so that $\mathfrak{S}(\mathfrak{N}, \theta)^*$ is a split Lie algebra, and $\mathfrak{L}_{\mathfrak{K}} \cong (\mathfrak{S}(\mathfrak{N}, \theta)^*)_{\mathfrak{K}} \cong \mathfrak{S}(\mathfrak{N}_{\mathfrak{K}}, \theta)^*$. Let $\{V_\sigma\}$ be the group of Galois semi-automorphisms of $(\mathfrak{N}_{\mathfrak{K}}, \theta)$ associated with a canonical embedding of $(\mathfrak{N}, \theta)$ in $(\mathfrak{N}_{\mathfrak{K}}, \theta)$. Then the V_σ map $\mathfrak{S}(\mathfrak{N}_{\mathfrak{K}}, \theta)$, $\mathfrak{S}(\mathfrak{N}_{\mathfrak{K}}, \theta)'$ into themselves, and induce a group of Galois semi-automorphisms $\{V_\sigma^*\}$ of $\mathfrak{S}(\mathfrak{N}_{\mathfrak{K}}, \theta)^*$.

Let $\{U_\sigma^*\}$ be a group of Galois semi-automorphisms of $\mathfrak{S}(\mathfrak{N}_{\mathfrak{K}}, \theta)^*$ associated with $\mathfrak{L}$ and a $\mathfrak{K}$-isomorphism of $\mathfrak{L}_{\mathfrak{K}}$ onto $\mathfrak{S}(\mathfrak{N}_{\mathfrak{K}}, \theta)^*$; let $Z_\sigma^* = U_\sigma^{*-1} V_\sigma^*$ be the associated 1-cocycle, with values in $\mathfrak{A}^* = \mathrm{Aut}(\mathfrak{S}(\mathfrak{N}_{\mathfrak{K}}, \theta)^*)$. The construction of the $\{V_\sigma^*\}$ above defines the action of $\mathfrak{G}$ on $\mathfrak{A}^*$ in such a way that the isomorphism $A \to A^*$ of $\mathfrak{A} = \mathrm{Aut}(\mathfrak{N}_{\mathfrak{K}}, \theta)$ onto $\mathfrak{A}^*$, described in the Corollary, is a $\mathfrak{G}$-isomorphism. Hence there is a unique 1-cocycle $\sigma \to Z_\sigma$ on $\mathfrak{G}$ with values in $\mathfrak{A}$ inducing Z_σ^*. By Th. 2.1, the fixed elements $\mathfrak{M}$ of $\mathfrak{N}_{\mathfrak{K}}$ under $\{V_\sigma Z_\sigma^{-1}\}$, together with the restriction η to $\mathfrak{M}$ of θ (acting in $\mathfrak{N}_{\mathfrak{K}}$), constitute an $\mathfrak{F}$-form $(\mathfrak{M}, \eta)$ of $(\mathfrak{N}_{\mathfrak{K}}, \theta)$ with associated cohomology class that of $\{Z_\sigma\}$. Now the $\{V_\sigma Z_\sigma^{-1}\}$ induce on $\mathfrak{S}(\mathfrak{N}_{\mathfrak{K}}, \theta)^*$ the $\{U_\sigma^*\}$ as the group of Galois semi-automorphisms associated with the canonical identification of $(\mathfrak{S}(\mathfrak{M}, \eta)^*)_{\mathfrak{K}}$ with $\mathfrak{S}(\mathfrak{N}_{\mathfrak{K}}, \theta)^*$; hence the associated 1-cocycle on $\mathfrak{G}$ with values in $\mathfrak{A}^*$ is $\{U_\sigma^{*-1} V_\sigma^*\} = \{Z_\sigma^*\}$. By Th. 2.1, $\mathfrak{L}$ is isomorphic with $\mathfrak{S}(\mathfrak{M}, \eta)^*$. We have thus proved the first assertion of the following:

Theorem 4.3.1. Let $\mathfrak{F}$ be a field of characteristic $\neq 2, 3$, and let $\mathfrak{L}$ be a normal simple Lie algebra over $\mathfrak{F}$ of type A, B_n $(n \geqq 2)$, C_n

($n \geqq 3$), or D_n ($n \geqq 5$). Then $\mathfrak{L}$ is isomorphic to $\mathfrak{S}(\mathfrak{M}, \eta)^*$, where $(\mathfrak{M}, \eta)$ is a normal simple involutorial algebra over $\mathfrak{F}$ of the same type as is $\mathfrak{L}$. Conversely each such $\mathfrak{S}(\mathfrak{M}, \eta)^*$ is a normal simple Lie algebra of the same type as $(\mathfrak{M}, \eta)$, and each isomorphism of the Lie algebra $\mathfrak{S}(\mathfrak{M}, \eta)^*$ onto $\mathfrak{S}(\mathfrak{P}, \zeta)^*$ is induced by an isomorphism of $(\mathfrak{M}, \eta)$ onto $(\mathfrak{P}, \zeta)$.

In view of earlier remarks, it remains only to prove the last assertion. Let φ be an isomorphism of $\mathfrak{S}(\mathfrak{M}, \eta)^*$ onto $\mathfrak{S}(\mathfrak{P}, \zeta)^*$, and let $\mathfrak{K}/\mathfrak{F}$ be a Galois extension such that both $(\mathfrak{M}_{\mathfrak{K}}, \eta)$ and $(\mathfrak{P}_{\mathfrak{K}}, \zeta)$ are split; let $(\mathfrak{N}, \theta)$ be the split involutorial simple algebra over $\mathfrak{F}$ of the same type as $(\mathfrak{M}, \eta)$ and $(\mathfrak{P}, \zeta)$ (these necessarily have the same type, by the fact that $\mathfrak{S}(\mathfrak{M}_{\mathfrak{K}}, \eta)^*$ and $\mathfrak{S}(\mathfrak{P}_{\mathfrak{K}}, \zeta)^*$ are split and isomorphic). Let $\sigma \to Z_\sigma$ and $\sigma \to Y_\sigma$, respectively, be cocycles on the Galois group $\mathfrak{G}$ with values in $\mathfrak{A} = \mathrm{Aut}(\mathfrak{N}_{\mathfrak{K}}, \theta)$, associated with $(\mathfrak{M}, \eta)$ and $(\mathfrak{P}, \zeta)$, respectively, and with the respective isomorphisms μ, ν of $(\mathfrak{M}_{\mathfrak{K}}, \eta)$ resp. $(\mathfrak{P}_{\mathfrak{K}}, \zeta)$ onto $(\mathfrak{N}_{\mathfrak{K}}, \theta)$. By Th. 2.1, we have an automorphism $A^* \in \mathfrak{A}^* = \mathrm{Aut}(\mathfrak{S}(\mathfrak{N}_{\mathfrak{K}}, \theta)^*)$ with $Z_\sigma^* = A^{*-1} Y_\sigma^* A^{*\sigma}$, where the notations are as before, viz. $A^* = \nu^{*-1} \varphi^{-1} \mu^*$, where φ is regarded as extended to an isomorphism of $\mathfrak{S}(\mathfrak{M}_{\mathfrak{K}}, \eta)^*$ onto $\mathfrak{S}(\mathfrak{P}_{\mathfrak{K}}, \zeta)^*$. Now there is $A \in \mathfrak{A}$ inducing A^*, and $(A^{-1} Y_\sigma A^\sigma)^* = Z_\sigma^*$ for all $\sigma \in \mathfrak{G}$; hence $A^{-1} Y_\sigma A^\sigma = Z_\sigma$ by the Corollary, and $(\mathfrak{M}, \eta)$ and $(\mathfrak{P}, \zeta)$ are isomorphic by Th. 2.1. In fact, the proof of Th. 2.1 showed that the restriction ψ to $\mathfrak{M}$ of $\mu A^{-1} \nu^{-1}$ is an isomorphism of $(\mathfrak{M}, \eta)$ onto $(\mathfrak{P}, \zeta)$; on $\mathfrak{S}(\mathfrak{M}_{\mathfrak{K}}, \eta)^*$ we have $\psi^* = \mu^* (A^*)^{-1} (\nu^*)^{-1} = \varphi$. Hence $\psi^* = \varphi$ on $\mathfrak{S}(\mathfrak{M}, \eta)^*$, and the proof is complete.

One can be somewhat more explicit in the description of normal simple involutorial algebras $(\mathfrak{M}, \eta)$ over $\mathfrak{F}$, hence in the description of normal simple Lie algebras of the types treated in this section. If $\mathfrak{M}$ is a simple algebra, then either: (a) $\mathfrak{M}$ is normal simple over $\mathfrak{F}$, or: (b) $\mathfrak{M}$ is normal simple over $\mathfrak{E} = \mathfrak{F}(\xi)$, the center of $\mathfrak{M}$ and a quadratic extension of $\mathfrak{F}$, where ξ may be so chosen that $\xi^\eta = -\xi$. In case (a), $\mathfrak{M}$ may be identified with the ring of all endomorphisms of a finite-dimensional vector space $\mathfrak{V}$ over a central division algebra $\mathfrak{D}$ over $\mathfrak{F}$, $\mathfrak{D}$ admitting an involution $\alpha \to \alpha'$, and the involution η may be identified with the adjoint mapping of endomorphisms of $\mathfrak{V}$ relative to a hermitian or anti-hermitian scalar product in $\mathfrak{V}$. Moreover (cf. [18, p. 114]) if $\mathfrak{D} \neq \mathfrak{F}$, $\mathfrak{S}(\mathfrak{M}, \eta)$ for the anti-hermitian case is isomorphic with an $\mathfrak{S}(\mathfrak{M}, \eta')$ for the hermitian case, by changing the involution in $\mathfrak{D}$ and the scalar product on $\mathfrak{V}$. In this case classification of the Lie algebras depends upon classifying the central involutorial division algebras over $\mathfrak{F}$, and upon classifying the hermitian forms in finite-dimensional vector spaces over these algebras (as well as quadratic forms over $\mathfrak{F}$—the case $\mathfrak{D} = \mathfrak{F}$). (For a detailed exposition of this material and that of the next paragraphs, cf. [234, Chap. 10; 5, Chap. 10; 226, Chap. IV; 218, Chap. 2].)

In case (b), $\mathfrak{M}$ may be identified with the ring of all endomorphisms of $\mathfrak{V}$ over $\mathfrak{D}$, where $\mathfrak{D}$ is now a central division algebra over $\mathfrak{C}$, $\mathfrak{D}$ possessing an involution inducing on $\mathfrak{C}$ the restriction of η, and η may be identified with the adjoint mapping relative to a hermitian or anti-hermitian form on $\mathfrak{V}$ over $\mathfrak{D}$; for purposes of studying $\mathfrak{S}(\mathfrak{M}, \eta)$ the hermitian case is again sufficient, so that the problem of classifying Lie algebras of this type ("type $A_{\rm II}$") may be regarded as solved once one has classified all involutorial division algebras $\mathfrak{D}$ over $\mathfrak{F}$ whose center is a quadratic extension of $\mathfrak{F}$ moved by the involution, and all hermitian scalar products in vector spaces over such algebras.

Finally, if $\mathfrak{M}$ is not simple ("type $A_{\rm I}$") we have the case $\mathfrak{M} = \mathfrak{B} \oplus \mathfrak{B}^\eta$, $\mathfrak{B}$ normal simple over $\mathfrak{F}$. Thus $\mathfrak{B}$ is a full ring of endomorphisms as above. As in the split case, we have an identification of $\mathfrak{S}(\mathfrak{M}, \eta)$ with $\mathfrak{B}$, considered as a Lie algebra, and $\mathfrak{M} = \mathfrak{B} \oplus \mathfrak{B}^\eta$ and $\mathfrak{P} = \mathfrak{C} \oplus \mathfrak{C}^\zeta$ are isomorphic as algebras with involution if and only if $\mathfrak{B}$ and $\mathfrak{C}$ are isomorphic or anti-isomorphic. The classification of the simple algebras $\mathfrak{S}(\mathfrak{M}, \eta)^*$ thus depends mainly upon the classification of central division algebras over $\mathfrak{F}$, and of hermitian scalar products.

§ 4. Derivation algebras of alternative and Jordan algebras

The passage from $(\mathfrak{M}, \eta)$ to $\mathfrak{S}(\mathfrak{M}, \eta)^*$ of the last section is an example of a process which has been treated in a more general context by Jacobson, who has given it the name of "forming a class of *derived* algebras" [237]. His theory provides a general setting in which determination of forms of a derived algebra can be equated with determination of forms of the original algebra. Rather than attempting a summary of this theory here, or trying to adapt the theory to the general formalism of § 2, we give some further examples of the process, which yield all algebras of types G_2 and F_4, as well as others of § 3.

a) Let $\mathfrak{C}$ be a normal simple *alternative*, but not associative algebra over $\mathfrak{F}$, i.e., the subalgebra of $\mathfrak{C}$ generated by any two elements is associative. As associated derived Lie algebra we take the algebra $\mathfrak{D}(\mathfrak{C})$ of derivations of $\mathfrak{C}$. Now it is known [7, 251, 341, 344, 425] that $\mathfrak{C}$ is a Cayley–Dickson algebra, that is, an 8-dimensional algebra carrying a non-singular symmetric bilinear form (x, y) satisfying $(xy, xy) = (x, x)(y, y)$ for all $x, y \in \mathfrak{C}$. The mapping $x \to 2(x, 1)1 - x$ is an involution η in $\mathfrak{C}$, and every element satisfies $x^2 - 2(x, 1)x + (x, x)1 = 0$. When $\mathfrak{C}$ contains isotropic vectors with respect to the bilinear form, then $\mathfrak{C}$ is determined to within isomorphism, and is the split Cayley algebra of Chapter III, § 6. This situation can always be achieved by quadratic extension of the base field. Thus if $\mathfrak{K}$ is a field (of characteristic $\neq 2$), and if $C_{\mathfrak{K}}$ consists of all algebras over $\mathfrak{K}$ which are isomorphic with split Cayley algebras, $C_{\mathfrak{F}}$ (for $\mathfrak{F}$ a subfield of $\mathfrak{K}$) of all $\mathfrak{F}$-forms of members of $C_{\mathfrak{K}}$, $S_{\mathfrak{F}}$ of all split Cayley algebras over $\mathfrak{F}$, then the conditions a) and b) of § 1 are satisfied, and c) holds if $\mathfrak{K}$ is algebraically closed. Furthermore, by Chapter III, § 6, when $\mathfrak{K}$ has characteristic $\neq 2, 3$ the mapping $A \to A^*$ of $\mathrm{Aut}(\mathfrak{C})$ onto $\mathrm{Aut}(\mathfrak{D}(\mathfrak{C}))$, assigning to A the automorphism $A^*: D \to A^{-1}DA$, is an isomorphism for $\mathfrak{C}$ in $C_{\mathfrak{K}}$, and $\mathfrak{D}(\mathfrak{C})$ is the split Lie algebra of type G_2. One now proceeds by formal analogy with arguments of § 3 to prove the following

Theorem IV.4.1 (Jacobson [214], Barnes [21]). Let $\mathfrak{F}$ be a field of characteristic $\neq 2, 3$, and let $\mathfrak{L}$ be a normal simple Lie algebra of type G_2 over $\mathfrak{F}$. Then $\mathfrak{L}$ is isomorphic to $\mathfrak{D}(\mathfrak{C})$, for $\mathfrak{C}$ a Cayley–Dickson algebra over $\mathfrak{F}$. Conversely, each such $\mathfrak{D}(\mathfrak{C})$ is a normal simple Lie algebra of type G_2, and each isomorphism of $\mathfrak{D}(\mathfrak{C}_1)$ onto $\mathfrak{D}(\mathfrak{C}_2)$ is induced by an isomorphism of $\mathfrak{C}_1$ onto $\mathfrak{C}_2$.

B) Let $\mathfrak{J}$ be an exceptional normal simple Jordan algebra over $\mathfrak{F}$ (characteristic $\neq 2, 3$); that is, $\mathfrak{J}$ is commutative, satisfies the identity $x^2(yx) = (x^2y)x$, and is not isomorphic with any subspace of an associative $\mathfrak{F}$-algebra with the operation $x \cdot y = \frac{1}{2}(xy + yx)$ (a *special* Jordan algebra). Then each element of $\mathfrak{J}$ satisfies an identity $x^3 - t(x)x^2 + s(x)x - n(x)1 = 0$, where $t(x), s(x), n(x)$ are forms of degrees $1, 2, 3$ respectively. Now either the polynomial $X^3 - t(x)X^2 + s(x)X - n(x)$ is irreducible whenever x and 1 are linearly independent (here $\mathfrak{J}$ is a *Jordan division algebra*), or $\mathfrak{J}$ is isomorphic with the set of 3 by 3 matrices over a Cayley–Dickson algebra $\mathfrak{C}$ which are hermitian under a certain involution (the operation being symmetrized matrix multiplication $\frac{1}{2}(xy + yx)$). Moreover, if $\mathfrak{C}$ is split, this second (*reduced*) $\mathfrak{J}$ is determined to within isomorphism, and may be identified with the "split exceptional Jordan algebra" of Chapter III, § 6. (For the preceding, see [6, 8, 10, 48, 232, 238, 342, 344].) In the first case, a cubic extension of the base field leads to a reduced Jordan algebra; then a quadratic extension splits the associated Cayley–Dickson algebra. Thus if $C_{\mathfrak{F}}$ is the family of $\mathfrak{F}$-forms of the split exceptional Jordan algebra over $\mathfrak{K}$, and $S_{\mathfrak{F}}$ the family of split exceptional Jordan algebras over $\mathfrak{F}$, the conditions a) and b) of § 1 hold, and c) holds if $\mathfrak{K}$ is algebraically closed. In Chapter III, § 6, we have seen that the mapping $A \to A^*$, where A^* sends $D \in \mathfrak{D}(\mathfrak{J})$ into $A^{-1}DA$, is an isomorphism of the automorphism groups $\mathrm{Aut}(\mathfrak{J})$ and $\mathrm{Aut}(\mathfrak{D}(\mathfrak{J}))$ in the split case, in which case $\mathfrak{D}(\mathfrak{J})$ is classical simple of type F_4. A straightforward analogy with § 3 now proves the

Theorem IV.4.2 (Tomber [400], Barnes [21]). Let $\mathfrak{F}$ be a field of characteristic $\neq 2, 3$, and let $\mathfrak{L}$ be a normal simple Lie algebra of type F_4 over $\mathfrak{F}$. Then $\mathfrak{L}$ is isomorphic to $\mathfrak{D}(\mathfrak{J})$ for $\mathfrak{J}$ an exceptional normal simple Jordan algebra over $\mathfrak{F}$. Conversely, each such $\mathfrak{D}(\mathfrak{J})$ is a normal simple Lie algebra of type F_4, and each isomorphism of $\mathfrak{D}(\mathfrak{J}_1)$ onto $\mathfrak{D}(\mathfrak{J}_2)$ is induced by an isomorphism of $\mathfrak{J}_1$ onto $\mathfrak{J}_2$.

C) Descriptions of a similar nature can be given, using the same formalism, for forms of the algebras of types discussed in § 3. If $(\mathfrak{M}, \eta)$ is a normal simple involutorial algebra, and $\mathfrak{D}(\mathfrak{M}, \eta)$ the Lie algebra of derivations of $\mathfrak{M}$ commuting with η, the derived algebra $\mathfrak{D}(\mathfrak{M}, \eta)'$ is normal simple when $(\mathfrak{M}, \eta)$ is of type A_n $(n \geqq 1)$, B_n $(n \geqq 2)$, C_n $(n \geqq 3)$, or D_n $(n \geqq 4)$, and $\mathfrak{D}(\mathfrak{M}, \eta)'$ is a form of the classical Lie algebra of the same type. In the split cases, except for D_4, we again have an isomorphism $A \to A^*$ as in A), B) from $\mathrm{Aut}(\mathfrak{M}, \eta)$ onto $\mathrm{Aut}(\mathfrak{D}(\mathfrak{M}, \eta)')$. (The essentials here are Chapter III, § 6, Chapter IV, § 3, and the fact that any derivation of a normal simple associative algebra is inner [218].) The conclusion then is that, *subject to the same restrictions on types as in* § 3, *every normal simple Lie algebra of a given type over* $\mathfrak{F}$ *is the algebra* $\mathfrak{D}(\mathfrak{M}, \eta)'$, *where* $(\mathfrak{M}, \eta)$ *is a normal simple involutorial algebra of the same type, and conversely; any isomorphism between* $\mathfrak{D}(\mathfrak{M}, \eta)'$ *and* $\mathfrak{D}(\mathfrak{N}, \theta)'$ *is induced by an isomorphism between* $(\mathfrak{M}, \eta)$ *and* $(\mathfrak{N}, \theta)$.

A somewhat closer analogy with B) is found if one considers algebras $\mathfrak{D}(\mathfrak{J})'$, where $\mathfrak{J}$ runs over normal simple Jordan algebras of suitable types (in the classification of such algebras). For characteristic zero, the conclusions here are to be found in [401]; once again the essentials involve only making the identifications of $\mathfrak{D}(\mathfrak{J})'$ with the split Lie algebras when $\mathfrak{J}$ is split, proving that $\mathfrak{J}$ has a separable splitting field, and showing that $A \to A^*$ is an isomorphism of $\mathrm{Aut}(\mathfrak{J})$ onto $\mathrm{Aut}(\mathfrak{D}(\mathfrak{J})')$ in the split case (cf. [209, 238]).

§ 5. Other types

A) D_4: Combining the considerations of § 3 and those of Chapter III, § 6, we see that if $\mathfrak{L}$ is an $\mathfrak{F}$-form of a split Lie algebra of type D_4 over $\mathfrak{K}$, a Galois splitting field for $\mathfrak{L}$, and if an associated cocycle on $\mathfrak{G} = \mathfrak{G}(\mathfrak{K}/\mathfrak{F})$ with values in $\mathrm{Aut}(\mathfrak{L}_{\mathfrak{K}})$ has values in a subgroup of index 3 in $\mathrm{Aut}(\mathfrak{L}_{\mathfrak{K}})$ containing the Chevalley group G, then $\mathfrak{L} \cong \mathfrak{S}(\mathfrak{M}, \eta)$, where $(\mathfrak{M}, \eta)$ is a normal simple involutorial algebra of type D_4 over F. Furthermore, any isomorphism between two $\mathfrak{F}$-algebras $\mathfrak{L}_1$, $\mathfrak{L}_2$ of type D_4 with this property relative to some common Galois splitting field $\mathfrak{K}$ and cocycles cohomologous within the same subgroup of index 3 in $\mathrm{Aut}(\mathfrak{L}_{\mathfrak{K}})$ is induced by an isomorphism of $(\mathfrak{M}_1, \eta_1)$ onto $(\mathfrak{M}_2, \eta_2)$. According as the subgroup in question of $\mathrm{Aut}(\mathfrak{L}_{\mathfrak{K}})$ is G or properly contains G, one says that $\mathfrak{L}$ is *of type* $D_{4\mathrm{I}}$ *or* $D_{4\mathrm{II}}$. That this is independent of the chosen Galois splitting field is not difficult to show [11, 236]. Likewise, if the associated cocycle has values in a subgroup of $\mathrm{Aut}(\mathfrak{L}_{\mathfrak{K}})$ of index 2 containing G, but if $\mathfrak{L}$ is not of type $D_{4\,\mathrm{I}}$, $\mathfrak{L}$ is *of type* $D_{4\,\mathrm{III}}$; in the remaining case, $\mathfrak{L}$ is *of type* $D_{4\,\mathrm{VI}}$.

All types can occur, and indeed as subalgebras of Lie algebras of type F_4; namely, let $\mathfrak{J}$ be an exceptional normal simple Jordan algebra over $\mathfrak{F}$ (cf. § 4), and let $\mathfrak{A}$ be a 3-dimensional semisimple associative subalgebra of $\mathfrak{J}$. Let $\mathfrak{L}$ be the Lie subalgebra of $\mathfrak{D}(\mathfrak{J})$ annihilating $\mathfrak{A}$. If $\mathfrak{A} \cong \mathfrak{F} \oplus \mathfrak{F} \oplus \mathfrak{F}$, then $\mathfrak{L}$ is of type $D_{4\,\mathrm{I}}$; if $\mathfrak{A} \cong \mathfrak{F} \oplus \mathfrak{E}$, $\mathfrak{E}$ a quadratic extension, $\mathfrak{L}$ is of type $D_{4\,\mathrm{II}}$; if $\mathfrak{A}$ is a cyclic cubic extension field of $\mathfrak{F}$, $\mathfrak{L}$ is of type $D_{4\,\mathrm{III}}$; if $\mathfrak{A}$ is a non-cyclic cubic extension field, $\mathfrak{L}$ is of type $D_{4\,\mathrm{VI}}$. Not all algebras of type $D_{4\,\mathrm{I}}$ or $D_{4\,\mathrm{II}}$ over a general field arise in this way [11, 236]; it seems to be an open question whether all algebras of type $D_{4\,\mathrm{III}}$ or $D_{4\,\mathrm{VI}}$ can be so realized. The most thorough study of these "Jordan D_4's" has been made by Allen [11] (cf. also [52, 227, 372]); for the question of isomorphisms between algebras of type $D_{4\,\mathrm{I}}$ or of type $D_{4\,\mathrm{II}}$, see Jacobson [236].

B) E_6, E_7, E_8: As with D_4, we may divide algebras of type E_6 into two classes, $E_{6\,\mathrm{I}}$ and $E_{6\,\mathrm{II}}$, according as an associated cocycle does or does not take values in the Chevalley group G associated with a split E_6. Neither of these classes has been described as fully as have those, other than $D_{4\,\mathrm{III}}$ and $D_{4\,\mathrm{VI}}$, discussed earlier in this chapter. Algebras of type $E_{6\,\mathrm{II}}$ do exist [131, 227, 233]. A class of algebras of type $E_{6\,\mathrm{I}}$ is provided by carrying out the procedure used to give a split E_6 in Chapter III, § 6, with the split exceptional Jordan algebra replaced by an arbitrary exceptional normal simple Jordan algebra; however, these do not exhaust the algebras of type $E_{6\,\mathrm{I}}$, as Jacobson has shown [227]. The most complete study to date of forms of E_6 and their isomorphisms is probably the thesis of Ferrar [131].

In the cases of E_7 and E_8, the descriptions known for the split algebras are not very simple; thus, even though the group $\mathrm{Aut}(\mathfrak{L}_{\mathfrak{K}})$ is the Chevalley group G when $\mathfrak{K}$ is a splitting field, this result does not yield a manageable list of forms for $\mathfrak{L}_{\mathfrak{K}}$. Some constructions due to Tits [345, 395, 397] which provide what seem to be all known algebras of types E_7 and E_8, are as follows:

1) Let $\mathfrak{A}$ be a Cayley–Dickson algebra over $\mathfrak{F}$, and let $\mathfrak{J}$ be a Jordan algebra which is a form over $\mathfrak{F}$ of a Jordan algebra (under $\frac{1}{2}(a\,b + b\,a)$) of 6 by 6 matrices of the form $g\,x$, where g is a fixed invertible skew-symmetric matrix, and where x runs over all skew-symmetric matrices.

2) Let $\mathfrak{A}$ be a normal simple associative algebra of dimension 4 over $\mathfrak{F}$, and let $\mathfrak{J}$ be an exceptional normal simple Jordan algebra over $\mathfrak{F}$.

3) Let $\mathfrak{A}$ be a Cayley–Dickson algebra, $\mathfrak{J}$ an exceptional normal simple Jordan algebra.

In each of the algebras $\mathfrak{A}$, $\mathfrak{J}$ above, a ("generic") trace is defined, which is a linear function; let $\mathfrak{A}_0$ resp. $\mathfrak{J}_0$ be its kernel. Let $\mathfrak{B}$ resp. $\mathfrak{D}$ be the Lie algebra of derivations of $\mathfrak{A}$ resp. $\mathfrak{J}$, and form the direct sum $\mathfrak{L} = \mathfrak{B} \oplus (\mathfrak{A}_0 \otimes \mathfrak{J}_0) \oplus \mathfrak{D}$. Introduce in $\mathfrak{L}$ a bilinear product $[x\,y]$ which is the ordinary Lie product when both factors are in $\mathfrak{B}$ or in $\mathfrak{D}$, such that $[\mathfrak{B}\,\mathfrak{D}] = 0 = [\mathfrak{D}\,\mathfrak{B}]$, such that $-[a \otimes c, B + D] = [B + D, a \otimes c] = aB \otimes c + a \otimes cD$ for $a \in \mathfrak{A}$, $c \in \mathfrak{J}$, $B \in \mathfrak{B}$, $D \in \mathfrak{D}$, and with $[a \otimes c, a' \otimes c'] = (c, c') \langle a, a' \rangle + (a^* a') \otimes (c^* c') + (a, a') \langle c, c' \rangle$, where $(a, a') = t_{\mathfrak{A}}(a\,a')$, $(c, c') = t_{\mathfrak{J}}(c\,c')$, $t_{\mathfrak{A}}$ resp. $t_{\mathfrak{J}}$ being the trace, where $a^* a' = a\,a' - (a, a')\, t_{\mathfrak{A}}$, $c^* c' = c\,c' - (c, c')\, t_{\mathfrak{J}}$, and where $\langle a, a' \rangle \in \mathfrak{B}$, $\langle c, c' \rangle \in \mathfrak{D}$ are defined as follows:

$$b\langle a, a' \rangle = \tfrac{1}{4}[b, [a', a]] - \tfrac{3}{4}((a\,a')\,b - a(a'b)),$$
$$d\langle c, c' \rangle = c(c'd) - c'(c\,d).$$

In cases 1) and 2), $\mathfrak{L}$ is a Lie algebra and a form of E_7; in case 3), $\mathfrak{L}$ is a form of E_8. Taking $\mathfrak{J}$ as in 2) or 3) and $\mathfrak{A}$ to be a separable associative algebra of dimension 2 over $\mathfrak{F}$, the construction gives forms of E_6, as it does when we take $\mathfrak{A}$ to be a Cayley-Dickson algebra and $\mathfrak{J}$ a form of the Jordan algebra of 3 by 3 matrices. The latter case provides "exceptional" forms of E_6.

§ 6. Finite fields

In case the field $\mathfrak{F}$ is finite, results of Lang on algebraic groups over finite fields [270] have been utilized by Hertzig [183] to study simple algebraic groups defined over $\mathfrak{F}$. We give here a slight generalization of the relevant result of Lang and apply it to the determination of all $\mathfrak{F}$-forms of classical simple algebras.

Let $\mathfrak{F}$ be the field of q elements, $\mathfrak{K}$ an extension of $\mathfrak{F}$ of finite degree d, and Ω an extension of $\mathfrak{K}$ which is algebraically closed and of infinite transcendency.

Let G be an irreducible algebraic subgroup of $GL(n, \Omega)$ which is defined over $\mathfrak{F}$, i.e., G is the intersection with $GL(n, \Omega)$ of the set of all n by n matrices over Ω which are the zeros of a prime ideal $\mathfrak{P}$ in the polynomial ring in n^2 variables (the matrix entries) over Ω, this ideal being generated by polynomials with coefficients in $\mathfrak{F}$. (Then $\mathfrak{P}$ is the totality of polynomials vanishing on G.) Now the map $\xi \to \xi^q$ is an automorphism of Ω with $\mathfrak{F}$ as fixed field, and its restriction to $\mathfrak{K}$ generates the Galois group $\mathfrak{G}$ of $\mathfrak{K}/\mathfrak{F}$. The map φ: $(\xi_{ij}) \to (\xi_{ij}^q)$ is thus an automorphism of $GL(n, \Omega)$ such that $G^\varphi \subseteq G$. By a *generic point of G over $\mathfrak{K}$* is meant an element $x \in G$ such that the polynomials over $\mathfrak{K}$ vanishing on G are exactly those vanishing at x. If $x = (\xi_{ij})$ is a generic point of G over $\mathfrak{K}$, and if $z = (\zeta_{ij})$ is any point of G, then there is a $\mathfrak{K}$-homomorphism of $\mathfrak{K}[x] = \mathfrak{K}[\xi_{ij}]$ onto $\mathfrak{K}[z] = \mathfrak{K}[\zeta_{ij}]$ mapping

ξ_{ij} onto ζ_{ij} for all i, j, and this mapping is an isomorphism if and only if z is a generic point of G over $\mathfrak{K}$, or if and only if the transcendency degrees over $\mathfrak{K}$ of $\mathfrak{K}[z]$ and $\mathfrak{K}[x]$ are equal (see [42, 271, 407] for these facts and for the existence of generic points). We denote by $G_{\mathfrak{K}}$ the group $G \cap GL(n, \mathfrak{K})$.

Lemma IV.6.1 (cf. LANG [270]). Let $c \in GL(n, \mathfrak{F})$ satisfy the condition $c^{-1} G c = G$. Then the image of G under the map $z \to c^{-1} z^{-1} c z^{\varphi}$ contains $G_{\mathfrak{K}}$.

For let x be a generic point of G over $\mathfrak{K}$, and let y be any point of $G_{\mathfrak{K}}$. Let $w = c^{-1} x c y (x^{\varphi})^{-1}$. Then w is a generic point of G over $\mathfrak{K}$; for $\mathfrak{K}(w) \subseteq \mathfrak{K}(x) = \mathfrak{K}(w, x)$, and $\mathfrak{K}(x) \subseteq \mathfrak{K}(w, x^{\varphi}) \subseteq \mathfrak{K}(w)\, \mathfrak{K}(x)^q$ (the subfield of $\mathfrak{K}(x)$ generated by $\mathfrak{K}(w)$ and q-th powers of elements of $\mathfrak{K}(x)$). Thus the only derivation of $\mathfrak{K}(x)$ vanishing on $\mathfrak{K}(w)$ is zero, so that $\mathfrak{K}(x)$ is (separably) algebraic over $\mathfrak{K}(w)$, by a result cited in Chapter III, § 4; hence w is a generic point of G over $\mathfrak{K}$. In particular, $v = c^{-1} x c (x^{\varphi})^{-1}$ is a generic point; hence there is a $\mathfrak{K}$-isomorphism of $\mathfrak{K}[v]$ onto $\mathfrak{K}[w]$ mapping v onto w. This mapping extends to a $\mathfrak{K}$-isomorphism η of $\mathfrak{K}(x)$ into Ω. Thus if $x = (\xi_{ij})$, $u = x^{\eta} = (\xi_{ij}^{\eta}) \in G$, and if $t = u^{-1} x$, then $c^{-1} t^{-1} c t^{\varphi} = c^{-1} x^{-1} c (c^{-1} u) c (u^{\varphi})^{-1} x^{\varphi} = c^{-1} \times$ $\times x^{-1} c v^{\eta} x^{\varphi} = c^{-1} x^{-1} c w x^{\varphi} = y$.

Now let $\mathfrak{L}$ be a split Lie algebra over an arbitrary field $\mathfrak{F}$, and let $\mathfrak{A} = \mathrm{Aut}(\mathfrak{L}_{\Omega})$. In terms of a basis for $\mathfrak{L}$, the matrices of $\mathfrak{A}$ constitute an $\mathfrak{F}$-closed subgroup of $GL(n, \Omega)$, also denoted by $\mathfrak{A}$. Since Ω is algebraically closed, the Chevalley groups G and G' of Chapter III relative to $\mathfrak{L}_{\Omega}$ coincide, so that G is generated by elements of $\mathfrak{A}$ of the form $E(\lambda e_{\alpha})$, $e_{\alpha} \in \mathfrak{L}$, $\lambda \in \Omega$. Relative to a basis for $\mathfrak{L}$ consistent with a classical Cartan decomposition, the matrices of the $E(\lambda e_{\alpha})$ constitute a homomorph of the additive group of Ω and have entries which are fixed polynomials in λ with coęfficients in $\mathfrak{F}$; moreover, at least one of these entries, the coefficient of e_{α} in the image under $E(\lambda e_{\alpha})$ of $h \in \mathfrak{H}$, $\alpha(h) \neq 0$, has the form $\beta \lambda$, $0 \neq \beta \in \mathfrak{F}$. It follows that $\mathfrak{E}(\alpha) = \{E(\lambda e_{\alpha}) \mid \lambda \in \Omega\}$ is defined over $\mathfrak{F}$, and hence, by [407], p. 79, that any finite product of these $\mathfrak{E}(\alpha)$'s is a variety defined over $\mathfrak{F}$; in fact, since $\mathfrak{E}(\alpha)$ is biregularly equivalent with the affine line, any finite product of the $\mathfrak{E}(\alpha)$'s is biregularly equivalent over $\mathfrak{F}$ with an affine space. Now let $\mathfrak{B}$ be such a product, with r factors; let $y_1, \ldots, y_r$ be independent generic points over $\mathfrak{F}$ for these factors, and consider the variety $\mathfrak{U}$ in $\Omega^{(r+1)n^2}$ with generic point $(y_1, \ldots, y_r, y_1 y_2 \ldots y_r)$. Then $\mathfrak{U}$ is clearly defined over $\mathfrak{F}$, and is biregularly equivalent with $\mathfrak{B}$ over $\mathfrak{F}$. By [407], Chapter I, § 7, and Chapter IV, § 3, $\mathfrak{U}$ has a projection $p(\mathfrak{U})$ on the last factor, which is a subvariety of Ω^{n^2} defined over $\mathfrak{F}$. If $\pi(\mathfrak{U})$ denotes the set-theoretic projection of $\mathfrak{U}$ on the last factor, $\pi(\mathfrak{U}) \subseteq G$ and $\pi(\mathfrak{U})$ is dense in $p(\mathfrak{U})$ in the Zariski topology of Ω^{n^2}. Since G is generated by the $\mathfrak{E}(\alpha)$, G is the union of the sets $\pi(\mathfrak{U})$ so formed for all finite products $\mathfrak{V}$ of the $\mathfrak{E}(\alpha)$.

Now $\dim p(\mathfrak{U}) \leq n^2$ for all $\mathfrak{U}$; choosing $\mathfrak{U}$ such that $\dim p(\mathfrak{U})$ is maximal, let $\mathfrak{V}$ be the product of varieties $\mathfrak{E}(\alpha)$ used in forming $\mathfrak{U}$, and let α be an arbitrary root. Let $\mathfrak{V}' = \mathfrak{E}(\alpha) \times \mathfrak{V}$, $\mathfrak{U}'$ defined as above for $\mathfrak{V}'$; then clearly $\pi(\mathfrak{U}) \subseteq \pi(\mathfrak{U}')$, so $p(\mathfrak{U}) \subseteq p(\mathfrak{U}')$. By the maximality of dimension of $p(\mathfrak{U})$ we have $p(\mathfrak{U}) = p(\mathfrak{U}')$, and it follows that $p(\mathfrak{U})$ contains all products in $GL(n, \Omega)$ of elements of the various $\mathfrak{E}(\alpha)$, hence that $G \subseteq p(\mathfrak{U})$. Thus if $q(X)$ is a polynomial over Ω vanishing on G, $q(X)$ vanishes on $\pi(\mathfrak{U})$, hence on $p(\mathfrak{U})$, and the ideal vanishing on G is generated by polynomials over $\mathfrak{F}$.

By the generation of G by the $\mathfrak{E}(\alpha)$, and by p. 123 of [71], G is closed in the Zariski topology of $GL(n, \Omega)$. It follows that $G = \bar{G} \cap GL(n, \Omega)$, $\bar{G}$ being the closure of G in Ω^{n^2}; but $\bar{G} = p(\mathfrak{U})$ by the density of G in $p(\mathfrak{U})$, from which we see that G is an irreducible algebraic subgroup of $GL(n, \Omega)$ defined over $\mathfrak{F}$.

If Γ is the group of graph automorphisms of $\mathfrak{L}$ (Chap. III, § 5) relative to a chosen basis and fundamental system, we denote again by Γ the group of extensions of these to $\mathfrak{L}_\Omega$. This is the group of graph automorphisms of $\mathfrak{L}_\Omega$, and we have $\mathfrak{A} = \Gamma G$, a semi-direct product with G invariant.

Now let $\mathfrak{F}$ and $\mathfrak{K}$ be finite fields as at the beginning of this section. The mapping $\varphi: (\xi_{ij}) \to (\xi_{ij}^q)$ in $GL(n, \Omega)$ sends $E(\lambda e_\alpha)$ onto $E(\lambda^q e_\alpha)$, hence maps G onto G. If $\sigma \in \mathfrak{G}(\mathfrak{K}/\mathfrak{F})$, the Galois group of $\mathfrak{K}$ over $\mathfrak{F}$, the semi-automorphism V_σ of $\mathfrak{L}_\mathfrak{K}$ over $\mathfrak{L}$ has the action $A \to A^\sigma = V_\sigma^{-1} A V_\sigma$ on $\mathrm{Aut}(\mathfrak{L}_\mathfrak{K})$; if A has matrix (α_{ij}), then A^σ has matrix (α_{ij}^q). In particular, the restriction of φ to the matrices of $G_\mathfrak{K}$ is the same as $A \to A^\sigma$, where $\sigma: \xi \to \xi^q$ is the canonical generator for G.

Now suppose $\mathfrak{M}$ is an $\mathfrak{F}$-form of a classical simple Lie algebra; let $\mathfrak{L}$ be the split algebra over $\mathfrak{F}$ of the same type, and let $\mathfrak{K}/\mathfrak{F}$ be a finite extension which splits $\mathfrak{M}$. Let $\mathfrak{A}_\mathfrak{K}$ be the automorphism group of $\mathfrak{L}_\mathfrak{K}$, and let $\tau \to Z_\tau$ be a 1-cocycle on $\mathfrak{G} = \mathfrak{G}(\mathfrak{K}/\mathfrak{F})$ with values in $\mathfrak{A}_\mathfrak{K}$. Write $Z_\tau = C_\tau X_\tau$, $C_\tau \in \Gamma$, $X_\tau \in G_\mathfrak{K}$. Since C_τ is the extension of an automorphism of $\mathfrak{L}$, we have $C_\tau^\theta = C_\tau$ for all $\theta \in \mathfrak{G}$. Thus $Z_{\tau\theta} = Z_\theta Z_\tau^\theta = C_\theta C_\tau (C_\tau^{-1} X_\theta C_\tau X_\tau^\theta)$. From the effect of θ on the generators of $G_\mathfrak{K}$ given in Chapter III, § 1 (cf. also Lemma III.2.7), we have $G_\mathfrak{K}^\theta = G_\mathfrak{K}$, so that $C_{\tau\theta} = C_\theta C_\tau$, $X_{\tau\theta} = C_\tau^{-1} X_\theta C_\tau X_\tau^\theta$. When σ is the generator $\xi \to \xi^q$ for G, $C = C_\sigma$ satisfies the hypotheses of Lemma 1, so that for some $W \in G$, $X_\sigma = C^{-1} W^{-1} C W^\varphi$, $Z_\sigma = W^{-1} C W^\varphi$, $Z_{(\sigma^k)} = Z_\sigma Z_\sigma^\sigma \ldots Z_\sigma^{(\sigma^{k-1})} = Z_\sigma Z_\sigma^\varphi \ldots Z_\sigma^{\varphi^{k-1}} = W^{-1} C^k W^{\varphi^k}$. If $d = [\mathfrak{K}:\mathfrak{F}]$, we have $Z_{(\sigma^d)} = Z_1 = 1 = W^{-1} C^d W^{\varphi^d}$, and $C^d = 1$ since $\tau \to C_\tau$ is a homomorphism with $\sigma \to C$. Thus $W^{\varphi^d} = W$, from which $W \in G_\mathfrak{K}$, $W^\varphi = W^\sigma$, and we have $Z_\tau = W^{-1} C_\tau W^\tau$ for all $\tau \in \mathfrak{G}$. Thus *every cocycle* $\tau \to Z_\tau$ *is cohomologous to a homomorphism* $\tau \to C_\tau$ *of* $\mathfrak{G}$ *into* Γ.

Now Γ is the symmetric group on 1, 2 or 3 letters. In the first case, every cocycle is cohomologous to the identity, and $\mathfrak{M}$ is split. In the second case, there are two possibilities if d is even, otherwise only one. In the third case, the fact that all elements of order 2 in $\mathfrak{S}_3$ are conjugate shows that all homomorphisms $\mathfrak{G} \to \Gamma$ with C of order 2 give cohomologous cocycles, and likewise all with C of order 3 are cohomologous. Thus there are at most three non-isomorphic Lie algebras over $\mathfrak{F}$ which are $\mathfrak{F}$-forms of $\mathfrak{L}_\mathfrak{K}$, according as C has order $1, 2, 3$. Moreover, if $A \in \mathfrak{A}_\mathfrak{K}$ has $A^{-1} C A^\sigma = D$, with $C, D \in \Gamma$, let $A = BY$, $B \in \Gamma$, $Y \in G_\mathfrak{K}$; then $(B^{-1} C B) Y^\sigma = D(D^{-1} Y D)$, and C and D are conjugate in Γ; thus no two of the above forms are isomorphic. By the existence of a common finite splitting field, it follows that the upper bounds given above for numbers of $\mathfrak{F}$-forms of $\mathfrak{L}_\mathfrak{K}$ are the maximum numbers of $\mathfrak{F}$-forms of $\mathfrak{L}_\Omega$. Our knowledge of Γ now gives the

Theorem IV.6.1. The number of forms of a classical simple Lie algebra over the finite field $\mathfrak{F}$ (characteristic $\neq 2, 3$) is as follows:

Types $A_1, B, C, G_2, F_4, E_7, E_8$:	one;
Types A_n $(n > 1)$, D_n $(n \geqq 5)$, E_6:	two;
Type D_4:	three.

If $\mathfrak{H} \subseteq \mathfrak{G}$ is the kernel of the homomorphism $\tau \to C_\tau$, and if $\mathfrak{E}$ is the fixed field of $\mathfrak{H}$, then the restriction to $\mathfrak{H}$ of the cocycle $\tau \to C_\tau$ is trivial, i.e., $\mathfrak{M}_\mathfrak{E}$ is split. Thus $\mathfrak{M}$ has a splitting field $\mathfrak{E}$ of degree 1, 2 or 3 over $\mathfrak{F}$, and the cocycle in $Z^1(\mathfrak{G}(\mathfrak{E}/\mathfrak{F}), \mathrm{Aut}\,\mathfrak{L}_\mathfrak{E}) = Z^1(\mathfrak{G}/\mathfrak{H}, \mathrm{Aut}\,\mathfrak{L}_\mathfrak{E})$ associated with $\mathfrak{M}$ is cohomologous with that induced by $\tau \to C_\tau$ $(C_\tau \in \Gamma \subseteq \mathrm{Aut}\,\mathfrak{L}_\mathfrak{E})$. These remarks enable one to give an explicit construction of all $\mathfrak{F}$-forms of classical simple Lie algebras: one chooses an extension $\mathfrak{E}$ of $\mathfrak{F}$ of degree 1, 2 or 3, and lets $\mathfrak{L}$ be the split algebra of a specified type over $\mathfrak{F}$, $\tau \to C_\tau$ an isomorphism of the Galois group $\mathfrak{G} = \mathfrak{G}(\mathfrak{E}/\mathfrak{F})$ into the group Γ associated with $\mathfrak{L}$. Let $\alpha_1, \ldots, \alpha_r$ be a fundamental system of roots of $\mathfrak{L}$ with respect to a classical Cartan subalgebra $\mathfrak{H}$, and let $\{e_\alpha, h_i\}$ be a basis for $\mathfrak{L}$ chosen as in Chapter II, § 8. The semi-automorphisms V_τ which define the action of $\mathfrak{G}$ on $\mathrm{Aut}(\mathfrak{L}_\mathfrak{E})$ send λe into $\lambda^\tau e$, where $e \in \mathfrak{L}$, $\lambda \in \mathfrak{E}$. Then $\tau \to U_\tau = V_\tau C_\tau^{-1}$ defines a group of Galois semi-automorphisms of $\mathfrak{L}_\mathfrak{E}$, and the fixed $\mathfrak{F}$-subalgebra is an $\mathfrak{F}$-form of $\mathfrak{L}_\mathfrak{E}$ with associated cocycle $\tau \to C_\tau$ (§ 2). Now C_τ sends e_α into a root-vector relative to $\mathfrak{H}_\mathfrak{E} = \mathfrak{H}_\mathfrak{E}^\tau$: denote this root by α^τ, and let $\alpha_i = \alpha_i'^{\tau}$, $1 \leqq i \leqq r$. Then U_τ sends λe_{α_i} onto $\lambda^\tau e_{\alpha_i'}$, and similarly for $\lambda e_{-\alpha_i}$; also, the effect on these generators for $\mathfrak{L}_\mathfrak{E}$ determines U_τ.

For type A_r with $r \geqq 2$, we have $\alpha_i^\sigma = \alpha_{r-i+1}$, where σ is of order 2 and generates $\mathfrak{G}(\mathfrak{E}/\mathfrak{F})$. Thus the $\mathfrak{F}$-form of $\mathfrak{L}_\mathfrak{E}$ with non-trivial cocycle is the set $\mathfrak{M}$ of fixed elements of $\mathfrak{L}_\mathfrak{E}$ under the semi-automorphism sending λh_i onto $\lambda^\sigma h_{r-i+1}$, $1 \leqq i \leqq r$, $\lambda[\ldots[e_{\alpha_i} e_{\alpha_{i+1}}]\ldots e_{\alpha_{i+j}}]$ onto $\lambda^\sigma[\ldots[e_{\alpha_{r-i+1}} e_{\alpha_{r-i}}]\ldots e_{\alpha_{r-i-j+1}}]$, and similarly for negative roots. In terms of our realization of $\mathfrak{L}_\mathfrak{E}$ as the $(r+1)$-rowed $\mathfrak{E}$-matrices of trace zero (modulo scalars), we obtain a realization of $\mathfrak{M}$ as $\mathfrak{S}(\mathfrak{B}, \eta)^*$, where $\mathfrak{B} = M_{r+1}(\mathfrak{E})$, and where η is the involution sending $(\beta_{ij}) \in \mathfrak{B}$ onto $s^{-1}\,{}^t(\beta_{ij}^\sigma)\, s$, with

$$s = \begin{pmatrix} 0 & & & -1 \\ & & +1 & \\ & & \cdot & \\ & \cdot & & \\ (-1)^{r+1} & & & 0 \end{pmatrix}.$$

Then $(\mathfrak{B}, \eta)$ is a normal simple involutorial $\mathfrak{F}$-algebra, and is $\mathfrak{F}$-isomorphic with $(\mathfrak{B}, (\beta) \to {}^t(\beta^\sigma))$. Hence *the only non-split algebras of type A over a finite field $\mathfrak{F}$ are those of the form $\mathfrak{S}(M_n(\mathfrak{E}), (\beta) \to {}^t(\beta^\sigma))^*$, where $\mathfrak{E}$ is the quadratic extension of $\mathfrak{F}$ and σ generates $\mathfrak{G}(\mathfrak{E}/\mathfrak{F})$.*

One can carry out a similar analysis in terms of the realizations of other types. One may also note that if s is the $2r$ by $2r$ matrix

$$s = \left(\begin{array}{cc|c} \begin{array}{c|c} 0 & I_{r-1} \\ \hline I_{r-1} & 0 \end{array} & & 0 \\ \hline & 0 & \begin{array}{cc} 1 & 0 \\ 0 & \beta \end{array} \end{array}\right),$$

where $-\beta$ is not a square in $\mathfrak{F}$, then for $r \geqq 4$ $(M_{2r}(\mathfrak{F}), (\alpha) \to s^{-1}\,{}^t(\alpha)\, s)$ is a normal simple involutorial algebra $(\mathfrak{M}, \eta)$ of first kind, not split, and hence that $\mathfrak{S}(\mathfrak{M}, \eta)^* = \mathfrak{S}(\mathfrak{M}, \eta)$ is a non-split algebra of type D_r, which is split by a quadratic extension of $\mathfrak{F}$. These algebras provide realizations of all types D_r except D_4 when $[\mathfrak{E}:\mathfrak{F}] = 3$. In this case, a realization is available as the Lie algebra of derivations of the (split) exceptional simple Jordan algebra $\mathfrak{J}$ over $\mathfrak{F}$ annihilating a cubic subfield [11, 227] (and § 5 above). Finally, the non-split E_6 over $\mathfrak{F}$ may be obtained from the split E_6, say $\mathfrak{L}$, over $\mathfrak{F}$, by taking the fixed elements of $\mathfrak{L}_\mathfrak{E}$ ($[\mathfrak{E}:\mathfrak{F}] = 2$) under the semi-automorphism $U_\sigma^-: \lambda D \to$ $\to -\lambda^\sigma D^*$, where $D \in \mathfrak{L}$, $\lambda \in \mathfrak{E}$, D^* is as in Chapter III, § 6, and where σ generates $\mathfrak{G}(\mathfrak{E}/\mathfrak{F})$ [131, 378, 392].

The process whereby the non-split algebras above are obtained is also applicable to forms of algebraic groups; even when $\mathfrak{F}$ is not finite, the construction is possible for an extension $\mathfrak{E}/\mathfrak{F}$ where $\mathfrak{G}(\mathfrak{E}/\mathfrak{F})$ is isomorphic to a subgroup of the group Γ for the split type being considered. The resulting forms have been variously named ("quasi-split", "semi-split", "twisted") by their investigators, chiefly TITS [392, 394] and STEINBERG [378]. The exceptional algebras D_4 and E_6 obtained by this process over finite fields yield new simple finite groups [378, 379], analogous to the group G' of CHEVALLEY in the split case. A refinement of the method for characteristics 2 and 3 has been shown by REE [330, 331] to yield further finite simple groups, including those of SUZUKI [57, 312, 386].

§ 7. On automorphism groups

In Chapter III we have seen that the automorphism group of a classical simple Lie algebra has a subgroup G of finite index whose commutator subgroup is simple. One may ask whether an analogous result holds for arbitrary forms of classical simple Lie algebras. Since the forms of algebras of type A include the derived algebras of central division algebras, one sees from § 3 that any such theory would be applicable to the quotient by its center of the multiplicative group of such a division algebra. It is known (e.g., see [226, p. 191]) that such a group can have a very complicated structure. Similar objections can be raised in the case of the Lie algebras associated with symmetric or hermitian scalar products of Witt index zero [18, 102, 112]. Now in these simple Lie algebras there is no element $x \neq 0$ with ad x nilpotent, since such an x is necessarily nilpotent as an element of the division algebra, or of the ring of linear transformations of the vector space admitting the scalar product. This is seen for characteristic p by $(\operatorname{ad} x)^{p^k} = 0 = \operatorname{ad}(x^{p^k})$ for some k, and for characteristic zero by embedding x in a 3-dimensional subalgebra as in [234, Chap. 3]. But there are no non-zero nilpotents in a division algebra or in the skew transformations of an anisotropic space. In an attempt to avoid these cases, we therefore restrict our attention to algebras we shall call *non-compact*, namely those which possess elements $x \neq 0$ with ad x nilpotent.

Except in the case of G_2, all root-vectors e_α of a classical simple Lie algebra $\mathfrak{L}$ satisfy $(\operatorname{ad} e_\alpha)^3 = 0$, so that the simple Chevalley group G' for such an algebra is a subgroup of the group of automorphisms generated by all $E(x) = \exp(\operatorname{ad} x)$, where $(\operatorname{ad} x)^3 = 0$. This is in fact the case for G_2 as well: one has $(\operatorname{ad} e_\alpha)^3 = 0$ for "long" roots α ($\pm\alpha_1$, $\pm(\alpha_1 + 3\alpha_2)$, $\pm(2\alpha_1 + 3\alpha_2)$, in the notations of Chap. II), and products of such $E(e_\alpha)$ for different Cartan decompositions yield the $E(e_\beta)$ for the remaining roots β. One can also verify in general that those $x \in \mathfrak{L}$ with $(\operatorname{ad} x)^3 = 0$ and ad x of minimal rank among such x are conjugate under G' to root-vectors e_α, where α is a long root in the cases B, C, F_4, G_2, and that the $E(e_\alpha)$ for long roots α in all (classical) Cartan decompositions suffice to generate G'. Thus the simple group G'

may be described as the group generated by all $E(x)$, $(\mathrm{ad}\, x)^3 = 0$ and $\mathrm{ad}\, x$ of minimal rank with this property. It does not seem to be known (even if a non-singular Killing form is assumed) whether every non-compact form of a classical simple Lie algebra contains elements $x \neq 0$ with $(\mathrm{ad}\, x)^3 = 0$. This is not hard to see in the case of characteristic zero, by embedding a non-zero nilpotent y in a 3-dimensional simple algebra $\{e, f, h\}$ and by taking x to be a characteristic vector belonging to the greatest characteristic root of $\mathrm{ad}\, h$ in $\mathfrak{L}$. It is also valid generally for algebras of types $A_n(p \nmid (n+1))$, B, C, D_n $(n > 4)$, and G_2, by appeal to the classification of these and their realizations as discussed earlier in this chapter. Because of the uncertainty as to the existence of such elements in general non-compact algebras, we introduce the term *anti-compact* to refer to a form of a classical simple Lie algebra containing elements $x \neq 0$ with $(\mathrm{ad}\, x)^3 = 0$.

Let $\mathfrak{L}$ be an anti-compact Lie algebra, and let $\mathfrak{U}$ be the group of automorphisms of $\mathfrak{L}$ generated by all $E(x)$, where $\mathrm{ad}\, x \neq 0$ is of minimal rank with $(\mathrm{ad}\, x)^3 = 0$. We may then conjecture that $\mathfrak{U}$ *is a simple group*. By identifying $\mathfrak{L}$ as one of our realizations, and by identifying elements $x \in \mathfrak{L}$ having the above property, we can hope to give a realization for $\mathfrak{U}$. For the cases A, B, C, D_n $(n > 4)$, one thus obtains an identification of $\mathfrak{U}$ with the quotient by its center of a classical group $SL_m(\mathfrak{D})$, $Sp_{2m}(\mathfrak{F})$, $\Omega_m(\mathfrak{F}, f)$, $T_m(\mathfrak{B}, g)$, in the notations of Dieudonné [112], where $\mathfrak{F}$ is the ground field, $\mathfrak{D}$ a finite-dimensional central division algebra over $\mathfrak{F}$, $\mathfrak{B}$ a finite-dimensional central involutorial division algebra over $\mathfrak{F}$, f a bilinear scalar product of positive Witt index, and g a hermitian scalar product of positive Witt index. Furthermore, m is bounded from below to an extent which guarantees that these quotients are simple [112]. The only non-compact form of G_2 is the split one, so $\mathfrak{U} = G'$ is simple in this case as well.

A general proof of our conjecture, when $\mathfrak{F}$ is *perfect, of characteristic* $\neq 2, 3, 5$, appears as a corollary of a theorem of Tits [398]: If Ω is a universal domain over $\mathfrak{F}$, a basis for the anti-compact Lie algebra $\mathfrak{L}$ over $\mathfrak{F}$ gives a set of matrices for the group $\mathrm{Aut}(\mathfrak{L}_\Omega)$ which are the zeros in $GL(n, \Omega)$ of a set of polynomials with coefficients in $\mathfrak{F}$. Letting G_0 be the component of the identity in the algebraic group consisting of these matrices for $\mathrm{Aut}(\mathfrak{L}_\Omega)$, we see that every automorphism of Ω over $\mathfrak{F}$ maps G_0 onto itself. But this condition guarantees that G_0 is defined over $\mathfrak{F}$ [42; 271, p. 74]. Now the Chevalley group G is of finite index in $\mathrm{Aut}(\mathfrak{L}_\Omega)$, since $\mathfrak{L}_\Omega$ is split, and the matrices for G relative to a suitable basis for $\mathfrak{L}_\Omega$ form a connected algebraic group, the component of the identity in the matrices of $\mathrm{Aut}(\mathfrak{L}_\Omega)$ relative to this basis. It follows that G and G_0 are conjugate in $GL(n, \Omega)$, and that $G_0 \cong G = G'$ is a simple group. Now let H be the subgroup of G_0 generated by all ***unipotent*** elements of G_0 ***rational over*** $\mathfrak{F}$, i.e., matrices of automorphisms of the form A_Ω, where $A \in \mathrm{Aut}(\mathfrak{L})$ is unipotent. By the theorem of TITS, the quotient of H by its center is a simple group.

Our group $\mathfrak{U}$ is evidently generated by elements of H, so is contained in H; since the set of generators is self-conjugate under automorphisms, $\mathfrak{U}$ is normal in H. For characteristics greater than 5, elements $x \in \mathfrak{L}$ with $(\mathrm{ad}\, x)^3 = 0$ can be embedded in 3-dimensional simple subalgebras; since an "opposite" element y to x in such a subalgebra can be chosen with $(\mathrm{ad}\, y)^3 = 0$, $\mathrm{Rank}(\mathrm{ad}\, y) = \mathrm{Rank}(\mathrm{ad}\, x)$ (cf. Chap. V, § 8), one sees that $\mathfrak{U}$ is non-abelian, hence coincides with H. An element of the center of $\mathfrak{U}$ must fix all $x \in \mathfrak{L}$ for which $\exp(\mathrm{ad}\, x)$ is one of our generators; under the hypotheses, this implies that a basis for $\mathfrak{L}$ is fixed (Chap. V, § 8), so that the center of $\mathfrak{U}$ is trivial, and $\mathfrak{U} = H$ is simple.

The theorem of TITS, in its general form, would enable the same conclusions to be drawn as to simplicity of $\mathfrak{U}$ when $\mathfrak{F}$ is not perfect if we knew that: 1) G_0 is defined over $\mathfrak{F}$; 2) each generator $E(x)$ of $\mathfrak{U}$ is contained in the unipotent radical of some parabolic subgroup of G_0 defined over $\mathfrak{F}$. The restrictions on the characteristic of $\mathfrak{F}$ are not those of TITS, but rather are imposed by the limitations of our theory of Lie algebras; in particular, the exclusion of the characteristic 5 is due to technical difficulties in finding appropriate 3-dimensional subalgebras.

In the case where $\mathfrak{L}$ is *semi-split* in the sense of § 6, but not split, our groups $\mathfrak{U}$ coincide with the simple groups of STEINBERG [378]. For perfect ground fields this is most easily seen by observing, as above, that the latter must coincide with the group H of TITS. In the general case, more direct methods establish the identity of these groups.

Chapter V

Comparison of the Modular and Non-modular Cases

§ 1. Solvable and nilpotent algebras

The famous theorem of Lie on solvable linear Lie algebras over an algebraically closed field of characteristic zero asserts that every such algebra has a common eigenvector or, equivalently, consists of triangular matrices when interpreted as matrices relative to a suitable basis. The common algebraic proofs of this theorem use an argument which infers the nilpotency of a matrix from the vanishing of the traces of certain polynomials in the matrix, especially its powers, as in [234, pp. 43—50, and 64, pp. 2-05, 2-06]. One may also deduce the result from the corresponding one (due to KOLCHIN [254]) for solvable connected linear algebraic groups over arbitrary algebraically closed fields, via the correspondence in characteristic zero between linear algebraic groups and their Lie algebras [71, 72]. The conclusion fails for modular fields, although some of the proofs referred to are still applicable when the degree of the matrices is less than the characteristic. Over a field $\mathfrak{F}$ of prime characteristic p, a 2-dimensional solvable Lie algebra $\mathfrak{L}$ of p by p matrices is spanned by

$$E = \begin{pmatrix} 0 & 1 & 0 & \cdot & \cdot & \cdot & 0 \\ 0 & 0 & 1 & 0 & \cdot & \cdot & 0 \\ \vdots & & & & & & \vdots \\ 0 & \cdot & \cdot & \cdot & 0 & & 1 \\ 1 & 0 & \cdot & \cdot & \cdot & & 0 \end{pmatrix}, \qquad F = \begin{pmatrix} 0 & & & & \\ & 1 & & & 0 \\ & & 2 & & \\ & 0 & & \ddots & \\ & & & & p-1 \end{pmatrix}.$$

These evidently have no common eigenvector; in fact, $\mathfrak{L}$ acts irreducibly [234, p. 52].

A corollary of Lie's theorem is the nilpotency of the derived algebra of a solvable Lie algebra over any non-modular field. Taking $\mathfrak{L}$ as above, acting in $\mathfrak{F}^p$, let $\mathfrak{M} = \mathfrak{L} \oplus \mathfrak{F}^p$ with the product $[A + x, B + y] = [AB] + (xB - yA)$ for $A, B \in \mathfrak{L}$, $x, y \in \mathfrak{F}^p$. Then $\mathfrak{M}' = \mathfrak{F}E + \mathfrak{F}^p$, $\mathfrak{M}^{(3)} = 0$, but $\mathfrak{M}^{(2)} = (\mathfrak{M}')' = \mathfrak{F}^p = [\mathfrak{M}', \mathfrak{M}^{(2)}]$, so that $\mathfrak{M}'$ is not nilpotent; thus this corollary of Lie's theorem also fails for prime characteristic [234, p. 53].

Applied following the conclusions of Th. I.5.2, Lie's theorem yields that each of the $\mathfrak{V}_\varphi$ of that theorem has a basis relative to which the matrix of each $T \in \mathfrak{L}$ (a nilpotent Lie subalgebra of $\mathfrak{E}(\mathfrak{V})$) has the form

$$\begin{pmatrix} \varphi(T) & & * \\ & \ddots & \\ 0 & & \varphi(T) \end{pmatrix}.$$

This would imply in particular that the weights of $\mathfrak{L}$ are linear functions. We give an example to show that the weights of $\mathfrak{L}$ are not, in general, linear; this will show that the conclusions of Lie's theorem must fail even for *nilpotent* Lie algebras. (It should be noted that when $\mathfrak{L}$ is a nilpotent Lie algebra of linear transformations of index at most p, the weights *are* linear, while $\mathfrak{L}$ need not be triangulable—[234, p. 53], where the result of the exercise mentioned is valid when the index of nilpotency does not exceed p, but fails in general.) Our example is as follows: Let E_{ij} denote the usual p by p matrix unit, and let $V_k = \sum_{\nu=0}^{k-1} (-1)^\nu \binom{k-1}{\nu} E_{p-\nu,\,k-\nu}$, $1 \leqq k \leqq p$. In particular, $V_p = I$, the identity matrix. The V_k span a commutative p-dimensional Lie algebra $\mathfrak{V}$, and $V_k^p = 0$ for $k < p$. Thus if $V = \sum \lambda_k V_k$, $V^p = \lambda_p^p I$. Let $U = E_{p1} + \sum_{\nu=1}^{p-1} E_{\nu,\,\nu+1}$. Then $U^p = I$ and $[V_k U] = V_{k+1}$, $k < p$. Thus the space $\mathfrak{L}$ spanned by $\mathfrak{V}$ and U is a $(p+1)$-dimensional Lie algebra of p by p matrices, and is nilpotent of index $p+1$. From (3) of Chapter I, § 3, we see that for $V = \sum \lambda_k V_k \in \mathfrak{V}$, $\mu \in \mathfrak{F}$, $(V + \mu U)^p = V^p + \mu^p U^p + \mu^{p-1} V (\mathrm{ad}\, U)^{p-1} = (\lambda_p^p + \mu^p + \lambda_1 \mu^{p-1}) I$. Thus the only weight φ of $\mathfrak{L}$ has

$$\varphi(V + \mu U) = \lambda_p + \mu + (\lambda_1 \mu^{p-1})^{\frac{1}{p}}, \tag{1}$$

and is not additive.

A general theory of irreducible representations for nilpotent Lie algebras over algebraically closed modular fields has been given by ZASSENHAUS [419]: One sees easily from the definition that each nilpotent Lie algebra $\mathfrak{L}$ has a *regular basis* $\{x_i\}$, i.e., for all i, j, $[x_i x_j]$ is a combination of the x_k with $k > \mathrm{Max}(i, j)$. Taking a fixed regular basis, and ϱ an irreducible representation of $\mathfrak{L}$, ZASSENHAUS showed that ϱ is determined to within equivalence by the values of its single weight at the x_i; moreover, these values may be chosen arbitrarily. The $\mathfrak{L}$-module associated with ϱ has dimension equal to a power of the characteristic p, and these dimensions are bounded. $\mathfrak{L}$ has a faithful irreducible representation if and only if its center is one-dimensional. The eigenvalue $\varphi(\sum \xi_i x_i)$ of $\varrho(\sum \xi_i x_i)$ is given in terms of the $\varphi(x_i)$

as a certain fixed combination of the $\varphi(x_i)^{p^{-j}}$, with coefficients which are homogeneous polynomials in the $\xi_i^{p^{-j}}$. In the example above, $U = x_1$, $V_i = x_{i+1}$ define a regular basis, with $\varphi(x_1) = 1 = \varphi(x_{p+1})$, $\varphi(x_i) = 0$ otherwise; the formula (1) may be regarded as a prototype for the formula of ZASSENHAUS. More recent investigations have been made by CURTIS [92], KANNO [244] and ZASSENHAUS [422], who have simplified the original proofs and have extended them somewhat. In particular, JACOBSON has shown [234, Chap. 2] that in an indecomposable nilpotent Lie algebra of linear transformations over an arbitrary field, the minimum polynomial of each element x is a power of a (monic) prime polynomial π_x; one may thus regard the mapping $x \to \pi_x$ as the unique weight of this Lie algebra. In this extended sense, the notion of a weight for a representation ϱ of a nilpotent Lie algebra $\mathfrak{L}$ is now clear. CURTIS [92] showed that there is a mapping of equivalence classes of irreducible representations of $\mathfrak{L}$ onto mappings of a fixed regular basis onto prime polynomials, the mapping associated with ϱ being the restriction to the regular basis of the unique weight of ϱ. Here the dimensions are evidently unbounded in general.

§ 2. Representations

The results of § 1 include the fact that a nilpotent irreducible modular Lie algebra of linear transformations need not be abelian. This contrasts with the situation in characteristic zero, where solvable irreducible linear Lie algebras are abelian; namely, JACOBSON has proved [210], [234, Chap. II] that a completely reducible linear Lie algebra of characteristic 0 is the direct sum of its center and a semisimple Lie algebra, the elements of the center being semisimple endomorphisms. These conditions characterize non-modular completely reducible linear Lie algebras. By way of contrast, JACOBSON has shown that *every modular Lie algebra has a faithful completely reducible representation and a faithful representation which fails to be completely reducible* [221; 234, Chap. VI]. One might hope to salvage part of the analogy with the non-modular case by considering only Lie algebras which are *restricted* and only restricted representations. Here HOCHSCHILD [191] has proved that *a restricted Lie algebra $\mathfrak{L}$ has all its restricted representations completely reducible if and only if $\mathfrak{L}$ is abelian and $x \to x^{[p]}$ is one-one*. Thus these conditions on $\mathfrak{L}$ are equivalent with semisimplicity of the finite-dimensional associative algebra $\mathfrak{U}(\mathfrak{L})$, the "*u*-algebra" of Chapter I, § 4, hence with the complete reducibility of the regular representation of $\mathfrak{U}(\mathfrak{L})$. Since this representation is faithful, it follows that *a restricted Lie algebra has a faithful representation which is not completely reducible except when $\mathfrak{L}$ is abelian and $x \to x^{[p]}$ is one-one*. It remains an open problem to determine the structure of those restricted Lie algebras having faithful completely reducible restricted representations; the abelian algebras in this class are those for which $x \to x^{[p]}$ is one-one.

The non-modular representation theory is most highly developed in the case of semisimple Lie algebras over an algebraically closed field or, more generally, for classical Lie algebras [54; 166; 411; 64, Exposés 17—20; 234, Chap. 7, 8]. A partial analogue of this program for classical modular Lie algebras, namely the study of irreducible (restricted) representations, has been carried out by CURTIS [95, 96]. In terms of these representations, one may describe the radical $\mathfrak{N}$ of the u-algebra $\mathfrak{U}(\mathfrak{L})$—it is the intersection of their kernels in $\mathfrak{U}(\mathfrak{L})$—and thus may give a condition $(\varrho(x) = 0$ for all $x \in \mathfrak{N})$ which is necessary and sufficient for the complete reducibility of a given restricted representation ϱ of $\mathfrak{L}$. To establish a condition of this type which can be readily checked for a given representation ϱ would require finding a fairly simple set of generators for the ideal $\mathfrak{N}$. The only case in which such generators are known is that of the classical 3-dimensional algebra A_1, where the author has shown in unpublished work [360], that if $\{e, f, h\}$ is a basis for $\mathfrak{L}$ with $[e\,h] = 2e$, $[f\,h] = -2f$, $[e\,f] = h$, then the elements $e^{p-1}(h+1)$, $(h+1)\,f^{p-1}$ of $\mathfrak{U}(\mathfrak{L})$ generate the ideal $\mathfrak{N}$. JACOBSON [228] had shown earlier that the ideal generated by e^{p-1}, f^{p-1} contains $\mathfrak{N}$, but is not equal to $\mathfrak{N}$. In unpublished work, CURTIS has given sufficient conditions for a restricted representation ϱ of a classical Lie algebra to be completely reducible, conditions which require at least that $\varrho(e_\alpha)^{p-1} = 0$ for all root-vectors e_α relative to a classical Cartan subalgebra. By the above, these conditions are not necessary.

The results of CURTIS on the irreducible restricted representations are as follows: Let $\mathfrak{L} = \mathfrak{H} + \sum_{\alpha \neq 0} \mathfrak{L}_\alpha$ be a classical Cartan decomposition of $\mathfrak{L}$, $\alpha_1, \ldots, \alpha_r$ a fundamental system of roots, and let $h_i \in [\mathfrak{L}_{\alpha_i}\, \mathfrak{L}_{-\alpha_i}]$, $\alpha_i(h_i) = 2$. Let ϱ be an irreducible restricted representation of $\mathfrak{L}$ in $\mathfrak{V}$. Then there is a unique one-dimensional subspace $\mathfrak{X}$ of $\mathfrak{V}$ such that $\mathfrak{X}\,\varrho(e_\alpha) = 0$ for all $e_\alpha \in \mathfrak{L}_\alpha$, $\alpha > 0$ in an ordering associated with the given fundamental system. For $x \in \mathfrak{X}$, one has $x\,\varrho(h_i) = \lambda_i\, x$, λ_i in the prime field $\boldsymbol{Z}_p$, and the linear function λ on $\mathfrak{H}$ with $\lambda(h_i) = \lambda_i$, $1 \leqq i \leqq r$, is called the *highest weight* of ϱ. Two irreducible restricted representations having the same highest weight are equivalent, and conversely. Given any linear function λ on $\mathfrak{H}$ such that $\lambda(h_i) \in \boldsymbol{Z}_p$ for all i, there is an irreducible restricted representation of $\mathfrak{L}$ with highest weight λ. Thus *there is a one-one correspondence between equivalence classes of irreducible restricted representations of $\mathfrak{L}$ and linear functions λ on $\mathfrak{H}$ satisfying $\lambda(h_i) \in \boldsymbol{Z}_p$ for all i.* The space $\mathfrak{V}$ is generated by $\mathfrak{X}$ and the

$\mathfrak{L}_{-\alpha_i}$, in the sense that $\mathfrak{V}$ is the smallest subspace of $\mathfrak{V}$ containing $\mathfrak{X}$ and stable under the $\varrho(f_i)$, $f_i \in \mathfrak{L}_{-\alpha_i}$. Proofs of these results are in [95]. In another paper [96], CURTIS has shown that under rather strong general conditions on the irreducible representation ϱ, the dimension of $\mathfrak{V}$ can be computed from the highest weight λ and the system of roots of $\mathfrak{L}$ by the formula of WEYL [411; 142; 64, Exposés 19, 20; 234, Chap. 8] for the non-modular case, suitably interpreted. NIELSEN [308] has given elements of $\mathfrak{U}(\mathfrak{L})$ which generate minimal right ideals affording all irreducible representations of $\mathfrak{L}$.

A procedure utilized by CURTIS in obtaining some of his results is a passage from the non-modular to the modular case. More recently, such techniques have been given a more comprehensive setting [332; 371; cf. also 78]: If $\mathfrak{L}_C$ is a complex semisimple Lie algebra, ϱ_C an irreducible representation of $\mathfrak{L}_C$ in $\mathfrak{V}_C$, then there is a basis for $\mathfrak{L}_C$ and one for $\mathfrak{V}_C$, the former consisting of root-vectors e_α and elements h_i as above and the latter consisting of weight-vectors, such that the former basis is a Chevalley basis for $\mathfrak{L}_C$ in the sense of Chapter II, § 3, and such that each $\varrho_C(x)$, $x \in \mathfrak{L}_Z$, the additive group generated by the $\{e_\alpha, h_i\}$, maps into itself the group $\mathfrak{V}_Z$ generated by the basis for $\mathfrak{V}_C$. (In fact, $\mathfrak{L}_Z$ is mapped into itself by all $\exp(\mathrm{ad}\, m\, e_\alpha)$, $m \in \boldsymbol{Z}$, and $\mathfrak{V}_Z$ is stable under all $\exp(\varrho_C(m\, e_\alpha))$.) If $\mathfrak{F}$ is a field, $\mathfrak{V} = \mathfrak{V}_Z \otimes_Z \mathfrak{F}$ thus affords a representation ϱ of $\mathfrak{L} = \mathfrak{L}_Z \otimes_Z \mathfrak{F}$, a classical Lie algebra over $\mathfrak{F}$ or one which becomes classical when reduced modulo its center. If $0 \neq v \in \mathfrak{V}_Z$ belongs to the highest weight of ϱ_C (with respect to a fixed admissible ordering of the roots) and if $\mathfrak{M}$ is a maximal ϱ-stable subspace of $\mathfrak{V}$ not containing $v \otimes 1$, then $\mathfrak{V}/\mathfrak{M}$ affords an irreducible restricted representation of $\mathfrak{L}$. Its highest weight λ has $\lambda(h_i \otimes 1) = \Lambda(h_i) \pmod p$, where Λ is the highest weight of ϱ_C. Using the mappings induced in $\mathfrak{V}$ by the exponentials above, one is able to obtain linear groups acting in $\mathfrak{V}$, or in $\mathfrak{V}/\mathfrak{M}$, which are analogous to the group G of CHEVALLEY (acting in $\mathfrak{L}$). For suitable representations (those whose weights generate an additive group containing all weights of all representations of $\mathfrak{L}_C$), one thus constructs a linear group G^* which admits each such linear group associated with $\mathfrak{L}$ as homomorphic image in a canonical way [95, 332]. For certain classical algebras $\mathfrak{L}$, the group G^* (associated with the modules $\mathfrak{V}/\mathfrak{M}$) has been identified as a special linear group, symplectic group, or as a spin group associated with a quadratic form of maximal index [95, 384].

For Lie algebras of characteristic zero, one has the result that the tensor product $(\varrho \otimes 1) + (1 \otimes \sigma)$ of two completely reducible representations, ϱ, σ is again completely reducible [234, Chap. 3]. This fails in the modular case, even for restricted representations of classical Lie algebras. Namely, consider the 3-dimensional algebra $\mathfrak{L}$ with basis

$\{e, f, h\}$ as above; then the irreducible restricted representation ϱ with highest weight $\lambda: \lambda(h) = p - 1$ has degree p, and a vector $x \neq 0$ belonging to λ has $x\,\varrho(f)^{p-1} \neq 0$, $x\,\varrho(f)^p = 0$. Thus the $x\,\varrho(f)^j$, $0 \leq j < p$, are linearly independent. Let σ be any non-zero irreducible restricted representation of $\mathfrak{L}$, with highest weight μ, and let $y \neq 0$ belong to μ; then $\mu(h) = k$, $0 < k < p$, and $y\,\sigma(f)^k \neq 0$, $y\,\sigma(f)^{k+1} = 0$. The representation $\tau = \varrho \otimes 1 + 1 \otimes \sigma$ is a restricted representation of $\mathfrak{L}$; by the remarks above, it is completely reducible only if $\tau(h)\,\tau(f)^{p-1} + \tau(f)^{p-1} = 0$. If we apply this transformation to $x \otimes y$, we obtain an expression $x\,\varrho(f)^{p-1} \otimes y\,\sigma(h) + \sum_{j=0}^{p-2} x\,\varrho(f)^j \otimes y_j$. Since the $x\,\varrho(f)^j$ are linearly independent, and since $y\,\sigma(h) = k\,y \neq 0$, this cannot be zero, and τ is not completely reducible.

The u-algebra $\mathfrak{U}(\mathfrak{L})$ of a restricted Lie algebra $\mathfrak{L}$ provides an interesting class of non-semisimple associative algebras. From Curtis' representation theory, it follows that all irreducible representations of $\mathfrak{U}(\mathfrak{L})$ are absolutely irreducible if $\mathfrak{L}$ is classical, hence that $\mathfrak{U}(\mathfrak{L})/\mathfrak{N}$ is separable in this case. If $e_1, \ldots, e_n$ is a basis for $\mathfrak{L}$, Berkson [29] has shown that the linear function $\lambda(\sum \alpha_{i_1 \ldots i_n} e_1^{i_1} \ldots e_n^{i_n}) = \alpha_{p-1, \ldots, p-1}$ makes $\mathfrak{U}(\mathfrak{L})$ into a Frobenius algebra; Schue [352] has shown that $\mathfrak{U}(\mathfrak{L})$ is a symmetric algebra whenever $\mathrm{Tr}(\mathrm{ad}\, x) = 0$ for all $x \in \mathfrak{L}$, thus in particular whenever $\mathfrak{L}$ is classical.

§ 3. Cohomology

The cohomology theory for Lie algebras may be regarded as beginning with the lemmas of Whitehead [412, 413], which have as consequences the complete reducibility of representations of a semisimple Lie algebra of characteristic zero and Levi's theorem on the existence of a subalgebra, isomorphic to the quotient by its radical, in a Lie algebra of characteristic zero [234, Chap. 3; 64, Exposés 4, 5; 47]. If φ is a representation of $\mathfrak{L}$ in $\mathfrak{M}$, we write $v\,x$ for $v\,\varphi(x)$, $v \in \mathfrak{M}$, $x \in \mathfrak{L}$; by a natural extension of the notions of the Whitehead lemmas, a k-linear alternating function on $\mathfrak{L}$ with values in $\mathfrak{M}$ is called a *cocycle* if it satisfies the functional equation $(\delta f)(x_0, \ldots, x_k) = 0$, where $(\delta f)(x_0, \ldots, x_k) = \sum_{i=0}^{k} (-1)^{i+k} f(x_0, \ldots, \hat{x}_i, \ldots, x_k)\, x_i + \sum_{i<j} (-1)^{i+j+1} \times f([x_i\, x_j], x_0, \ldots, \hat{x}_i, \ldots, \hat{x}_j, \ldots, x_k)$, where the circumflex indicates an omitted argument. The function f is a *coboundary* if there is a $(k-1)$-linear alternating function g on $\mathfrak{L}$ with values in $\mathfrak{M}$ such that $f = \delta g$. One verifies that $\delta(\delta g) = 0$, hence that a quotient group $H^k(\mathfrak{L}, \mathfrak{M})$ = cocycles/coboundaries may be defined for each k (0-coboundaries being defined to be 0). Thus $H^0(\mathfrak{L}, \mathfrak{M})$ = elements of $\mathfrak{M}$ annihilated

by $\mathfrak{L}$, and the Whitehead lemmas assert the vanishing of H^1 and H^2 whenever $\mathfrak{L}$ is semi-simple and non-modular.

From the fact that $H^1(\mathfrak{L}, \mathfrak{M}) = 0$ for all $\mathfrak{M}$ implies the complete reducibility of all $\mathfrak{L}$-modules, and from the existence of non-semi-simple modules for all modular $\mathfrak{L} \neq 0$ (§ 2), we see that *if $\mathfrak{L}$ is a modular Lie algebra such that $H^1(\mathfrak{L}, \mathfrak{M}) = 0$ for all $\mathfrak{L}$-modules $\mathfrak{M}$, then $\mathfrak{L} = 0$*. It appears likely that a similar conclusion holds with H^1 replaced by H^2, and perhaps by H^k for each $k < \dim \mathfrak{L}$. A simple example shows that it is not true that $H^2(\mathfrak{L}, \mathfrak{M}) = 0$ for all $\mathfrak{M}$ even when $\mathfrak{L}$ is classical (hence restricted) and when $\mathfrak{M}$ is an irreducible restricted $\mathfrak{L}$-module. Namely, let $\mathfrak{L}$ be the classical algebra $\{e, f, h\}$ of § 2, and let $\mathfrak{M}$ be the $(p-1)$-dimensional irreducible restricted $\mathfrak{L}$-module; then $\mathfrak{M}$ has basis $v_1, \ldots, v_{p-1}$, with

$$v_i f = v_{i+1}, \quad 1 \leq i < p-1, \qquad v_{p-1} f = 0;$$
$$v_i e = -(i-1)(p-i)\, v_{i-1}, \quad i > 1, \qquad v_1 e = 0;$$
$$v_i h = (p-2i)\, v_i, \quad 1 \leq i \leq p-1.$$

Let g be the alternating bilinear function on $\mathfrak{L}$ with $g(e, f) = 0 = g(f, h)$, $g(e, h) = v_{p-1}$. One verifies at once that $(\delta g)(e, f, h) = 0$, from which it follows that $\delta g = 0$. On the other hand, if k is any linear mapping of $\mathfrak{L}$ into $\mathfrak{M}$, $(\delta k)(e, h) = k(e)\, h - k(h)\, e - k([e\, h])$ is seen to lie in the subspace of $\mathfrak{M}$ spanned by the v_i, $i < p-1$. Thus $g \neq \delta k$, and $H^2(\mathfrak{L}, \mathfrak{M}) \neq 0$. (In fact, $H^2(\mathfrak{L}, \mathfrak{M})$ has dimension 2 over the base field $\mathfrak{F}$, and if $\mathfrak{N}$ is any other irreducible restricted module for this $\mathfrak{L}$, $H^2(\mathfrak{L}, \mathfrak{N}) = 0$.)

When the module $\mathfrak{M}$ is one-dimensional, with $v\, x = 0$ for all $v \in \mathfrak{M}$, $x \in \mathfrak{L}$, we may identify $\mathfrak{M}$ with the ground field $\mathfrak{F}$. If $\mathfrak{F} = \boldsymbol{R}$, the field of real numbers, and if $\mathfrak{L}$ is the Lie algebra of a compact connected Lie group $\mathfrak{G}$, Chevalley and Eilenberg [80] have shown that this cohomology of $\mathfrak{L}$ is isomorphic with the real cohomology of $\mathfrak{G}$. Thus the groups $H^k(\mathfrak{L}, \mathfrak{F})$ take on special interest; if $\mathfrak{L} = [\mathfrak{L}\, \mathfrak{L}]$, then $H^1(\mathfrak{L}, \mathfrak{F}) = 0$ since all 1-cocycles are zero. If $\mathfrak{L}$ carries a nonsingular symmetric associative bilinear form (x, y), and if all derivations of $\mathfrak{L}$ are inner (e.g., if $\mathfrak{L}$ has nonsingular Killing form), then every 2-cocycle f on $\mathfrak{L}$ with values in $\mathfrak{F}$ has the form $f(x, y) = (x\, S, y)$, where S is a linear transformation of $\mathfrak{L}$ skew with respect to (x, y); it further follows from the cocycle condition that S is a *derivation* of $\mathfrak{L}$, and from this that $f(x, y) = ([x\, z], y)$ for some fixed $z \in \mathfrak{L}$. That is, $f = \delta g$, where $g(x) = (x, z)$, and $H^2(\mathfrak{L}, \mathfrak{F}) = 0$ for such $\mathfrak{L}$. In fact, this reasoning shows that the mapping $S \to f$, where $f(x, y) = (x\, S, y)$, induces an isomorphism of $\mathfrak{S}\mathfrak{D}(\mathfrak{L})/(\mathrm{ad}\, \mathfrak{L})$, the algebra of "outer derivations" skew with respect to (x, y), onto $H^2(\mathfrak{L}, \mathfrak{F})$ whenever $\mathfrak{L}$ has a nonsingular symmetric associative bilinear form (x, y). In particular, for $\mathfrak{L}$

classical of type A_n, $p|(n+1)$, $H^2(\mathfrak{L}, \mathfrak{F})$ has dimension one (cf. § 5). If $\mathfrak{L}$ possesses a symmetric associative bilinear form (x, y), let $g(x, y, z) = (x, [y\,z])$; one verifies directly that g is a 3-cocycle with values in $\mathfrak{F}$. If (x, y) is non-singular, and if $g = \delta f$, there is a linear transformation T of $\mathfrak{L}$ such that $f(x, y) = (xT, y) = -(x, yT)$. Setting $g = \delta f$ yields $[y\,z] = [y\,z]T - [yT, z] - [y, zT]$ for all $y, z \in \mathfrak{L}$, or $[\mathrm{ad}\, z, T] = \mathrm{ad}\, z + \mathrm{ad}(zT)$ for all $z \in \mathfrak{L}$. Now suppose $\mathfrak{L} = [\mathfrak{L}\,\mathfrak{L}]$; it then follows from the non-singularity and associativity of (x, y) that the center of $\mathfrak{L}$ is zero, so that $\mathrm{ad}\,\mathfrak{L}$ is isomorphic to $\mathfrak{L}$. If, in addition, every derivation of $\mathfrak{L}$ is inner, the fact that $[\mathrm{ad}\, z, T] \in \mathrm{ad}\,\mathfrak{L}$ for all $z \in \mathfrak{L}$ implies that $[\mathrm{ad}\, z, T] = [\mathrm{ad}\, z, \mathrm{ad}\, t]$ for some $t \in \mathfrak{L}$, all $z \in \mathfrak{L}$, from which $zT = [z\,t] - z$ for all z. Substituting in δf gives $g(x, y, z) = (\delta f)(x, y, z) = -3g(x, y, z) + (x, [[y\,z]\,t] - [[y\,t]\,z] - [y[z\,t]])$, from which $4g(x, y, z) = 0$ by the Jacobi identity. If the characteristic of $\mathfrak{F}$ is not 2, this is absurd. In particular, we see that for characteristic $\neq 2$, if $\mathfrak{L} \neq 0$ has nonsingular Killing form, then $H^3(\mathfrak{L}, \mathfrak{F}) \neq 0$. For semi-simple algebras of characteristic zero, this is a familiar result [80, 266].

To discuss the cohomology in terms of general theory, one starts with a projective resolution of $\mathfrak{F}$, regarded as a trivial $\mathfrak{L}$-module, or right $\mathfrak{U}'(\mathfrak{L})$-module, where $\mathfrak{U}'(\mathfrak{L})$ is the universal associative algebra of $\mathfrak{L}$. That is, one constructs an exact sequence

$$0 \longleftarrow \mathfrak{F} \underset{\varepsilon}{\longleftarrow} X_0 \underset{\partial_1}{\longleftarrow} X_1 \underset{\partial_2}{\longleftarrow} \cdots,$$

where the X_i are projective $\mathfrak{U}'(\mathfrak{L})$-modules and ε, ∂_i are $\mathfrak{U}'(\mathfrak{L})$-homomorphisms. With this resolution and with a given $\mathfrak{L}$-module $\mathfrak{M}$ one may associate the additive groups $\mathfrak{M}_i = \mathrm{Hom}(X_i, \mathfrak{M})$ of $\mathfrak{U}'(\mathfrak{L})$-homomorphisms of X_i into $\mathfrak{M}$. Composition with the maps ∂_i gives rise to a complex

$$\mathfrak{M}_0 \xrightarrow{\delta_0} \mathfrak{M}_1 \xrightarrow{\delta_1} \mathfrak{M}_2 \xrightarrow{\delta_2} \cdots,$$

i.e., composition of successive δ_i is zero. The associated k-th cohomology group is $\mathrm{Ker}(\delta_k)/\mathrm{Im}(\delta_{k-1})$, for $k = 0$ simply $\mathrm{Ker}(\delta_0)$ being taken. It follows from general principles that the cohomology groups so obtained are independent of the particular projective resolution chosen, as are various operations on the cohomology [56]. The groups $H^k(\mathfrak{L}, \mathfrak{M})$ defined above in terms of functional equations may be identified with those obtained by this latter process from a particular free resolution of $\mathfrak{F}$ as $\mathfrak{U}'(\mathfrak{L})$-module [56, 64, 234, 266, 287].

This second point of view may be taken whenever one has an associative algebra $\mathfrak{A}$ over $\mathfrak{F}$ and an augmentation $\varepsilon: \mathfrak{A} \to \mathfrak{F}$ which is a homomorphism of $\mathfrak{A}$-modules (one takes $X_0 = \mathfrak{A}$) [56, 287]. In particular, if $\mathfrak{L}$ is a restricted Lie algebra, one may take $\mathfrak{A}$ to be the "u-algebra" $\mathfrak{U}(\mathfrak{L})$ of Chapter I, § 3, and use this procedure to define the *restricted cohomology groups* $H^k_*(\mathfrak{L}, \mathfrak{M})$, where $\mathfrak{M}$ is a restricted $\mathfrak{L}$-module. Interpretations of $H^k_*(\mathfrak{L}, \mathfrak{M})$ for $k = 1, 2, 3$ have been given by Hochschild [192—194, 198]: for $H^1_*(\mathfrak{L}, \mathfrak{M})$, in terms of extensions $0 \to \mathfrak{R} \to \mathfrak{E} \to \mathfrak{Q} \to 0$ of restricted $\mathfrak{L}$-modules, where $\mathfrak{M}$ is the space of all $\mathfrak{F}$-linear mappings $\mathfrak{Q} \to \mathfrak{R}$ with the action $(f\,x)\,(q) = f(q)\,x - - f(q\,x)$ for $f \in \mathfrak{M}$, $x \in \mathfrak{L}$, $q \in \mathfrak{Q}$; for $H^2_*(\mathfrak{L}, \mathfrak{M})$, in terms of extensions of restricted Lie algebras $0 \to \mathfrak{M} \to \mathfrak{E} \to \mathfrak{L} \to 0$, where $\mathfrak{M}$ has the structure of an abelian restricted Lie algebra with $x^{[p]} = 0$ for all $x \in \mathfrak{M}$; for $H^3_*(\mathfrak{L}, \mathfrak{M})$, in terms of extensions of restricted Lie algebras with non-abelian kernels. In particular, the vanishing of $H^1_*(\mathfrak{L}, \mathfrak{M})$ for all finite-dimensional $\mathfrak{M}$ implies the complete reducibility of all finite-dimensional restricted $\mathfrak{L}$-modules (i.e., the semisimplicity of $\mathfrak{U}(\mathfrak{L})$) which by results of Hochschild cited in § 2 implies that $\mathfrak{L}$ is abelian with one-one p-th power operator. Conversely, if the base field is perfect and $\mathfrak{L}$ has this structure, then so does $\mathfrak{L}_{\mathfrak{R}}$, for $\mathfrak{R}$ the algebraic closure of $\mathfrak{F}$, so that by [225] (or [234, Chap. V]) $\mathfrak{L}_{\mathfrak{R}}$ has a basis $\{e_i\}$ with $e_i^{[p]} = e_i$. It follows that $\mathfrak{U}(\mathfrak{L}_{\mathfrak{R}}) \cong \mathfrak{U}(\mathfrak{L})_{\mathfrak{R}}$ is a semisimple associative algebra, hence that $\mathfrak{U}(\mathfrak{L})$ is separable, and that every restricted $\mathfrak{L}$-module is a projective $\mathfrak{U}(\mathfrak{L})$-module. Thus $H^k_*(\mathfrak{L}, \mathfrak{M}) = 0$ for all $k > 0$ and all restricted $\mathfrak{L}$-modules $\mathfrak{M}$. One thus sees that the condition that the first cohomology groups be zero for all finite-dimensional modules is equivalent to: for characteristic zero, the semisimplicity of $\mathfrak{L}$ [47, 64, 234, 266]; for characteristic p and ordinary modules, $\mathfrak{L} = 0$; for characteristic p and restricted modules, $\mathfrak{L}$ abelian with one-one p-power map.

Although an interpretation for $H^1_*(\mathfrak{L}, \mathfrak{M})$ in terms of functional equations on $\mathfrak{L}$ has been given by Hochschild [192], similar interpretations for higher restricted cohomology seem to be available only in terms of functions on $\mathfrak{U}(\mathfrak{L})$ (i.e., from $\mathrm{Hom}_{\mathfrak{U}(\mathfrak{L})}(B(\mathfrak{U}(\mathfrak{L})), \mathfrak{M})$, where B is the "bar construction" of [287]). A resolution of $\mathfrak{F}$ which is $\mathfrak{U}(\mathfrak{L})$-free, and which appears rather efficient for computation has been found by May [302], in the more general context of graded restricted Lie algebras. He has used this resolution and connections with the cohomology of Hopf algebras to obtain extensive information on the homotopy groups of spheres [301]. In particular, May is able to use his resolution to show that $H^2_*(\mathfrak{L}, \mathfrak{F}) = 0$ only if $\mathfrak{L}$ is abelian with one-one p-power [300]; the converse (for perfect fields) holds by the above. On the other hand, $H^2(\mathfrak{L}, \mathfrak{F}) \neq 0$ if $\mathfrak{L}$ is abelian of dimension at least 2, while $H^2(\mathfrak{L}, \mathfrak{F}) = 0$ for most classical semisimple algebras $\mathfrak{L}$ by remarks above. (For further studies on graded Lie algebras, restricted or not, see [305, 336], and references therein; in [305], it is shown that Hopf algebras of the form $\mathfrak{U}(\mathfrak{L})$, where $\mathfrak{L}$ is a graded restricted Lie algebra, play an important part in the general study of Hopf algebras.)

If the p-power map of the restricted Lie algebra $\mathfrak{L}$ is one-one (indeed, if $x^{[p]} = 0$ implies $x = 0$, $x \in \mathfrak{L}$) and if the base field $\mathfrak{F}$ is algebraically closed, then $\mathfrak{L}$ is abelian [234, p. 196, ex. 14; 83]. Chwe has established the analogous result over arbitrary fields when $\mathfrak{L}$ is solvable [82]. In this paper, he gives relations between the ordinary and restricted cohomology of a restricted Lie algebra $\mathfrak{L}$ by connecting the restricted cohomology with relative ordinary cohomology with respect to the subalgebra $\mathfrak{S}$ of $\mathfrak{U}'(\mathfrak{L})$ generated by all $x^p - x^{[p]}$, $x \in \mathfrak{L}$.

§ 4. Known simple Lie algebras

In Chapter IV, we have given what seems to be the current state of knowledge concerning simple modular Lie algebras which have analogues of characteristic zero. Except for small prime characteristics, the determination of all forms of simple classical Lie algebras has reached the same degree of completeness in the modular and non-modular cases. While these exhaust the normal simple Lie algebras of characteristic zero, it will be seen from the following examples that the situation is radically different for modular Lie algebras.

A. The Jacobson–Witt algebras $\mathfrak{W}_n$

Let $\mathfrak{F}$ be a field of prime characteristic p. Let $\mathfrak{F}[X] = \mathfrak{F}[X_1, \ldots, X_n]$ be the polynomial ring in the indeterminates X_i and let $\mathfrak{A}_n$ be the quotient of $\mathfrak{F}[X]$ by the ideal generated by the X_i^p, $1 \leqq i \leqq n$. Then $\mathfrak{A}_n$ is a commutative associative $\mathfrak{F}$-algebra with unit, of dimension p^n, and with basis consisting of monomials in $x_1, \ldots, x_n$, the images of the X_i, of degree at most $p-1$ in each x_i, and with $x_i^p = 0$. *The algebra $\mathfrak{W}_n$ is the Lie algebra of derivations of $\mathfrak{A}_n$* as $\mathfrak{F}$-algebra; $\mathfrak{W}_1$ is known as the *Witt algebra* [66, 419]. A more general case, in which the generators for the ideal to be factored from $\mathfrak{F}[X]$ have the form $X_i^p - \xi_i$, has been studied by Jacobson [217]; this algebra is a form of $\mathfrak{W}_n$, as one sees by adjoining $\xi_i^{1/p}$ to $\mathfrak{F}$ and replacing X_i by $X_i - \xi_i^{1/p}$. Thus we refer to $\mathfrak{W}_n$ *as the split Jacobson–Witt algebra of order n* over $\mathfrak{F}$.

The algebra $\mathfrak{W}_n$ is a simple Lie algebra except when it is 2-dimensional. Being a derivation algebra, $\mathfrak{W}_n$ is necessarily restricted. A Cartan subalgebra $\mathfrak{H}$ for $\mathfrak{W}_n$ is obtained as the set of derivations of $\mathfrak{A}_n$ preserving each of the one-dimensional spaces $\mathfrak{F} x_i$, $1 \leqq i \leqq n$. This $\mathfrak{H}$ is commutative, and its roots are linear $\mathfrak{F}$-valued functions on $\mathfrak{H}$; in fact, they may be described as follows:

Let $H \in \mathfrak{H}$, $x_i H = \lambda_i(H)\, x_i$ all i; then

$$\Big(\prod_i x_i^{\nu_i}\Big) H = \Big(\sum_i \nu_i \lambda_i(H)\Big) \Big(\prod_i x_i^{\nu_i}\Big).$$

Now $\mathfrak{W}_n$ has a basis consisting of derivations $D_{j,(\nu)}$ sending x_k into $\delta_{jk} \prod_i x_i^{\nu_i}$, and $[D_{j,(\nu)} H]$ sends x_k into $\delta_{jk}\left(\sum_i \nu_i \lambda_i(H) - \lambda_k(H)\right) \prod x_i^{\nu_i}$. Thus $D_{j,(\nu)}$ belongs to the root $\sum_i (\nu_i - \delta_{ij}) \lambda_i$, and *the roots relative to* $\mathfrak{H}$ *are the members of the additive group generated by the* λ_i. *Each root space has dimension* n; for $n > 1$ and $p > 3$, these last remarks represent a substantial departure from the situation treated in Chapter II. Every derivation of $\mathfrak{W}_n$ is an inner derivation, and for $p \neq 2, 3$, the mapping sending $A \in \mathrm{Aut}(\mathfrak{A}_n)$ onto $\sigma_A \in \mathrm{Aut}(\mathfrak{W}_n)$: $D\sigma_A = A^{-1} DA$, is an isomorphism of $\mathrm{Aut}(\mathfrak{A}_n)$ onto $\mathrm{Aut}(\mathfrak{W}_n)$. In constrast with the classical case, consideration of a normal series for $\mathrm{Aut}(\mathfrak{A}_n)$ does not lead to simple groups distinctively associated with $\mathfrak{W}_n$. What one obtains, even for $\mathfrak{A}_n$ in the more general form considered by JACOBSON, is a normal series for $\mathrm{Aut}(\mathfrak{A}_n)$, whose only nonabelian factor is a full linear group over a certain subfield of $\mathfrak{A}_n$. Any isomorphism between the derivation algebras of two algebras $\mathfrak{A}_n$ (in the general form) is realized by an isomorphism between the associative algebras. Any form of $\mathfrak{W}_n$ which is split by a separable extension, or by a purely inseparable extension of exponent one, is the derivation algebra of a general $\mathfrak{A}_n$ (if $p > 3$). These facts are to be found in [217], with details.

B. Some simple subalgebras of $\mathfrak{W}_n$

A large number of simple Lie algebras of prime characteristic (and all the known simple restricted Lie algebras not included in Chap. II) may be realized as certain distinguished subalgebras of Jacobson–Witt algebras. (This is not so surprising since JACOBSON (unpublished) and, independently, JU. I. MANIN [292] have proved that every restricted Lie algebra can be embedded in a Jacobson–Witt algebra.) A unified treatment which provides most known simple Lie subalgebras of $\mathfrak{W}_n$ has been given by M. S. FRANK [314, 135]. She shows that, if $\mathfrak{F}$ is a field of prime characteristic, $\mathfrak{M}$ a Lie subalgebra of the matrix algebra $\mathfrak{M}_n(\mathfrak{F})$, $\mathfrak{A}_n$ as in A) above, $\mathfrak{L} = \mathfrak{M} \otimes_{\mathfrak{F}} \mathfrak{A}_n$, a Lie algebra over $\mathfrak{F}$ with $[m_1 \otimes a_1, m_2 \otimes a_2] = [m_1 m_2] \otimes a_1 a_2$, then $\mathfrak{D}(\mathfrak{M}) = \mathfrak{L}^*$ is a Lie subalgebra of $\mathfrak{W}_n$, where $\mathfrak{L}^*$ is defined as follows: Let $\mathfrak{L}$ be any subspace of $\mathfrak{X}_n \equiv \mathfrak{M}_n(\mathfrak{F}) \otimes_{\mathfrak{F}} \mathfrak{A}_n \cong \mathfrak{M}_n(\mathfrak{A}_n)$; if $D \in \mathfrak{W}_n$, define $\sigma(D)$ to be the element of $\mathfrak{X}_n$ corresponding to the matrix $\left(\frac{\partial}{\partial x_j}(x_i D)\right)$ in $\mathfrak{M}_n(\mathfrak{A}_n)$ under the canonical isomorphism indicated above. Then $\mathfrak{L}^*$ is the inverse image of $\mathfrak{L}$ under the mapping $D \to \sigma(D)$. Here $\frac{\partial}{\partial x_j}$ may be interpreted as the derivation denoted in A) by $D_{j,(0)}$. Conjugate subalgebras of $\mathfrak{M}_n(\mathfrak{F})$ give rise to isomorphic subalgebras of $\mathfrak{W}_n$, but isomorphic subalgebras need not do so. The algebra $\mathfrak{M}$ is closed under p-th powers if and only if $\mathfrak{D}(\mathfrak{M})$ is. Mrs. FRANK gives simple conditions on $\mathfrak{D}(\mathfrak{M})$ which guarantee that the derived series of $\mathfrak{D}(\mathfrak{M})$ terminates in a normal simple Lie algebra, provided that $\mathfrak{M}$ is absolutely irreducible as subalgebra of $\mathfrak{M}_n(\mathfrak{F})$. Application of the procedure yields simple Lie subalgebras of $\mathfrak{W}_n$ in the following cases:

a) The class $\mathfrak{S}_n$ [133, 9]

Let $\mathfrak{M}$ be the Lie algebra of matrices of trace zero in $\mathfrak{M}_n(\mathfrak{F})$, $n > 1$. For $n \geqq 3$, $\mathfrak{D}(\mathfrak{M})^{(1)} = \mathfrak{D}(\mathfrak{M})^{(2)}$, and this is a simple restricted subalgebra of $\mathfrak{W}_n$, denoted by $\mathfrak{S}_n$; its dimension is $(n-1)(p^n-1)$. For $n = 2$, $\mathfrak{D}(\mathfrak{M})^{(1)} \neq \mathfrak{D}(\mathfrak{M})^{(2)} = \mathfrak{D}(\mathfrak{M})^{(3)}$, and, for $p > 2$, $\mathfrak{D}(\mathfrak{M})^{(2)}$ is a simple restricted subalgebra of $\mathfrak{W}_2$, of dimension $p^2 - 2$. We denote this algebra by $\mathfrak{S}_2$.

b) The class $\mathfrak{V}_n$ [9, 33, 356]

Let $\mu_1, \ldots, \mu_n$ be nonzero elements of $\mathfrak{F}$, and let $\mathfrak{M}$ be the Lie subalgebra of $\mathfrak{M}_{2n}(\mathfrak{F})$ consisting of matrices (α) with $(\alpha)\,({}^t(\mu)) = (\mu)\,({}^t(\alpha))$, where ${}^t(\alpha)$ is the transpose of (α), and where (μ) is the skew-symmetric matrix of degree $2n$

$$\left(\begin{array}{ccc|ccc} & & & \mu_1 & & 0 \\ & 0 & & & \ddots & \\ & & & 0 & & \mu_n \\ \hline -\mu_1 & & 0 & & & \\ & \ddots & & & 0 & \\ 0 & & -\mu_n & & & \end{array}\right).$$

Clearly $\mathfrak{M}$ is closed under p-th powers; hence $\mathfrak{D}(\mathfrak{M})$ is a restricted subalgebra of $\mathfrak{W}_{2n}$. Conjugation of $\mathfrak{M}$ by

$$\left(\begin{array}{ccc|ccc} \mu_1^{-1} & & 0 & & & \\ & \ddots & & & 0 & \\ 0 & & \mu_n^{-1} & & & \\ \hline & & & 1 & & 0 \\ & 0 & & & \ddots & \\ & & & 0 & & 1 \end{array}\right)$$

carries $\mathfrak{M}$ onto the set of the matrices satisfying the same conditions for $\mu_1 = \cdots = \mu_n = 1$; hence the resulting algebra $\mathfrak{D}(\mathfrak{M})$ is isomorphic with that obtained for all $\mu_i = 1$. One has $\mathfrak{D}(\mathfrak{M}) \neq \mathfrak{D}(\mathfrak{M})^{(1)} = \mathfrak{D}(\mathfrak{M})^{(2)}$ for $p > 2$ or $n > 1$, and $\mathfrak{D}(\mathfrak{M})^{(1)}$ is a simple restricted subalgebra $\mathfrak{V}_{n,(\mu)}$ of dimension p^{2n-2} of $\mathfrak{W}_{2n}$ under these conditions. By the above, all $\mathfrak{V}_{n,(\mu)}$ are isomorphic with $\mathfrak{V}_{n,(1)}$, which we denote simply by $\mathfrak{V}_n$. (For another interpretation of these algebras and this isomorphism, cf. [343].)

c) *The class* $\mathfrak{R}_n$ [134, 135]

Taking the basis for $\mathfrak{A}_n$ consisting of the monomials $x_1^{\nu_1} \ldots x_n^{\nu_n}$, $0 \leqq \nu_i < p$, and ordering these by a lexicographic ordering of the n-tuples $(\nu_1, \ldots, \nu_n)$, the matrices of the regular representation of $\mathfrak{A}_n$ constitute a commutative subalgebra $\mathfrak{B}$ of $\mathfrak{M}_{p^n}(\mathfrak{F})$. Let $\mathfrak{M}$ be the normalizer (as Lie algebras) of $\mathfrak{B}$ in $\mathfrak{M}_{p^n}(\mathfrak{F})$, a restricted subalgebra. Then $\mathfrak{D}(\mathfrak{M})$ is a restricted subalgebra of $\mathfrak{W}_{p^n}$ of dimension p^{2n+1}. *If p does not divide $n+2$, then $\mathfrak{D}(\mathfrak{M})$ is simple; if $p \mid (n+2)$, then $\mathfrak{D}(\mathfrak{M})^{(1)}$ is simple, of dimension $p^{2n+1} - 1$.* We denote the simple algebra corresponding to the index n by $\mathfrak{R}_n$.

Mrs. FRANK has also shown how the four "great classes" of split classical simple algebras may be obtained from the construction $\mathfrak{M} \to \mathfrak{D}(\mathfrak{M})$ as restricted subalgebras of Jacobson–Witt algebras. *Together with their forms and those of the classical algebras, the above exhaust the known normal simple restricted Lie algebras defined for "almost all" characteristics*; for the special case $p = 2$ see also [346].

As subalgebras of Jacobson–Witt algebras, the algebras from which one starts in obtaining the algebras $\mathfrak{S}_n$ and $\mathfrak{V}_n$ as members of the derived series may be described as follows: For $\mathfrak{S}_n$, one takes all derivations D of $\mathfrak{A}_n$ such that $\sum_i \frac{\partial a_i}{\partial x_i} = 0$, where $a_i = x_i D$ [9, 133]; for $\mathfrak{V}_n$, one starts with all derivations D of $\mathfrak{A}_{2n}$ such that, for $1 \leqq i, j \leqq n$, $\frac{\partial a_i}{\partial x_{n+j}} = \frac{\partial a_j}{\partial x_{n+i}}$, $\frac{\partial a_{n+i}}{\partial x_j} = \frac{\partial a_{n+j}}{\partial x_i}$, $\frac{\partial a_i}{\partial x_j} + \frac{\partial a_{n+j}}{\partial x_{n+i}} = 0$ [9]. In each case, these are a set of homogeneous linear conditions on the $\frac{\partial a_i}{\partial x_j}$. A single homogeneous linear condition involving both the a_i and their partial derivatives gives

d) *The class* $\mathfrak{T}_n$ [9]

Let $\mathfrak{L}$ be the set of all derivations D of $\mathfrak{A}_n$ such that $\sum_i \frac{\partial a_i}{\partial x_i} = \sum_i a_i$, where $a_i = x_i D$ as above. Then $\mathfrak{L}$ is a simple Lie subalgebra of $\mathfrak{W}_n$ if $p > 2$ and $n > 2$; it has dimension $(n-1)\,p^n$ over $\mathfrak{F}$, and has been denoted as $\mathfrak{T}_n$ by ALBERT and FRANK [9]. The algebras $\mathfrak{T}_n$ are not restricted subalgebras of $\mathfrak{W}_n$, *nor do they admit any structure of restricted Lie algebras compatible with their structure as Lie algebras* [33, 356].

Another means of distinguishing and generalizing interesting subalgebras of $\mathfrak{W}_n$ has been given by JENNINGS and REE [243, 325, 329]. These Lie algebras are described generally by REE [325] as follows: Let $\mathfrak{F}$ be a field of prime characteristic p, $\mathfrak{A}$ a finite-dimensional commutative associative $\mathfrak{F}$-algebra with unit. With the action $D \cdot a = D R_a$, the derivation algebra $\mathfrak{D}(\mathfrak{A})$ becomes a right $\mathfrak{A}$-module. Ree considers Lie subalgebras $\mathfrak{L}$ of $\mathfrak{D}(\mathfrak{A})$ which are free as $\mathfrak{A}$-submodules. He shows that $\mathfrak{L}$ is simple only if $\mathfrak{A}$ is completely primary, in which case a commutative basis $\{D_i\}$ can be chosen for $\mathfrak{L}$ as $\mathfrak{A}$-module, along with elements $x_i \in \mathfrak{A}$ such that $x_i D_j = \delta_{ij}$. Assuming this is the case, and that $\mathfrak{F}$ is algebraically closed, make the further assumption that the only common eigenvectors of all the D_i are units, and that the only vectors annihilated by all D_i are elements of $\mathfrak{F}$; then one may assume that $\mathfrak{A}$ is an algebra $\mathfrak{A}_m$ and that all the D_i are nilpotent. These are the *generalized Witt algebras* of ZASSENHAUS [419] and KAPLANSKY [246]. By selecting subalgebras $\mathfrak{L}(D_i, b_i)$ in terms of conditions of the form $\sum_i a_i D_i = \sum_i a_i b_i$ on the derivation $D = \sum_i D_i \cdot a_i \in \mathfrak{L}$, where $b_i \in \mathfrak{A}$ satisfy $b_i D_j = b_j D_i$ for all i, JENNINGS and REE [243] generalize some of the algebras of ALBERT and FRANK [9]. In particular, for all $b_i = 0$ one obtains a generalization of the algebras $\mathfrak{S}_n$: the derived algebra is simple of dimension $n(p^m - 1)$, where there are $n + 1$ of the D_i and where $\mathfrak{A}$ is of the form $\mathfrak{A}_m (1 \leqq n < m)$. For $b_i \in \mathfrak{F}$, not all zero, if there is $c \neq 0$ in $\mathfrak{A}$ with $cD_i = b_i c$, all i then the derived algebra is simple for $1 < n < m$. For $n = 1$, $p > 2$, $m > 1$, the second derived algebra $\mathfrak{L}(D_i, b_i)^{(2)}$ is simple of dimension $p^m - 2$. For b_i as above, in the case where there is no $c \neq 0$ with $cD_i = b_i c$, all i, the algebra $\mathfrak{L}(D_i, b_i)$ is itself simple for $n < m$, provided only that $p > 2$ or $n > 1$; its dimension is $n p^m$, which coincides with that of a generalized Witt algebra. There appears to be no discussion of this coincidence in the literature.

C. Algebras defined by finite groups of functions

a) *Generalized Witt algebras* [246, 325]

Let $\mathfrak{F}$ be a field of prime characteristic p, $\mathfrak{I}$ a set of m elements, $\mathfrak{G}$ a finite additive group of functions on $\mathfrak{I}$ with values in $\mathfrak{F}$, and suppose that the only $\mathfrak{F}$-valued function λ on $\mathfrak{I}$ such that $\sum_{i \in \mathfrak{I}} \lambda(i)\, \sigma(i) = 0$ for all $\sigma \in \mathfrak{G}$ is the zero-function ("$\mathfrak{G}$ is *total*"). Then $\mathfrak{G}$ has order p^n. Let $\mathfrak{L}$ be a vector space over $\mathfrak{F}$ with basis the set $\mathfrak{G} \times \mathfrak{I}$, and define a bilinear product in $\mathfrak{L}$ by $[(\sigma, i), (\tau, j)] = \tau(i)\,(\sigma + \tau, j) - \sigma(j)\,(\sigma + \tau, i)$ for basis elements. With this product, $\mathfrak{L}$ becomes a Lie algebra, called by Ree a generalized Witt algebra, of dimension $m p^n$; $\mathfrak{L}$ is simple if $m > 1$ or $p > 2$. As remarked in B above, REE has identified generalized Witt algebras (over algebraically closed fields) with subalgebras of Jacobson–Witt algebras.

b) *Another generalization of the Witt algebra* [9]

Let $\mathfrak{F}$, $\mathfrak{I}$, $\mathfrak{G}$ be as above, and let $m = 1$. Then we may identify $\mathfrak{G} \times \mathfrak{I}$ with $\mathfrak{G} \times \{1\}$; let h be a homomorphism of $\mathfrak{G}$ into the additive group of $\mathfrak{F}$. Let $\mathfrak{L}$ be a vector space over $\mathfrak{F}$ with $\mathfrak{G} \times \{1\}$ as basis, and define a bilinear product in $\mathfrak{L}$ by $[(\sigma, 1), (\tau, 1)] = \{\sigma(1)(h(\tau)+1) - \tau(1)(h(\sigma)+1)\}(\sigma+\tau, 1)$. (For $h = 0$, this is the algebra of 1) with $m = 1$.) *It is a Lie algebra and is simple if $p > 2$; its dimension is p^n.*

c) The algebras of Block [33]

Let $\mathfrak{G} = \mathfrak{G}_0 + \cdots + \mathfrak{G}_m$, the direct sum of $m+1$ finite abelian groups. For $0 \leqq i \leqq m$, let f_i be a skew-symmetric biadditive function on $\mathfrak{G}_i$ with values in the modular field $\mathfrak{F}$, and let f be the biadditive function on $\mathfrak{G}$ whose restriction to $\mathfrak{G}_i \times \mathfrak{G}_i$ is f_i and which vanishes on $\mathfrak{G}_i \times \mathfrak{G}_j$ for $i \neq j$. Let $\delta = \delta_0 + \cdots + \delta_m$, $\delta_i \in \mathfrak{G}_i$, and suppose $\delta_0 = 0$, $\delta_i \neq 0$ for all $i > 0$. Suppose that for each i, there exist additive functions g_i, h_i on $\mathfrak{G}_i$, with $f(\alpha_i, \beta_i) = g_i(\alpha_i)\, h_i(\beta_i) - g_i(\beta_i)\, h_i(\alpha_i)$, $g_i(\delta_i) = 0$. Let $\mathfrak{L}$ be a vector space over $\mathfrak{F}$ with basis $\{u_\alpha\}$ in one-one correspondence $u_\alpha \leftrightarrow \alpha$ with elements of $\mathfrak{G}$, and define a product in $\mathfrak{L}$ by bilinearity and $[u_\alpha u_\beta] = \sum_{i=0}^{m} f(\alpha_i, \beta_i)\, u_{\alpha+\beta-\delta_i}$. Then $\mathfrak{L}$ is a Lie algebra over $\mathfrak{F}$, and u_0 clearly is central. The cosets v_α, modulo this one-dimensional ideal, of all u_α, $\alpha \neq \delta$, form a basis for an ideal in the quotient. Block has denoted this ideal by $\mathfrak{L}(\mathfrak{G}, \delta, f)$, and has shown that $\mathfrak{L}(\mathfrak{G}, \delta, f)$ can only be simple if all f_i are non-singular and if $\mathfrak{G}$ is an elementary p-group, where p is the characeristic of $\mathfrak{F}$. If $\mathfrak{G}$ has order p^n and $\mathfrak{G} \neq \mathfrak{G}_0$, then $\mathfrak{L}(\mathfrak{G}, \delta, f)$ has dimension $p^n - 2$; if $\mathfrak{G} = \mathfrak{G}_0$, the dimension is $p^n - 1$. *Simplicity of $\mathfrak{L}(\mathfrak{G}, \delta, f)$ follows from the non-singularity of f and any of the following*: a) $0 \neq \mathfrak{G}_1 \neq \mathfrak{G}$; b) $\mathfrak{G} = \mathfrak{G}_0$, $n > 1$; c) $\mathfrak{G} = \mathfrak{G}_1$, $n > 1$, $p > 2$. The cases b) and c) had previously been investigated by Albert and Frank [9; see also 310].

Quite recently, Mrs. Frank has further generalized the procedure described under B above [136]. Her construction appears to give explicit realizations, as subalgebras of algebras $\mathfrak{W}_n$, of all the simple algebras listed here. (See also [265] for other realizations.)

D. Isomorphisms among known simple algebras

Unless otherwise indicated, we exclude characteristics 2 and 3 from consideration here, even though most of our assertions remain valid for these cases when meaningful. The question of isomorphisms among classical simple algebras has been dealt with in Chapter IV. In particular, the results of Chapter III show that the automorphism group of a split classical simple algebra has no non-trivial solvable invariant subgroup; on the other hand, the group of automorphisms of $\mathfrak{W}_n$ induced by all automorphisms A of $\mathfrak{A}_n$ such that $x_i A \equiv x_i \bmod (x_1, \ldots, x_n)^2$ for each i is such a non-trivial subgroup of $\operatorname{Aut}(\mathfrak{W}_n)$. Thus *there are no isomorphisms between Jacobson–Witt algebras and classical simple algebras.*

Dimension and characteristic considerations *rule out isomorphisms between Jacobson–Witt algebras and algebras of the classes* $\mathfrak{S}_n$ *and* $\mathfrak{V}_n$, *as well as* $\mathfrak{R}_n$ *for* $p \mid (n+2)$. We have seen in Chapter II, § 10 and Chapter IV, § 1 that every Cartan subalgebra of a classical simple Lie algebra over an algebraically closed field is classical. On the other hand, the intersection with $\mathfrak{S}_n$ of the Cartan subalgebra $\mathfrak{H}$ of $\mathfrak{W}_n$ given in A above is contained in a Cartan subalgebra of dimension $(n-1)(p-1)$, relative to which root-spaces have dimension $(n-1)p \neq 1$. Hence $\mathfrak{S}_n$ $(n \geqq 3)$ *cannot be classical.* In [33], Block has shown that the algebras $\mathfrak{S}_n$ $(n > 3)$ cannot admit a nonsingular symmetric associative bilinear form, whereas the algebras $\mathfrak{V}_n$ do admit such a form; thus *there can be no isomorphisms between algebras* $\mathfrak{S}_n$ *and algebras* $\mathfrak{V}_m$ *except perhaps for* $n = 2, 3$. For $n = 3$, dimension considerations rule out isomorphism; for $n = 2$ and $m = 1$, the dimensions of $\mathfrak{S}_n$ and $\mathfrak{V}_m$ coincide, and in fact they are seen from B to arise from the same subalgebra $\mathfrak{M}$ of $\mathfrak{M}_2(\mathfrak{F})$; thus $\mathfrak{S}_2$ *and* $\mathfrak{V}_1$ *are isomorphic.*

The algebras $\mathfrak{V}_n$ have abelian Cartan subalgebras with root-spaces of dimension p^n, from which it follows as above that *no algebra* $\mathfrak{V}_n$ *is classical.* For $n \not\equiv -2 \pmod{p}$, the dimension of $\mathfrak{R}_n$ coincides with that of $\mathfrak{W}_m$ only if $m = p^k$, $n = \frac{1}{2}(p^k + k - 1)$; *this is the only case of possible isomorphism between* $\mathfrak{R}_n$ *and* $\mathfrak{W}_m$, *and it is not known whether the algebras are actually isomorphic in this case. For* $n \not\equiv -2 \pmod{p}$, *dimension considerations show there can be no isomorphism between* $\mathfrak{R}_n$ *and any* $\mathfrak{V}_m$ *or* $\mathfrak{S}_m$. In this case, Mrs. Frank [135] has displayed a Cartan decomposition for $\mathfrak{R}_n$ with root-spaces of dimension p^n: it follows as above that $\mathfrak{R}_n$ *is not classical.* This is even true for $n \equiv -2 \pmod{p}$, again by a Cartan decomposition found in [135]. For $n \equiv -2 \pmod{p}$, dimension considerations *rule out isomorphism between* $\mathfrak{R}_n$ *and any* $\mathfrak{V}_m$. There remains the question of isomorphisms $\mathfrak{R}_n \cong \mathfrak{S}_m$, where by $\mathfrak{S}_2 \cong \mathfrak{V}_1$ we may assume $m \geqq 3$. In this case one sees easily that one must have $m \equiv 2 \pmod{p^m}$, which is absurd if $m > 2$. Thus *there are no isomorphisms* $\mathfrak{R}_n \cong \mathfrak{S}_m$.

The above discussion completes our treatment of the existence of isomorphisms among known simple restricted Lie algebras. This is not to say that the procedure of Jennings and Ree or that of Mrs. Frank yields no new restricted Lie algebras not in the above list, but rather that these are the only ones whose identity has been so well established as to make them distinguishable from classical Lie algebras and from one another. The same kind of qualification applies in the remainder of this section.

By the fact that the algebras $\mathfrak{T}_n$ do not admit the structure of restricted Lie algebras, they cannot be isomorphic with any restricted Lie algebras; hence *they are new with respect to the algebras discussed above under this subsection.* As for the generalized Witt algebras of C, a), it is not hard to see that the *only restricted generalized Witt algebras are the algebras* $\mathfrak{W}_n$. There is a dimension-coincidence of certain generalized Witt algebras and the algebras $\mathfrak{T}_n[(n-1)\,p^n]$; it has not been established whether there are any isomorphisms at these dimensions. *Apart from the* $\mathfrak{W}_n$ *and perhaps these* $\mathfrak{T}_n$, *there are no algebras isomorphic to generalized Witt algebras among those discussed above in this subsection.* Except for certain cases (treated below), the algebras of JENNINGS and REE introduced in B have not been discussed with respect to restrictedness or isomorphism, except to show that they are distinct from the algebras $\mathfrak{R}_n$ [135]; their dimensions are not new in general, and they include the algebras $\mathfrak{S}_n$. For the simple algebras of ALBERT and FRANK of C, b), one sees easily that restrictedness is possible if and only if the algebra is $\mathfrak{W}_1$, the original Witt algebra; otherwise (and for $h \neq 0$) the question of isomorphism with others in our list of non-restricted algebras is not settled. In the case of the algebras $\mathfrak{L}(\mathfrak{G}, \delta, f)$ of BLOCK (C, c)), BLOCK has shown [33] that $\mathfrak{L}(\mathfrak{G}, \delta, f)$ is restricted if and only if $\mathfrak{G}_0 = 0$ and $\mathfrak{G}_i$ has order p^2 for $1 \leq i \leq m$, in which case $\mathfrak{L}(\mathfrak{G}, \delta, f)$ is an algebra $\mathfrak{B}_{m,(\mu)}$ of B, b), hence is isomorphic with $\mathfrak{B}_m$. He has also remarked in a footnote that the algebras $\mathfrak{L}(\mathfrak{G}, \delta, f)$ include the simple algebras of JENNINGS and REE of dimensions $p^n - 1$, $p^n - 2$, the latter being exactly the $\mathfrak{L}(\mathfrak{G}, \delta, f)$ with $\mathfrak{G} = \mathfrak{G}_1$. By studying their derivation algebras, he has obtained necessary conditions in terms of the groups $\mathfrak{G}$ for two algebras $\mathfrak{L}(\mathfrak{G}, \delta, f)$ of the same dimension to be isomorphic. In the non-restricted case, the only possible algebras in our list which can be isomorphic with $\mathfrak{L}(\mathfrak{G}, \delta, f)$ are seen from the dimensions to be those of JENNINGS and REE having the same dimension.

§ 5. Derivations

It is well known (cf. Chap. I, § 8) that all derivations of a Lie algebra with nonsingular Killing form are inner. In particular, all derivations of a semisimple Lie algebra of characteristic zero are inner. This conclusion fails, even for simple Lie algebras, in prime characteristic; probably the simplest counterexample is the classical simple algebra $\mathfrak{L}$ of type A_n, $p \mid (n+1)$. Here $\mathfrak{L}$ may be regarded as the derived algebra of $\mathfrak{M} = \mathfrak{M}_n(\mathfrak{F})/\mathfrak{F}\,1$, and the centralizer of $\mathfrak{L}$ in $\mathfrak{M}$ is zero; it follows that any inner derivation of $\mathfrak{M}$ induces a derivation in $\mathfrak{L}$, and that not all of these are inner since $\mathfrak{L} \neq \mathfrak{M}$. Except for (perhaps) $p = 5$ and type E_8, these are the only classical simple algebras having outer derivations, and over an algebraically closed field, they are the only classical simple Lie algebras (of characteristic $p > 3$) not having a representation with nonsingular trace form. One thus has the result, due to BLOCK [35], that if $\mathfrak{L}$ is semisimple and has such a representation, then every derivation of $\mathfrak{L}$ is inner; Block's proof uses the full classification theory, and no direct proof is known to the author. Further simple algebras with outer derivations are the $\mathfrak{L}(\mathfrak{G}, \delta, f)$ of BLOCK [33] and the $\mathfrak{R}_n$ of Mrs. FRANK [135]. For characteristic 2, it is even possible for the Lie algebra of outer derivations of a simple restricted Lie algebra to be simple [346].

If $\mathfrak{L}$ is a Lie algebra of characteristic zero, and $\mathfrak{S}$ its radical, it is well known that $\mathfrak{S}$ is a *characteristic ideal*, i.e., $\mathfrak{S}D \subseteqq \mathfrak{S}$ for all derivations D of $\mathfrak{L}$; in fact $\mathfrak{S}D \subseteqq \mathfrak{N}$, the maximal nilpotent ideal of $\mathfrak{L}$ (cf. [234], p. 74). This conclusion fails for prime characteristic, as the following example shows: The algebra $\mathfrak{A}_n$ of § 4 has radical $\mathfrak{N}$ of codimension 1, generated by the x_i; let $\mathfrak{M}$ be any simple Lie algebra over the field $\mathfrak{F}$ and let $\mathfrak{L} = \mathfrak{M} \otimes_{\mathfrak{F}} \mathfrak{A}_n$. One may regard $\mathfrak{L}$ as a Lie algebra over $\mathfrak{A}_n$, obtained by extending the base ring, hence as a Lie algebra over $\mathfrak{F}$, and one easily sees that the nilpotent ideal $\mathfrak{M} \otimes \mathfrak{N}$ is the radical $\mathfrak{S}$ of $\mathfrak{L}$. The mappings $1 \otimes D$, $D \in \mathfrak{W}_n$, are derivations of $\mathfrak{L}$, and it is clear that $\mathfrak{S}$ is not preserved by these; in fact, $\mathfrak{L}$ *contains no proper characteristic ideal.* A prototype for this class of examples is to be found in [354], where some conclusions are drawn as to the structure of Lie algebras in terms of their characteristic ideals.

In dealing with restricted Lie algebras $\mathfrak{L}$, there are several natural reasons for considering derivations D satisfying $x^p D = (xD)\,(\operatorname{ad} x)^{p-1}$; such derivations have been called *restricted* by JACOBSON [215], who has shown that they are the derivations which extend to derivations of the u-algebra $\mathfrak{U}(\mathfrak{L})$, and include the inner derivations. From the point of view of general "algebras" defined by polynomial mappings, he has also shown that they are the appropriate mappings to be regarded as derivations in restricted Lie algebras [237]. In any restricted Lie algebra $\mathfrak{L}$, one sees from $\binom{p-1}{j} \equiv (-1)^j \bmod p$ that $\operatorname{ad}(x^p D) = \operatorname{ad}\big((xD)\,(\operatorname{ad} x)^{p-1}\big)$ for all $x \in \mathfrak{L}$ and all derivations D; thus if $\mathfrak{L}$ has center zero, all derivations are restricted. On the other hand, if $\mathfrak{L}$ is abelian and $x \to x^p$ is a semilinear automorphism of $\mathfrak{L}$, the only restricted derivation of $\mathfrak{L}$ is zero, while all linear transformations are derivations.

If D is a derivation of $\mathfrak{L}$, it is customary to refer to a *constant* of D as an element $x \in \mathfrak{L}$ with $xD = 0$. Let $\mathfrak{L}_D$ be the space of D-constants. For Lie algebras $\mathfrak{L}$ of characteristic zero, JACOBSON [224] has proved that if $\mathfrak{L}_D = 0$ for some D, then $\mathfrak{L}$ is nilpotent, and the same conclusion holds if both $\mathfrak{L}$ and D are *restricted* of prime characteristic. In fact, he establishes these results with D replaced by a nilpotent Lie algebra of (restricted) derivations, and $\mathfrak{L}_D$ by their common constants.

An example of DIXMIER and LISTER [216] shows that it is possible for a nilpotent Lie algebra $\mathfrak{L}$ to have $\mathfrak{L}_D \neq 0$ for every derivation D.

Clearly, a derivation D with $\mathfrak{L}_D = 0$ cannot be inner; thus a partial converse to the results above on nilpotency when $\mathfrak{L}_D = 0$ is given by the existence of non-inner derivations of a nilpotent Lie algebra [234, p. 29, 399]. In proofs of nilpotency under conditions like the above, one may assume the ground field algebraically closed, and consider the decomposition of $\mathfrak{L}$ into subspaces $\mathfrak{L}_\lambda$ corresponding to the characteristic roots λ of D; Jacobson's device is to use the fact that $[\mathfrak{L}_\lambda \mathfrak{L}_\mu] \subseteqq \mathfrak{L}_{\lambda+\mu}$ to apply his theory of "weakly closed sets of linear transformations" to $\mathfrak{M} = U_\lambda \mathfrak{L}_\lambda$ to show that if $\operatorname{ad} x$ is nilpotent for each $x \in \mathfrak{M}$, then the same holds for all $x \in \mathfrak{L}$. When all $\lambda \neq 0$, the desired conclusions follow if it is not the case that $\mu, \mu + \lambda, \mu + 2\lambda, \ldots$ are all characteristic roots of D for any characteristic root λ.

Over a perfect ground field $\mathfrak{F}$, one has the well-known Chevalley decomposition $T = S + N$ of an endomorphism of a vector space into semisimple and nilpotent parts, which may be written as polynomials in T without constant term. In the algebraically closed case, these correspond to the diagonal and off-diagonal parts of the Jordan canonical matrix for T. If D is a derivation of $\mathfrak{L}$ over $\mathfrak{F}$, and if $\mathfrak{K}$ is the algebraic closure of $\mathfrak{F}$, one sees from the Jordan form and the relation $[\mathfrak{L}_\lambda \mathfrak{L}_\mu] \subseteqq \mathfrak{L}_{\lambda+\mu}$ that the semisimple and nilpotent parts of $D_\mathfrak{K}$, acting in $\mathfrak{L}_\mathfrak{K}$, are again derivations. Since these are $S_\mathfrak{K}$, $N_\mathfrak{K}$, where S and N are the semisimple and nilpotent parts of D, it follows that S and N are derivations of $\mathfrak{L}$, or that the derivation algebra $\mathfrak{D}(\mathfrak{L})$ is "almost algebraic" (Jacobson) or "splittable" (Malcev). This conclusion remains valid for restricted derivations of a restricted Lie algebra. A general result of this type for restricted Lie algebras follows in § 7.

If $\mathfrak{L}$ has center zero, $\mathfrak{L}$ may be regarded as an ideal (the inner derivations) in $\mathfrak{D}(\mathfrak{L})$; then if D is central in $\mathfrak{D}(\mathfrak{L})$, xD is central in $\mathfrak{L}$ for each $x \in \mathfrak{L}$, so that $D = 0$: $\mathfrak{D}(\mathfrak{L})$ has center zero. Then, repeating, one has the *derivation tower* $\mathfrak{L} \subseteqq \mathfrak{D}(\mathfrak{L}) \subseteqq \mathfrak{D}(\mathfrak{D}(\mathfrak{L})) \subseteqq \cdots$ Schenkman [348] has proved that the derivation tower is stable after finitely many steps.

Analogous to the study of Lie algebras $\mathfrak{L}$ with some $\mathfrak{L}_D = 0$ is the study of Lie algebras $\mathfrak{L}$ having an automorphism A without fixed points $\neq 0$. If A has finite prime period, then Jacobson [224] has shown that $\mathfrak{L}$ is nilpotent (the analogue of the much deeper theorem of J. Thompson [388] for finite groups); Jacobson also shows that $\mathfrak{L}$ is nilpotent if no characteristic root of A is a root of unity. It has been shown by Kreknin [267] that if A has finite period, then $\mathfrak{L}$ is solvable (cf. also [264] for special cases); for $\mathfrak{L}$ a Lie *ring* and A of prime period, nilpotency has been proved by G. Higman [188]. For classical simple algebras and for the Jacobson–Witt algebras, every automorphism has a fixed point $\neq 0$, and the dimension of the space of fixed points has been studied by JACOBSON [235], by SMITH [370], and by LISSNER [285]. In the case of characteristic zero, the existence of an automorphism without fixed points $\neq 0$ implies solvability [43, 318]. Using the results of KREKNIN, WINTER has recently extended this result to the case of prime characteristic [414].

§ 6. Extension of the base field

By analogy with the theory of associative algebras, it seems natural to call a Lie algebra $\mathfrak{L}$ over $\mathfrak{F}$ *separable* if $\mathfrak{L}_\mathfrak{K}$ is semisimple for each field extension $\mathfrak{K}$ of $\mathfrak{F}$. In particular, a separable $\mathfrak{L}$ is itself semi-simple. It is clear from Chapter I, § 9 that if $\mathfrak{L}$ has nonsingular Killing form, then $\mathfrak{L}$ is separable. In general, let $\mathfrak{L}$ be semisimple, and let $\mathfrak{M}$ be the sum of all the minimal ideals of $\mathfrak{L}$. By finite-dimensionality, $\mathfrak{M}$ is the direct sum of finitely many minimal ideals of $\mathfrak{L}$; the centralizer $\mathfrak{N}$ of $\mathfrak{M}$ in $\mathfrak{L}$ is an ideal, so that $\mathfrak{N} \cap \mathfrak{M} = 0$ by the semisimplicity of $\mathfrak{L}$. On the other hand, if $\mathfrak{N} \neq 0$, $\mathfrak{N}$ contains a non-zero minimal ideal not contained in $\mathfrak{M}$; hence $\mathfrak{N} = 0$. Thus the adjoint representation of $\mathfrak{L}$ in $\mathfrak{M}$ is faithful and completely reducible. Let $\mathfrak{L}^*$ denote the associative enveloping algebra of the set of representing transformations; then $\mathfrak{L}^*$ is a semisimple associative subalgebra of $\mathfrak{E}(\mathfrak{M})$. If $\mathfrak{K}$ is an extension field of $\mathfrak{F}$, then $\mathfrak{M}_\mathfrak{K}$ is an ideal in $\mathfrak{L}_\mathfrak{K}$, whose centralizer is zero, and $(\mathfrak{L}^*)_\mathfrak{K}$ is naturally identified with the enveloping algebra of the restriction to $\mathfrak{M}_\mathfrak{K}$ of the elements of $\operatorname{ad}\mathfrak{L}_\mathfrak{K}$. If $(\mathfrak{L}^*)_\mathfrak{K}$ is semisimple, then $\mathfrak{M}_\mathfrak{K}$ is a completely reducible module for $\mathfrak{L}_\mathfrak{K}$ under the adjoint representation. Now let $\mathfrak{B}$ be an ideal of $\mathfrak{L}_\mathfrak{K}$, $[\mathfrak{B}\,\mathfrak{B}] = 0$; if $\mathfrak{B} \neq 0$, we may assume $\mathfrak{B}$ to be a minimal ideal. Now $\mathfrak{B} \cap \mathfrak{M}_\mathfrak{K} = 0$ or $\mathfrak{B} \subseteqq \mathfrak{M}_\mathfrak{K}$. The former is impossible for any ideal $\mathfrak{B} \neq 0$ since $[\mathfrak{B}\,\mathfrak{M}_\mathfrak{K}] \neq 0$; in the latter case, $\mathfrak{M}_\mathfrak{K} = \mathfrak{B} \oplus \mathfrak{C}$, where $\mathfrak{C}$ is an ideal in $\mathfrak{L}_\mathfrak{K}$, and $[\mathfrak{B}\,\mathfrak{M}_\mathfrak{K}] = [\mathfrak{B}\,\mathfrak{B}] = 0$, also a contradiction. Thus if $(\mathfrak{L}^*)_\mathfrak{K}$ is semisimple, $\mathfrak{L}_\mathfrak{K}$ is semisimple. In particular, if $\mathfrak{K}$ is a separable extension of $\mathfrak{F}$, it follows from associative theory (e.g., [5, 100, 218]) that $\mathfrak{L}_\mathfrak{K}$ is semisimple, and hence, that if $\mathfrak{F}$ is perfect, every semisimple Lie algebra over $\mathfrak{F}$ is separable.

If one adopts the more stringent definition of semisimplicity which requires that $\mathfrak{L}$ be a direct sum of minimal nonabelian ideals, one has $\mathfrak{M} = \mathfrak{L}$ in the above setting, and $(\mathfrak{L}^*)_\mathfrak{K}$ becomes the enveloping algebra of $\operatorname{ad}\mathfrak{L}_\mathfrak{K}$. In this case, the semisimplicity of $(\mathfrak{L}^*)_\mathfrak{K}$, as associative algebra, is equivalent with the complete reducibility of the adjoint representation of $\mathfrak{L}_\mathfrak{K}$, hence with the semisimplicity, in this stricter sense, of $\mathfrak{L}_\mathfrak{K}$. Once again, one sees that this notion of semisimplicity is also preserved under separable extension of the base field.

On the other hand, if $\mathfrak{F}$ is not perfect, let $\mathfrak{L}$ be a Lie algebra over $\mathfrak{F}$ of the form $\mathfrak{M} \otimes_\mathfrak{F} \mathfrak{K}$, where $\mathfrak{M}$ is a normal simple Lie algebra over $\mathfrak{F}$ and $\mathfrak{K}$ is a purely inseparable extension field of $\mathfrak{F}$, $\mathfrak{K} \neq \mathfrak{F}$. Then, as a Lie algebra over $\mathfrak{K}$, $\mathfrak{L}$ is normal simple and $\mathfrak{L}$ is simple as a Lie algebra over $\mathfrak{F}$. (For if $\mathfrak{N} \neq 0$ is an ideal, let $\mathfrak{E}$ be the subset of elements $\beta \in \mathfrak{K}$ such that $\mathfrak{M} \otimes \beta \subseteqq \mathfrak{N}$; if $\beta \in \mathfrak{E}$, $\alpha \in \mathfrak{K}$, then $[\mathfrak{M} \otimes \beta, \mathfrak{M} \otimes \alpha] = [\mathfrak{M}\,\mathfrak{M}] \otimes \otimes\, \beta\alpha = \mathfrak{M} \otimes \beta\alpha \subseteqq \mathfrak{N}$, and $\mathfrak{E}$ is seen to be an ideal in $\mathfrak{K}$, hence either $\mathfrak{K}$ or 0. If $\mathfrak{E} = \mathfrak{K}$, then $\mathfrak{N} = \mathfrak{L}$. But $\mathfrak{E} \neq 0$, since if $e_1, \ldots, e_n$ are a basis for $\mathfrak{M}$ over $\mathfrak{F}$, and if $\sum e_i \otimes \alpha_i$ $(\alpha_i \in \mathfrak{K})$ is in $\mathfrak{N}$ with, say, $\alpha_1 \neq 0$, we know by the normal simplicity of $\mathfrak{M}$ that for each $x \in \mathfrak{M}$, there is T in the enveloping algebra $(\mathrm{ad}\,\mathfrak{M})^*$ such that $x = e_1 T$, $e_i T = 0$ for $i > 1$. Now $T \otimes I$ is in the $\mathfrak{F}$-enveloping algebra of $\mathrm{ad}\,\mathfrak{L}$, from which $\mathfrak{N}$ contains all $x \otimes \alpha_1$, $x \in \mathfrak{M}$; thus $0 \neq \alpha_1 \in \mathfrak{E}$, so that $\mathfrak{E} = \mathfrak{K}$, $\mathfrak{N} = \mathfrak{L}$.) Now $\mathfrak{L}_{\mathfrak{K}}$ may be identified with $\mathfrak{M} \otimes_{\mathfrak{F}} (\mathfrak{K} \otimes_{\mathfrak{F}} \mathfrak{K})$, regarded as $\mathfrak{K}$-algebra by the $\mathfrak{K}$-action on $\mathfrak{K} \otimes_{\mathfrak{F}} \mathfrak{K}$. Since the commutative associative algebra $\mathfrak{K} \otimes_{\mathfrak{F}} \mathfrak{K}$ has a radical $\mathfrak{R} \neq 0$, $\mathfrak{M} \otimes_{\mathfrak{F}} \mathfrak{R}$ is a non-zero nilpotent ideal in $\mathfrak{L}_{\mathfrak{K}}$, and the simple algebra $\mathfrak{L}$ is not separable.

§ 7. Cartan subalgebras

We have defined a Cartan subalgebra $\mathfrak{H}$ of a Lie algebra $\mathfrak{L}$ as a nilpotent subalgebra which is its own normalizer. Over algebraically closed fields of characteristic zero, CHEVALLEY [67; cf. also 72, Chap. VI; 234, Chap. IX] showed that any two Cartan subalgebras are conjugate under the group of automorphisms of $\mathfrak{L}$ generated by all $\exp(\mathrm{ad}\,x)$, $x \in \mathfrak{L}$, $\mathrm{ad}\,x$ nilpotent. It follows from Chapter I, § 9 that all Cartan subalgebras of a Lie algebra over an arbitrary non-modular field have the same dimension.

For the existence of Cartan subalgebras (Chap. I, § 6), we have referred to a proof which requires a sufficiently large base field and then displays a Cartan subalgebra as the sum of the spaces annihilated by powers of $\mathrm{ad}\,x$, where x is a regular element. For characteristic zero, every Cartan subalgebra is of this form. In the algebraically closed case, this follows from the conjugacy of Cartan subalgebras, and the general result then follows from the observation that, for infinite ground fields, if $\mathfrak{H}_{\mathfrak{K}}$ contains a regular element of $\mathfrak{L}_{\mathfrak{K}}$ then $\mathfrak{H}$ contains a regular element of $\mathfrak{L}$. All these conclusions fail, even for simple Lie algebras over algebraically closed fields, in the modular case. Namely, BLOCK has shown [33] that the simple algebra $\mathfrak{L}(\mathfrak{G}, \delta, f)$ of § 4, C with $\mathfrak{G} = \mathfrak{G}_0$ of order $p^{\binom{m+1}{2}}$ possesses Cartan subalgebras of each of the dimensions $p^k - 1$, $k = 1, 2, \ldots, m-1$. On the other hand, we have seen in Chapter III, § 4, Chapter IV, § 1, that all Cartan subalgebras are conjugate in classical Lie algebras over algebraically closed fields. This result has the consequences of commutativity and equality of dimensions for Cartan subalgebras of forms of these algebras, as well as the conclusion that every Cartan subalgebra of a form of such an algebra over an infinite field contains a regular element. Over general fields, we have seen in Chapter III, § 4 that *classical* Cartan subalgebras of classical semisimple Lie algebras are conjugate.

One may also seek an analogue of the theorem of Chevalley asserting conjugacy of Cartan subalgebras in *solvable* Lie algebras over arbitrary fields of characteristic zero [72, Chap. VI]. Such an analogue does not exist in general; namely, let $\mathfrak{M}$ be the $(p+2)$-dimensional solvable Lie algebra of § 1 whose derived algebra is not nilpotent. The centralizers in $\mathfrak{M}$ of F resp. $E+F$ are Cartan subalgebras of dimensions 2 resp. 1, hence cannot be conjugate, nor can the Cartan subalgebra containing F contain a regular element. There is, however, an interesting class of solvable algebras in which all Cartan subalgebras are conjugate. Namely, let $\mathfrak{L}^\omega$ be the intersection of the lower central series of $\mathfrak{L}$, an ideal in $\mathfrak{L}$, and suppose that $\mathfrak{H}$ is a Cartan subalgebra of $\mathfrak{L}$. By [234, p. 39], and Lemma I.6.1, $\mathfrak{L} = \mathfrak{H} \oplus \mathfrak{L}_1$, where $\mathfrak{L}_1$ is the sum of the spaces $\mathfrak{L}(\operatorname{ad} h)^n$, $n = \dim \mathfrak{L}$, h running over $\mathfrak{H}$. Moreover, since $\mathfrak{L}(\operatorname{ad} h)^{n+1} = \mathfrak{L}(\operatorname{ad} h)^n$ and $[\mathfrak{L}_1 \mathfrak{H}] \subseteqq \mathfrak{L}_1$, we have $[\mathfrak{L}_1 \mathfrak{H}] = \mathfrak{L}_1$. It follows that $\mathfrak{L}_1(\operatorname{ad}\mathfrak{H})^n = \mathfrak{L}_1 \subseteqq \mathfrak{L}^{n+1} = \mathfrak{L}^\omega$. Now suppose $\mathfrak{L}^\omega$ is *abelian*; then $[\mathfrak{L}_1 \mathfrak{L}_1] = 0$, $\mathfrak{L}^k = \mathfrak{H}^k \oplus \mathfrak{L}_1(\operatorname{ad}\mathfrak{H})^{k-1} = \mathfrak{H}^k \oplus \mathfrak{L}_1$, and $\mathfrak{L}^n = \mathfrak{L}_1$. Thus $\mathfrak{L}_1 = \mathfrak{L}^\omega$, and we have $\mathfrak{L} = \mathfrak{H} \oplus \mathfrak{L}^\omega$ for every Cartan subalgebra $\mathfrak{H}$, provided $\mathfrak{L}^\omega$ is abelian. In particular, *all Cartan subalgebras of* $\mathfrak{L}$ *have the same dimension*; *in fact, they are all conjugate by automorphisms of* $\mathfrak{L}$ *of the form* $I + \operatorname{ad} a$, $a \in \mathfrak{L}^\omega$ (each such map is an automorphism since $\mathfrak{L}^\omega$ is an abelian ideal). To see the last assertion, first assume $\mathfrak{F}$ is infinite, and let h_0 be a regular element of $\mathfrak{L}$, $\mathfrak{H}_0$ the Cartan subalgebra containing h_0, and $\mathfrak{H}$ any Cartan subalgebra of $\mathfrak{L}$. By the above, $[\mathfrak{L}^\omega \mathfrak{H}_0] = \mathfrak{L}^\omega$, and by Th. I.6.1, $[\mathfrak{L}^\omega h_0] = \mathfrak{L}^\omega$. Since $\mathfrak{L} = \mathfrak{H} \oplus \mathfrak{L}^\omega$, we have $h_0 = h + x$, $x \in \mathfrak{L}^\omega$, $h \in \mathfrak{H}$. Now $x = [a\, h_0]$, $a \in \mathfrak{L}^\omega$; thus $h_0(I + \operatorname{ad} a) = h_0 - x = h$ is a regular element in $\mathfrak{H}$, so that $\mathfrak{H}$ is the space annihilated by sufficiently high powers of $\operatorname{ad} h$. It follows that $\mathfrak{H}_0(I + \operatorname{ad} a) = \mathfrak{H}$, and the proof for infinite base fields is complete since automorphisms $I + \operatorname{ad} a$, $a \in \mathfrak{L}^\omega$, form a group. Now let $\mathfrak{F}$ be arbitrary, and let $\mathfrak{K}$ be an infinite field containing $\mathfrak{F}$. Let $\mathfrak{H}_1$, $\mathfrak{H}_2$ be two Cartan subalgebras of $\mathfrak{L} = \mathfrak{H}_1 + \mathfrak{L}^\omega = \mathfrak{H}_2 + \mathfrak{L}^\omega$. Then $(\mathfrak{H}_i)_\mathfrak{K}$ is a Cartan subalgebra of $\mathfrak{L}_\mathfrak{K}$, $\mathfrak{L}^\omega_\mathfrak{K} = (\mathfrak{L}^\omega)_\mathfrak{K}$ is abelian, and there is $a \in (\mathfrak{L}^\omega)_\mathfrak{K}$ with $(\mathfrak{H}_1)_\mathfrak{K}(I + \operatorname{ad} a) = (\mathfrak{H}_2)_\mathfrak{K}$. Letting $\{e_i\}$ be a basis for $\mathfrak{H}_1$, $\{f_j\}$ for $\mathfrak{L}^\omega$, and letting $\{\varphi_k\}$ be a basis for the subspace of the dual space $\mathfrak{L}^*$ annihilating $\mathfrak{H}_2$, this says that the system of linear equations $\{\varphi_k(e_i) + \sum_j \xi_j\, \varphi_k([e_i\, f_j]) = 0\}$ with coefficients in $\mathfrak{F}$ has a solution (ξ_j) (viz. $a = \sum \xi_j f_j$) in $\mathfrak{K}$; hence it has a solution (β_j) in $\mathfrak{F}$, and $b = \sum \beta_j f_j$ satisfies $\mathfrak{H}_1(I + \operatorname{ad} b) \subseteqq \mathfrak{H}_2$. By equality of the dimensions, this completes the proof (cf. [43]).

In general if $\mathfrak{L}^\omega$ is abelian, and if $\mathfrak{H}$ is any subalgebra such that $\mathfrak{L} = \mathfrak{H} \oplus \mathfrak{L}^\omega$ (vector space direct sum), then $\mathfrak{H}$ is a Cartan subalgebra. For $\mathfrak{H}^\omega \subseteqq \mathfrak{L}^\omega \cap \mathfrak{H} = 0$ shows that $\mathfrak{H}$ is nilpotent. Now $\mathfrak{L} = \mathfrak{L}_0 \oplus \mathfrak{L}_1$, where $\mathfrak{L}_0 \supseteqq \mathfrak{H}$ is the intersection of the kernels of all $(\operatorname{ad} h)^n$, $h \in \mathfrak{H}$, $n = \dim \mathfrak{L}$, and where $\mathfrak{L}_1$ is as above. It suffices to show $\mathfrak{L}_0 = \mathfrak{H}$, or that $\mathfrak{L}_1 = \mathfrak{L}^\omega$. By the above, $\mathfrak{L}_1 \subseteqq \mathfrak{L}^\omega$, since $\mathfrak{L}_1 = [\mathfrak{L}_1 \mathfrak{H}]$; from $\mathfrak{L} = \mathfrak{H} + \mathfrak{L}^\omega$, $[\mathfrak{L}^\omega \mathfrak{L}^\omega] = 0$, we have $\mathfrak{L}(\operatorname{ad}\mathfrak{H})^n = \mathfrak{L}^\omega(\operatorname{ad}\mathfrak{H})^n = \mathfrak{L}^\omega(\operatorname{ad}\mathfrak{L})^n = \mathfrak{L}^\omega$. Now for each $h \in \mathfrak{H}$, $\operatorname{ad} h$ acts nilpotently in $\mathfrak{L}_0$; by Engel's theorem, there is a basis for $\mathfrak{L}_0$ relative to which all $\operatorname{ad} h$, $h \in \mathfrak{H}$, have properly (upper) triangular matrices. Thus $\mathfrak{L}_0(\operatorname{ad}\mathfrak{H})^n = 0$, and $\mathfrak{L}^\omega = \mathfrak{L}(\operatorname{ad}\mathfrak{H})^n = \mathfrak{L}_0(\operatorname{ad}\mathfrak{H})^n + \mathfrak{L}_1(\operatorname{ad}\mathfrak{H})^n = \mathfrak{L}_1(\operatorname{ad}\mathfrak{H})^n \subseteqq \mathfrak{L}_1$. This proves our assertion. (These considerations and the reduction to the case where $\mathfrak{L}^\omega$ is abelian in the theorem below were suggested to the author by D. J. Winter.)

Theorem V.7.1. Let $\mathfrak{L}$ be a solvable Lie algebra over an arbitrary field $\mathfrak{F}$. Then $\mathfrak{L}$ has a Cartan subalgebra.

First we suppose $\mathfrak{L}^\omega$ abelian, in which case the above shows that it is enough to establish that the exact sequence $0 \to \mathfrak{L}^\omega \to \mathfrak{L} \to \mathfrak{M} \to 0$ splits, where $\mathfrak{M} = \mathfrak{L}/\mathfrak{L}^\omega$, or that the cohomology class in $H^2(\mathfrak{M}, \mathfrak{L}^\omega)$ of any cocycle associated with this extension is zero, where the action of $\mathfrak{M}$ on $\mathfrak{L}^\omega$ is induced by the adjoint representation of $\mathfrak{L}$. The existence of a Cartan subalgebra and earlier remarks show that this is indeed the case if $\mathfrak{F}$ is infinite. In general, the result then follows by the same kind of "descent" used above to show conjugacy of Cartan subalgebras.

Next let $\mathfrak{L} \neq 0$ be an arbitrary solvable Lie algebra, and suppose the theorem true for solvable algebras of lower dimension. Now $\mathfrak{L}$ has an abelian ideal $\mathfrak{A} \neq 0$, and $\mathfrak{M} = \mathfrak{L}/\mathfrak{A}$ is solvable, of lower dimension. Let $\mathfrak{H}'$ be a Cartan subalgebra of $\mathfrak{M}$; then $\mathfrak{H}' = \mathfrak{R}/\mathfrak{A}$, where $\mathfrak{R}$ is a subalgebra of $\mathfrak{L}$ containing $\mathfrak{A}$. We distinguish two cases:

1) $\mathfrak{R} = \mathfrak{L}$. Then $\mathfrak{L}/\mathfrak{A}$ is nilpotent, or $\mathfrak{L}^\omega \subseteqq \mathfrak{A}$ is abelian. This case has been treated above.

2) $\mathfrak{R} \neq \mathfrak{L}$. Then $\mathfrak{R}$ is solvable, of lower dimension than is $\mathfrak{L}$, so has a Cartan subalgebra $\mathfrak{H}$. We claim $\mathfrak{H}$ is a Cartan subalgebra of $\mathfrak{L}$. For since $\mathfrak{H}' = \mathfrak{R}/\mathfrak{A}$ is nilpotent, $\mathfrak{R}^\omega \subseteqq \mathfrak{A}$ is abelian, and $\mathfrak{R} = \mathfrak{H} \oplus \mathfrak{R}^\omega$. Let φ be the canonical homomorphism of $\mathfrak{L}$ onto $\mathfrak{L}/\mathfrak{A}$. From $\mathfrak{R}^\omega = \mathfrak{R}^\omega(\operatorname{ad}\mathfrak{H})^n$ as above, we have $\varphi(\mathfrak{R}^\omega) \subseteqq \varphi(\mathfrak{R})^{n+1} = (\mathfrak{H}')^{n+1} = 0$ (for $n = \dim \mathfrak{L}$, say). Thus $\varphi(\mathfrak{H}) = \varphi(\mathfrak{R}) = \mathfrak{H}'$. Hence if $x \in \mathfrak{L}$, $[x\,\mathfrak{H}] \subseteqq \mathfrak{H}$, we have $[\varphi(x), \varphi(\mathfrak{H})] = [\varphi(x), \mathfrak{H}'] \subseteqq \mathfrak{H}'$, so that $x \in \mathfrak{R}$, and therefore $x \in \mathfrak{H}$. This completes the proof.

We next propose to demonstrate the existence of Cartan subalgebras in all *restricted* Lie algebras; it is clearly sufficient to treat the case of finite ground fields. More generally, we assume the ground field $\mathfrak{F}$ to be *perfect* and of prime characteristic p. First we consider the Jordan–Chevalley decomposition in restricted Lie algebras over $\mathfrak{F}$.

Let $\mathfrak{L}$ be a restricted Lie algebra over $\mathfrak{F}$. We denote by x^p the image of $x \in \mathfrak{L}$ under the p-power operation of $\mathfrak{L}$, by x^{p^k} the image of x under the k-th iterate of $x \to x^p$ (with $x^{p^0} = x$), and by $\langle x \rangle$ the smallest restricted subalgebra of $\mathfrak{L}$ containing x, i.e., the space of linear combinations of the x^{p^k}, $k = 0, 1, 2, \ldots$ The free restricted Lie algebra on one generator X is simply the subalgebra $\langle X \rangle$ of the polynomial ring $\mathfrak{F}[X]$, regarded as a commutative restricted Lie algebra with the associative p-power for polynomials. We call $x \in \mathfrak{L}$ *semisimple* if $x \in \langle x^p \rangle$, *nilpotent* if $x^{p^k} = 0$ for some k. Assuming $\mathfrak{L}$ finite-dimensional, $\langle x \rangle$ is finite-dimensional for each $x \in \mathfrak{L}$. Let $\mathfrak{K}$ be the kernel of the homomorphism of $\langle X \rangle$ onto $\langle x \rangle$ sending X onto x.

Lemma V.7.1. $\mathfrak{K} = \langle q(X) \rangle$, where $0 \neq q(X)$ is of minimal degree in $\mathfrak{K}$. If x is semisimple, the coefficient of X in $q(X)$ is different from zero.

Since $\mathfrak{K} \neq 0$, we may choose $q(X)$ as indicated. If $r(X) \in \mathfrak{K}$ has degree p^n, $r(X) \notin \langle q(X) \rangle$, and $r(X)$ is of minimal degree in $\mathfrak{K}$ with the latter property, let $q(X)$ have degree p^m. Then $n \geqq m$, and for suitable $\lambda \in \mathfrak{F}$, $r(X) - \lambda\, q(X)^{p^{n-m}} \in \mathfrak{K}$ has lower degree than has $r(X)$, and cannot be in $\langle q(X) \rangle$. This contradiction establishes the first assertion. For the second, the condition that x be semisimple amounts to saying that $\mathfrak{K}$ contains some $r(X)$ in which the coefficient of X is not zero. Thus $r(X) = \lambda_0\, q(X) + \lambda_1\, q(X)^p + \cdots$, $\lambda_i \in \mathfrak{F}$. By comparing terms in X, we have the lemma.

Lemma V.7.2. Let $\mathfrak{L}$ and $\mathfrak{F}$ be as above. Then the following are equivalent: (i) Every $x \in \mathfrak{L}$ is semisimple. (ii) No $x \neq 0$ in $\mathfrak{L}$ is nilpotent.

For if $x \in \mathfrak{L}$ is semisimple, then $\langle x \rangle = \langle x^p \rangle = \langle x^{p^k} \rangle$ for each k; thus $x^{p^k} = 0$ implies $x = 0$. Conversely suppose only $0 \in \mathfrak{L}$ is nilpotent; let $x \in \mathfrak{L}$ and let $q(X)$ be as above. If $q(X)$ has no term in X, then since $\mathfrak{F}$ is perfect $q(X) = s(X)^p$, $s(X) \in \langle X \rangle$ of lower degree than $q(X)$. Thus $s(x)^p = 0$, $s(x) = 0$, and $s(X) \in \mathfrak{K}$, contrary to the choice of $q(X)$ as generator for $\mathfrak{K}$. It follows that a term in X is present in $q(X)$, hence that x is semisimple.

Lemma V.7.3. Let $\mathfrak{L}$ and $\mathfrak{F}$ be as above. Let $x \in \mathfrak{L}$ be semisimple. Then $\langle x \rangle$ consists of semisimple elements.

For if $r(X) \in \langle X \rangle$ is such that $r(x)$ is nilpotent, we have $r(X)^{p^k} \in \langle q(X) \rangle$ for some k. By examining terms of degree less than p^k in X in a representation $r(X)^{p^k} = \lambda_0\, q(X) + \lambda_1\, q(X)^p + \cdots$, we see by Lemma 1 that $\lambda_i = 0$ for all $i < k$, i.e., $r(X)^{p^k} \in \langle q(X)^{p^k} \rangle$. Since $\mathfrak{F}$ is perfect, $r(X)^{p^k} = u(X)^{p^k}$, $u(X) \in \langle q(X) \rangle$, from which $r(X) = u(X) \in \langle q(X) \rangle$ and $r(x) = 0$.

Lemma V.7.4. If $x, y \in \mathfrak{L}$, $[x\, y] = 0$, and if x and y are nilpotent (resp. semisimple), so is every element of the restricted subalgebra $\langle x \rangle + \langle y \rangle$.

The assertions follow at once from the fact that the conclusions hold for the formal direct sum $\langle x\rangle \oplus \langle y\rangle$, and from the fact that nilpotency resp. semisimplicity of elements is preserved under homomorphisms of restricted Lie algebras.

Theorem V.7.2. Let $x \in \mathfrak{L}$, a restricted Lie algebra of finite dimension over a perfect field $\mathfrak{F}$. Then there exist elements $s, n \in \langle x\rangle$, s semisimple and n nilpotent, such that $x = s + n$. If $y \in \mathfrak{L}$ is semisimple, $z \in \mathfrak{L}$ nilpotent, $[y\, z] = 0$, $x = y + z$, then $y = s$ and $z = n$.

Once the existence of s and n has been proved, the rest is clear; for $[s\, y] = 0 = [z\, n]$, so that $s - y = z - n$ is at once semisimple and nilpotent by Lemma 4. But this element must then be zero, e.g., by the proof of Lemma 2. Now let $0 \neq q(X) \in \langle X\rangle$, $q(x) = 0$. If $q(X) \notin \langle X^p\rangle$, then x is semisimple and we are done. Thus we may assume $q(X) = \sum_{i=0}^{m} \alpha_i X^{p^{k+i}}$, $k > 0$, $\alpha_0 \neq 0$. We argue by induction on k to show the existence of s and n, having our assertion if $k = 0$. Let $\lambda_i \in \mathfrak{F}$, $\lambda_i^{p^k} = \alpha_i$, $0 \leqq i \leqq m$; let $r(X) = \sum_{i=0}^{m} \lambda_i X^{p^i}$, and let $u = r(x)$, so that $u \in \langle x\rangle$ satisfies $u^{p^k} = q(x) = 0$. Let $w = x - \lambda_0^{-1} u$, $t(X) = \sum_{i=0}^{m} \lambda_i^{p^{k-1}} X^{p^{k-1+i}}$. Then $t(w) = t(x) - t(\lambda_0^{-1} u) = t(x) - u^{p^{k-1}}$, since $u^{p^k} = 0$; but $t(x) = u^{p^{k-1}}$, so that $t(w) = 0$. Since $\lambda_0 \neq 0$, we have by induction that $w = s + n_1$, s semisimple, n_1 nilpotent, s and n_1 in $\langle w\rangle \subseteqq \langle x\rangle$. Hence we take $n = n_1 + \lambda_0^{-1} u$ and have $x = s + n$ as required.

If $\mathfrak{L}$ is an associative algebra over $\mathfrak{F}$, regarded as a restricted Lie algebra, then $x \in \mathfrak{L}$ is semisimple in our sense only if x satisfies a semisimple polynomial (no repeated roots), so has semisimple minimum polynomial as an element of the associative algebra. Conversely, if x has semisimple minimum polynomial, then $\mathfrak{F}[x]$ is without nilpotent elements, so that $\langle x\rangle \subseteqq \mathfrak{F}[x]$ consists of semisimple elements. Thus the above proof also shows the existence and uniqueness of the Jordan–Chevalley decomposition for linear transformations of finite-dimensional vector spaces over perfect fields of prime characteristic.

Theorem V.7.3. Let $\mathfrak{L}$ be a restricted Lie algebra over a perfect field $\mathfrak{F}$. Let $\mathfrak{H}$ be a subalgebra of $\mathfrak{L}$ which is maximal with respect to the properties: 1) $\mathfrak{H}$ is commutative; 2) each $h \in \mathfrak{H}$ is semisimple. Then the normalizer $\mathfrak{N}(\mathfrak{H})$ is a Cartan subalgebra.

For let $\mathfrak{N} = \mathfrak{N}(\mathfrak{H})$, a restricted subalgebra of $\mathfrak{L}$; now we have $[\mathfrak{N}\,\mathfrak{H}] = 0$, since $h \in \mathfrak{H}$ implies that $\operatorname{ad} h$ is semisimple as linear transformation of $\mathfrak{L}$, and $\mathfrak{N}(\operatorname{ad} h)^2 = 0$ since $\mathfrak{H}$ is commutative; thus $\mathfrak{N}(\operatorname{ad} h) = 0$. Now let $x \in \mathfrak{N}$, $x = s + n$ as in Th. 2. Then $\mathfrak{F}\, s + \mathfrak{H}$ is commutative and consists of semisimple elements, so that $s \in \mathfrak{H}$ by maximality of $\mathfrak{H}$. Therefore every $x \in \mathfrak{N}$ has its semisimple part in $\mathfrak{H}$. Now consider $\mathrm{ad}\,x$, acting in $\mathfrak{N}$; we have $(\mathrm{ad}\,s) \mid \mathfrak{N} = 0$; hence $(\mathrm{ad}\,x) \mid \mathfrak{N} = (\mathrm{ad}\,n) \mid \mathfrak{N}$ is nilpotent, from which $\mathfrak{N}$ is a nilpotent subalgebra by Engel's theorem. Relative to $\mathfrak{H}$, $\mathfrak{L} = \mathfrak{L}_0 \oplus \mathfrak{L}_1$, the *Fitting decomposition* of [234, p. 39], where $\mathfrak{L}_0 = \mathfrak{N}$ since all $\mathrm{ad}\,h$ $(h \in \mathfrak{H})$ are semisimple, and where $[\mathfrak{L}_0\, \mathfrak{L}_1] \subseteqq \mathfrak{L}_1$. If $u = x + v$ normalizes $\mathfrak{N}$, where $x \in \mathfrak{N}$, $v \in \mathfrak{L}_1$, we have $[v\, \mathfrak{H}] \subseteqq \mathfrak{L}_1 \cap \mathfrak{N} = 0$, and $v \in \mathfrak{N}$, hence $v = 0$. Thus $\mathfrak{N}$ is a Cartan subalgebra.

Corollary. Every restricted Lie algebra has a Cartan subalgebra.

For restricted Lie algebras $\mathfrak{L}$ over perfect fields which have (restricted) representations with nonsingular trace form, the algebra $\mathfrak{H}$ above is itself a Cartan subalgebra, as we shall now show. Conversely, in such an algebra which is its own derived algebra every Cartan subalgebra $\mathfrak{H}$ satisfies these conditions, as is seen by applying Chapter II, § 1 to $\mathfrak{L}_\Omega$ and $\mathfrak{H}_\Omega$, where Ω is an algebraically closed extension of $\mathfrak{F}$. We prove our assertion in a slightly more general form:

Theorem V.7.4. Let $\mathfrak{L}$ be a restricted Lie algebra over a perfect field $\mathfrak{F}$. Let (x, y) be a nonsingular symmetric associative bilinear form on $\mathfrak{L}$ such that for all $x, y \in \mathfrak{L}$ with $[x\, y] = 0$, y nilpotent, we have $(x, y) = 0$. Let $\mathfrak{H}$ be as in Th. 3. Then $\mathfrak{H}$ is a Cartan subalgebra.

For let $\mathfrak{N}$ be as in Th. 3, $\mathfrak{L} = \mathfrak{N} \oplus \mathfrak{L}_1$, $\mathfrak{L}_1 = [\mathfrak{L}_1\, \mathfrak{H}]$. Thus if $x \in \mathfrak{N}$, $y \in \mathfrak{L}_1$, we have $y = \Sigma[h_i\, z_i]$, $h_i \in \mathfrak{H}$, and $(x, y) = \Sigma([x\, h_i], z_i) = 0$. Therefore $\mathfrak{N}$ *is a nonsingular subspace* of $\mathfrak{L}$ with respect to the form. Next let $h \in \mathfrak{H}$, $y \in \mathfrak{N}$, $y = s + n$ as before ($s \in \mathfrak{H}$, n nilpotent), and assume $(h, y) \neq 0$. Since $[h\, n] = 0$, we have $(h, n) = 0$ by the hypothesis, hence $(h, s) \neq 0$. It follows that $\mathfrak{H}$ *is a nonsingular subspace.*

Next we show that $\mathfrak{N}$ is commutative; if not, let $0 \neq y \in [\mathfrak{N}\, \mathfrak{N}]$, y central in $\mathfrak{N}$, $y = s + n$ as above. Then n is central nilpotent in $\mathfrak{N}$, from which $(n, \mathfrak{N}) = 0$ by assumption, and $y = s \in \mathfrak{H}$. Hence there is $h \in \mathfrak{H}$ with $(h, s) \neq 0$. But $(\mathfrak{H}, [\mathfrak{N}\, \mathfrak{N}]) = ([\mathfrak{H}\, \mathfrak{N}], \mathfrak{N}) = 0$, so this is impossible, and $\mathfrak{N}$ is commutative. Therefore $\mathfrak{N}$ contains no nonzero nilpotent elements (since $\mathfrak{N}$ is nonsingular) and $\mathfrak{N} = \mathfrak{H}$. This completes the proof. For analogues of Theorems 3 and 4 in characteristic zero, see [43].

§ 8. Nilpotent elements and special subalgebras

For characteristic zero, one has the result of Morozov and Jacobson that if E is a nilpotent linear transformation belonging to a completely reducible Lie algebra $\mathfrak{L}$ of endomorphisms of the finite-dimensional vector space $\mathfrak{V}$, then there are elements $F, H \in \mathfrak{L}$, with F nilpotent and H semisimple, such that $[E\,H] = 2E$, $[F\,H] = -2F$, $[E\,F] = H$

(cf. [234, pp. 98—100]). In particular, one may apply this result to ad $\mathfrak{L}$, when $\mathfrak{L}$ is semisimple, to embed every $e \in \mathfrak{L}$ with ad$e$ nilpotent as a root-vector in a split 3-dimensional simple subalgebra of $\mathfrak{L}$. We consider the latter problem for prime characteristics, where $\mathfrak{L}$ is assumed to be a form of a classical simple Lie algebra. (This is the case connected with the group $\mathfrak{U}$ of Chap. IV, § 7.) By considering the classification in Chapter IV, one may proceed as in [167 and 230] to show that the analogous result holds for types A, B, C and D_n $(n > 4)$. In fact, it holds for all characteristics other than two and without finiteness restriction on the dimension over its center of the division algebra involved in Chapter IV, § 3 in the description of a realization of each of these classes. This has been shown in unpublished work of the author [361], where canonical forms for certain nilpotent linear transformations have been given; knowledge of these forms may then be applied to identify the group $\mathfrak{U}$ with a classical group as indicated in Chapter IV, § 7 [362]. A similar method has been applied to G_2 to yield the result in that case for characteristic $\neq 2, 3$. No generally applicable method for establishing this result seems to have been developed. For characteristics other than 2, 3, 5, and for elements e with $(\mathrm{ad}\, e)^3 = 0$, the result does hold generally, as we now show.

Lemma V.8.1. Let $\mathfrak{L}$ be a Lie algebra with nonsingular Killing form over a field of characteristic not two. Let $x \in \mathfrak{L}$, ad$x$ nilpotent. Then there exist $y, z \in \mathfrak{L}$, $z = [x\, y]$, $[x\, z] = 2x$.

For it is possible to choose $Y, Z \in \mathfrak{E}(\mathfrak{L})$, the full algebra of linear transformations, with $Z = [\mathrm{ad}\, x, Y]$, $[\mathrm{ad}\, x, Z] = 2\,\mathrm{ad}\, x$, as in [234, p. 100]. Let $(U, V) = Tr(UV)$ on $\mathfrak{E}(\mathfrak{L})$, so that (U, V) is nonsingular on $\mathfrak{E}(\mathfrak{L})$ and ad $\mathfrak{L}$ is a nonsingular subspace. Thus $Z = \mathrm{ad}\, z + Z_0$, $Y = \mathrm{ad}\, y + Y_0$, where $y, z \in \mathfrak{L}$ and $Y_0, Z_0 \in (\mathrm{ad}\, \mathfrak{L})^{\perp}$. By the associativity of (U, V), $[\mathrm{ad}\, \mathfrak{L}, (\mathrm{ad}\, \mathfrak{L})^{\perp}] \subseteqq (\mathrm{ad}\, \mathfrak{L})^{\perp}$, from which $[x\, z] = 2x$, $[x\, y] = z$.

Lemma V.8.2. Let $\mathfrak{L}$ be a Lie algebra over a field of characteristic other than 2, 3, 5. Let $0 \neq e \in \mathfrak{L}$, $(\mathrm{ad}\, e)^3 = 0$, $2e = [e\, h]$ for some $h \in [\mathfrak{L}\, e]$. Then there is $f \in \mathfrak{L}$, $[e\, f] = h$, $[f\, h] = -2f$, and for such f we have $(\mathrm{ad}\, f)^3 = 0$.

This lemma is really a summary of certain results of JACOBSON (see also [50]). As in [234, p. 99], one finds that if $[u\, e] = 0$, then $u(\mathrm{ad}\, h)((\mathrm{ad}\, h) - 1)((\mathrm{ad}\, h) - 2) = 0 = u(\mathrm{ad}\, h)(\mathrm{ad}\, e)$, hence that $(\mathrm{ad}\, h) + 2$ induces an automorphism (as vector space) of the kernel of ad$e$. Taking $v \in \mathrm{Ker}(\mathrm{ad}\, e)$ with $v((\mathrm{ad}\, h) + 2) = y((\mathrm{ad}\, h) + 2) \in \mathrm{Ker}(\mathrm{ad}\, e)$, where $h = [y\, e]$, and letting $f = v - y$, we have $[e\, f] = h$, $[f\, h] = -2f$. Next one notes that if E, F, H are linear transformations of a vector space $\mathfrak{V}$ with $[EF] = H$, $[EH] = 2E$, $[FH] = -2F$, and $E^3 = 0$, then by

Lemma 1 of [228], $\prod_{j+1}^{5}(H+3-j)=0$; that is, H is semisimple with characteristic roots among 0, ± 1, ± 2. By the restriction on characteristics, ± 3 and ± 4 are not characteristic roots of H, and if $v\in V$, $vH=\lambda v$, then $vFH=(\lambda-2)\,vF$; from this it follows that $F^3=0$. Applied to $E=\mathrm{ad}\,e$, $F=\mathrm{ad}\,f$, $H=\mathrm{ad}\,h$, this completes the proof.

Combining the cases previously mentioned with the observations of Chapter II, § 10, we see that the conclusion of Lemma 1 is valid for all forms of classical simple Lie algebras of characteristic $\neq 2, 3, 5$. Hence we have the conclusion of Lemma 2 for all such algebras:

Lemma V.8.3. Let $\mathfrak{L}$ be a form of a classical simple Lie algebra over a field of characteristic not $2, 3, 5$. Let $e\in\mathfrak{L}$, $(\mathrm{ad}\,e)^3=0$. Then there exist $f, h\in\mathfrak{L}$ with $[e\,h]=2e$, $[e\,f]=h$, $[f\,h]=-2f$, and $(\mathrm{ad}\,f)^3=0$.

With the assumptions above, the results of [228] may be applied to study the adjoint representation on $\mathfrak{L}$ of $\mathfrak{T}=\mathfrak{F}e+\mathfrak{F}h+\mathfrak{F}f$. This representation is completely reducible, and its irreducible summands are of (at most) three types; 1) one-dimensional, annihilated by $\mathfrak{T}$; 2) two-dimensional, of the form $\mathfrak{F}v+\mathfrak{F}[v\,f]$, where $[[v\,f]\,f]=0$, $[v\,h]=v$, $[[v\,f]\,h]=-[v\,f]$, $[v\,e]=0$, $[[v\,f]\,e]=-v$; 3) three-dimensional, isomorphic to the adjoint representation of $\mathfrak{T}$. Let $\mathfrak{M}$, $\mathfrak{N}$, $\mathfrak{P}$ be respectively the sums of the irreducible $\mathfrak{T}$-submodules of types 3), 2), 1). Let $\mathfrak{L}_\lambda$, $\mathfrak{M}_\lambda$, etc. denote the subspace of $\mathfrak{L}$, $\mathfrak{M}$, etc. belonging to the characteristic root λ of $\mathrm{ad}\,h$; then $\mathfrak{M}=\mathfrak{M}_2+\mathfrak{M}_0+\mathfrak{M}_2$, $\mathfrak{N}=\mathfrak{N}_1+\mathfrak{N}_{-1}$, $\mathfrak{P}=\mathfrak{P}_0$, and $\mathfrak{L}=\mathfrak{M}\oplus\mathfrak{N}\oplus\mathfrak{P}$. Moreover, the dimensions of $\mathfrak{M}_2$, $\mathfrak{M}_0$, $\mathfrak{M}_{-?}$ are equal, as are those of $\mathfrak{N}_1$ and $\mathfrak{N}_{-1}$; the range of $\mathrm{ad}\,e$ is $\mathfrak{M}_0+\mathfrak{M}_2+\mathfrak{N}_1$, while that of $\mathrm{ad}\,f$ is $\mathfrak{M}_0+\mathfrak{M}_{-2}+\mathfrak{N}_{-1}$. Thus $\mathrm{Rank}(\mathrm{ad}\,f)=\mathrm{Rank}(\mathrm{ad}\,e)$. If $x\in\mathfrak{M}_2\cup\mathfrak{M}_{-2}$, then $(\mathrm{ad}\,x)^3=0$ from $[\mathfrak{L}_\lambda\,\mathfrak{L}_\mu]\subseteqq\mathfrak{L}_{\lambda+\mu}$. From this, we also have for $x\in\mathfrak{M}_2$, $\mathrm{Rank}(\mathrm{ad}\,x)=\dim[\mathfrak{M}_{-2}\,x]+\dim[\mathfrak{N}_{-1}\,x]+\dim[\mathfrak{L}_0\,x]\leqq\dim\mathfrak{M}_{-2}+\dim\mathfrak{N}_{-1}+\dim\mathfrak{M}_2=\mathrm{Rank}(\mathrm{ad}\,e)$. Thus if $e\neq 0$ is such that $(\mathrm{ad}\,e)^3=0$ and $\mathrm{ad}\,e$ has minimal rank with this property, the same holds for every $x\neq 0$ in $\mathfrak{M}_2\cup\mathfrak{M}_{-2}$.

Lemma V.8.4. $\mathfrak{L}=\mathfrak{M}+\mathfrak{N}+[\mathfrak{M}_2\,\mathfrak{M}_{-2}]+[\mathfrak{N}_1\,\mathfrak{N}_{-1}]$.

Since $\mathfrak{M}\neq 0$, it suffices by simplicity to show that the right-hand side, $\mathfrak{K}$, is an ideal in $\mathfrak{L}$, and for this to show $[\mathfrak{L}_0\,\mathfrak{K}]\subseteqq\mathfrak{K}$, or $[\mathfrak{M}_0\,\mathfrak{K}]\subseteqq\mathfrak{K}$, $[\mathfrak{P}\,\mathfrak{K}]\subseteqq\mathfrak{K}$. Again, it suffices to show $[\mathfrak{M}_0\,\mathfrak{K}_0]\subseteqq\mathfrak{K}$, $[\mathfrak{P}\,\mathfrak{K}_0]\subseteqq\mathfrak{K}$, where $\mathfrak{K}_0=\mathfrak{M}_0+[\mathfrak{M}_2\,\mathfrak{M}_{-2}]+[\mathfrak{N}_1\,\mathfrak{N}_{-1}]$. That $[\mathfrak{L}_0[\mathfrak{M}_2\,\mathfrak{M}_{-2}]]$, $[\mathfrak{L}_0[\mathfrak{N}_1\,\mathfrak{N}_{-1}]]$ are contained in $[\mathfrak{M}_2\,\mathfrak{M}_{-2}]$, $[\mathfrak{N}_1\,\mathfrak{N}_{-1}]$, respectively, follows from the Jacobi identity. Thus it remains to show $[\mathfrak{M}_0\,\mathfrak{M}_0]\subseteqq\mathfrak{K}$, $[\mathfrak{M}_0\,\mathfrak{P}]\subseteqq\mathfrak{K}$. Now let x, $y\in\mathfrak{M}_0$, $x=[u\,f]$, $y=[v\,f]$, u, $v\in\mathfrak{M}_2$; then $[x\,y]=[[[u\,f]\,v]\,f]-[[[u\,f]\,f]\,v]\in[\mathfrak{L}\,f]+[\mathfrak{M}_{-2}\,\mathfrak{M}_2]\subseteqq\mathfrak{K}$. Finally, if $x=[u\,f]\in\mathfrak{M}_0$, $z\in\mathfrak{P}$, then $[x\,z]=[[u\,f]\,z]=[[u\,z]\,f]\in[\mathfrak{L}\,f]\subseteqq\mathfrak{K}$, and we are done.

We refer to automorphisms of $\mathfrak{L}$ of the form $\exp(\operatorname{ad} x)$, where $x \neq 0$, $(\operatorname{ad} x)^3 = 0$, and $\operatorname{ad} x$ is of minimal rank with this property, as *transvections* of $\mathfrak{L}$, and to the elements x of this kind as *pre-transvections* ("ptv" for short). The terminology is suggested by the identifications indicated in Chapter IV, § 7. Clearly the automorphisms of $\mathfrak{L}$ map ptvs onto ptvs, and we have seen above that all non-zero elements of $\mathfrak{M}_2 \cup \mathfrak{M}_{-2}$ are ptvs when e is a ptv. Hence, assuming e to be a ptv, we see that $\mathfrak{M}_2 + \mathfrak{M}_{-2}$ is spanned by ptvs; we wish to show this is the case for $\mathfrak{L}$ itself. Now let $u \in \mathfrak{N}_1$; then $\mathfrak{L}(\operatorname{ad} u)^3 \subseteq \mathfrak{N}_1 + \mathfrak{M}_2$, $\mathfrak{L}(\operatorname{ad} u)^4 \subseteq \mathfrak{M}_2$, from which $[\mathfrak{L}(\operatorname{ad} u)^3, \mathfrak{L}(\operatorname{ad} u)^4] = 0$, while $\mathfrak{L}(\operatorname{ad} u)^5 = 0$. It follows that $\exp(\operatorname{ad} u)$ is an automorphism of $\mathfrak{L}$, and the same holds if $u \in \mathfrak{N}_{-1}$.

Lemma V.8.5. $\mathfrak{L}$ is spanned by pre-transvections.

From $x(\exp(\operatorname{ad} y)) \equiv [x y] \pmod{\mathfrak{M}_2 + \mathfrak{M}_{-2}}$, where $x \in \mathfrak{M}_2$, $y \in \mathfrak{M}_{-2}$, and from $\mathfrak{M}_0 = [\mathfrak{M}_2 f] \subseteq [\mathfrak{M}_2 \mathfrak{M}_{-2}]$, we see, by the fact that $\mathfrak{M}_2 + \mathfrak{M}_{-2}$ is spanned by ptvs, that $\mathfrak{M} + [\mathfrak{M}_2 \mathfrak{M}_{-2}]$ is spanned by ptvs. Now let $u \in \mathfrak{N}_1$; then $f(\exp(\operatorname{ad} \lambda u))$ is a ptv for each $\lambda \in \mathfrak{F}$, and $f(\exp(\operatorname{ad} \lambda u)) \equiv \lambda [f u] + \frac{1}{2}\lambda^2 [[f u] u] + \frac{1}{6}\lambda^3 [[[f u] u] u] \pmod{\mathfrak{M} + [\mathfrak{M}_2 \mathfrak{M}_{-2}]}$. Since $\mathfrak{F}$ has at least seven elements, we see by varying λ that the space $\mathfrak{R}$ spanned by the ptvs contains $[f u]$, $[[f u] u]$, $f(\operatorname{ad} u)^3$ for each $u \in \mathfrak{N}_1$. As u runs over $\mathfrak{N}_1$, $[f u]$ runs over $\mathfrak{N}_{-1}$, so that $\mathfrak{N}_{-1} \subseteq \mathfrak{R}$; similarly, $\mathfrak{N}_1 \subseteq \mathfrak{R}$. Finally, let $v \in \mathfrak{N}_1$, $w \in \mathfrak{N}_{-1}$; then $w = [f u]$ for some $u \in \mathfrak{N}_1$. By the above, $\mathfrak{R}$ contains $f(\operatorname{ad}(u + v))^2 - f(\operatorname{ad} u)^2 - f(\operatorname{ad} v)^2 = [[f u] v] + [[f v] u] = 2[[f u] v] + [f[v u]]$; but $[f[v u]] \in [\mathfrak{M}_{-2} \mathfrak{M}_2] \subseteq \mathfrak{R}$, so that $[w v] = [[f u] v] \in \mathfrak{R}$, $[\mathfrak{N}_1 \mathfrak{N}_{-1}] \subseteq \mathfrak{R}$, and $\mathfrak{R} = \mathfrak{L}$ by Lemma 4.

From Lemma 5 and the fact that $\mathfrak{L}$ is normal simple it follows at once that the group $\mathfrak{U}$ of automorphisms of $\mathfrak{L}$ generated by the transvections acts irreducibly in $\mathfrak{L}$ and that its centralizer consists of scalar transformations.

A form of a classical simple Lie algebra cannot contain elements $x \neq 0$ with $(\operatorname{ad} x)^2 = 0$; for algebras with nonsingular Killing form, this is seen at once from $-2(\operatorname{ad} x)(\operatorname{ad} y)(\operatorname{ad} x) = (\operatorname{ad} x)^2(\operatorname{ad} y) - 2(\operatorname{ad} x)(\operatorname{ad} y)(\operatorname{ad} x) + (\operatorname{ad} y)(\operatorname{ad} x)^2 = \operatorname{ad}(y(\operatorname{ad} x)^2) = 0$ for such x, a fact which implies $((\operatorname{ad} x)(\operatorname{ad} y))^2 = 0$, hence $(x, y) = 0$ for all y. For the remaining cases (except E_8 when $p = 5$) the assertion may be verified by use of the classification. Lie algebras containing such elements x have been called *strongly degenerate* by KOSTRIKIN [261—263], who has investigated restricted Lie algebras for this property. In particular, he has shown that the only strongly degenerate simple restricted Lie algebra of dimension less than $2p$ is the Witt algebra [261].

The embedding of nilpotent elements in split three-dimensional algebras may be regarded as a special case of a general problem as to embedding subalgebras of nilpotent elements ("*nil* subalgebras") of semisimple algebras in a canonical way in the algebras, i.e., as to conjugacy of every nil subalgebra with a subalgebra of a canonically chosen maximal nil subalgebra. This amounts to the conjugacy of maximal nil subalgebras under the automorphism group. Analogous questions for algebraic groups have affirmative answers; over algebraically closed fields, the conjugacy of maximal solvable connected subgroups ("Borel subgroups") of a connected algebraic group was established by BOREL [42], and each maximal connected subgroup consisting of unipotent elements is the set of unipotent elements of a Borel subgroup [42, 79]. Over general fields, one may see [45] for an indication as to corresponding results. For algebraic Lie algebras of characteristic zero, in particular for semisimple Lie algebras, these results are reflected in conjugacy of the corresponding maximal solvable resp. nil subalgebras. A direct proof of conjugacy of maximal nil subalgebras of semisimple Lie algebras of characteristic zero has been given by MOSTOW (unpublished). To the writer's knowledge, no results of this type have been proved for characteristic p, even for classical semisimple Lie algebras. In view of the non-triangulability of some solvable algebras (§ 1) it is clear that conjugacy of maximal solvable subalgebras is not to be expected; one can be more optimistic about maximal triangulable and nil subalgebras in the classical cases. They should be of the form $\mathfrak{H} + \sum_{\alpha>0} \mathfrak{L}_\alpha$ resp. $\sum_{\alpha>0} \mathfrak{L}_\alpha$ for some admissible ordering of the roots.

Chapter VI

Related Topics

§ 1. Nilpotent groups and Lie algebras

The restricted Burnside problem

Let $\mathfrak{G}$ be a group, and let $\mathfrak{G} = \mathfrak{G}_1 \supseteqq \mathfrak{G}_2 \supseteqq \cdots$ be its lower central series: $\mathfrak{G}_{i+1} = (\mathfrak{G}, \mathfrak{G}_i)$, the group generated by all commutators $(x, y) = x^{-1} y^{-1} x y$, where $x \in \mathfrak{G}$, $y \in \mathfrak{G}_i$. Then each $\mathfrak{G}_i$ is normal in $\mathfrak{G}$ and $\mathfrak{G}_i/\mathfrak{G}_{i+1}$ is abelian. Let $\mathfrak{L} = \sum_{i=1}^{\infty} \oplus \mathfrak{G}_i/\mathfrak{G}_{i+1}$, writing the group operation additively. One sees by induction that if $x \in \mathfrak{G}_i$, $y \in \mathfrak{G}_j$, then $(x, y) \in \mathfrak{G}_{i+j}$, and it follows from the identities $(x, y\, z) = (x, z)\,(x, y)^z$, $(xy, z) = (x, z)^y\,(y, z)$ (where $u^v = v^{-1}\, u\, v$) that the coset of $(x, y) \bmod \mathfrak{G}_{i+j+1}$ depends only on the coset of $x \bmod \mathfrak{G}_{i+1}$ and that of $y \bmod \mathfrak{G}_{j+1}$. Thus one obtains a pairing $(x\, \mathfrak{G}_{i+1}, y\, \mathfrak{G}_{j+1}) \to (x, y)\, \mathfrak{G}_{i+j+1}$ from $\mathfrak{L}_i \times \mathfrak{L}_j$ into $\mathfrak{L}_{i+j}$, where $\mathfrak{L}_k = \mathfrak{G}_k/\mathfrak{G}_{k+1}$. Using the identities $(u, v)\,((u, v), w) = (u, v)^w = (w, (v, u))\,(u, v)$ to substitute in the above, one sees that this pairing is biadditive; it is then extended to a biadditive product $[x\, y]$ on $\mathfrak{L}$ in the only possible manner. From $(u, u) = 1$ and $(u, v) = (v, u)^{-1}$ in $\mathfrak{G}$ it follows that $[x\, x] = 0$ for all $x \in \mathfrak{L}$. From the identity $(x^y, (y, z))\,(y^z, (z, x))\,(z^x, (x, y)) = 1$ in $\mathfrak{G}$ one obtains the Jacobi identity in $\mathfrak{L}$; thus $\mathfrak{L}$ is a Lie ring. If $\mathfrak{G}$ is finite and if $\mathfrak{G}_k = \mathfrak{G}_{k+1}$, then $\mathfrak{L}$ is finite of order $(\mathfrak{G}:\mathfrak{G}_k)$; conversely, if $\mathfrak{L}$ is finite of order $|\mathfrak{L}|$, then $\mathfrak{G}_k = \mathfrak{G}_{k+1}$ for some k, and $|\mathfrak{L}| = (\mathfrak{G}:\mathfrak{G}_k)$.

In particular if $x^m = 1$ for all $x \in \mathfrak{G}$, then $m\, x = 0$ for all $x \in \mathfrak{L}$, and $\mathfrak{L}$ may be regarded as a Lie algebra over the ring $\boldsymbol{Z}_m$ of integers modulo m. The case where this observation seems to be especially useful is that where m is a prime p; in this case the fact that $\mathfrak{L}$ is a Lie algebra over the field $\boldsymbol{Z}_p$ is supplemented by the less trivial one that $\mathfrak{L}$ satisfies the $(p-1)$-th *Engel condition*: $(\operatorname{ad} x)^{p-1} = 0$ for all $x \in \mathfrak{L}$ (cf. [186; 161, Chap. 18]). Now Kostrikin [259] has shown the following: *Let $\mathfrak{L}$ be a Lie algebra over $\boldsymbol{Z}_p$ and suppose that $\mathfrak{L}$ satisfies the n-th Engel condition for some $n \leqq p$; then $\mathfrak{L}$ is locally nilpotent,* i.e., *the subalgebra generated by any finite set of elements of $\mathfrak{L}$ is nilpotent.*

If $x^p = 1$ for all $x \in \mathfrak{G}$, and if $\mathfrak{G}$ has r generators, one is in the situation of the "Burnside problem" for prime exponent: Is $\mathfrak{G}$ finite? A negative answer to this question has been announced by P. S. Novikov [309] for $p > 72$ and $r \geqq 2$; positive answers are known for general r with $p = 2, 3$ [161], and with the non-prime exponents 4 [338] and 6 [160]. Since $\mathfrak{L}$ is generated by $\mathfrak{L}_1$, which is generated by the cosets modulo $\mathfrak{G}_2$ of a set of generators for $\mathfrak{G}$, one may apply Kostrikin's theorem to conclude that $\mathfrak{L}$ is a nilpotent Lie algebra, hence that $\mathfrak{G}_k = \mathfrak{G}_{k+1}$ for some k. Since $\mathfrak{L}$ is finite, the finiteness of $\mathfrak{G}/\mathfrak{G}_k$ also follows. The method of the Lie ring wipes out $\mathfrak{G}_k$ if $\mathfrak{G}_k = \mathfrak{G}_{k+1}$, and $\mathfrak{G}_\infty = \bigcap\limits_k \mathfrak{G}_k$ in every case; thus one can only expect this method to yield information about $\mathfrak{G}/\mathfrak{G}_\infty$.

Now suppose $\mathfrak{G}$ has r generators, and that $x^m = 1$ for all $x \in \mathfrak{G}$; suppose further that $\mathfrak{G}$ is *nilpotent* (which is equivalent with finiteness of $\mathfrak{G}$ when m is prime). Suppose one has a bound $l(m, r)$ in terms of m and r for the order of the Lie ring $\mathfrak{L}$ associated with the group $\mathfrak{H} = \mathfrak{F}/\mathfrak{F}^m$, where $\mathfrak{F}$ is the free group on r generators and where $\mathfrak{F}^m$ is the subgroup generated by m-th powers of elements of $\mathfrak{F}$. Then $\mathfrak{H}_k = \mathfrak{H}_{k+1}$ for some $k = k(m, r)$, so that $\mathfrak{G}_k = \mathfrak{G}_{k+1} = 1$ by nilpotency and the fact that $\mathfrak{G}$ is a homomorphic image of $\mathfrak{H}$; in fact, $\mathfrak{G}$ is a homomorphic image of $\mathfrak{H}/\mathfrak{H}_k$, so has nilpotency class at most k and order at most $(\mathfrak{H}:\mathfrak{H}_k) \leqq l(m, r)$. For $m = p$, a prime, the theorem of Kostrikin establishes the existence of the bound $l(p, r)$. This is an affirmative solution to the "*restricted Burnside problem*" for prime exponent: *Among all finite groups $\mathfrak{G}$ on r generators and having prime period p, there is one, $\mathfrak{H}/\mathfrak{H}_k$, having all others as homomorphic images.* Thus $l(p, r)$ is a bound for the orders of such groups $\mathfrak{G}$. Some admissible values for $l(m, r)$ and bounds on the nilpotency class of $\mathfrak{G}$ are found in [161], especially for $r = 2$.

One may conjecture that *solvable* groups $\mathfrak{G}$ satisfying $x^m = 1$ will be somewhat more accessible than general groups in the context of the restricted Burnside problem; in fact, the result of Feit and Thompson [130] shows that if $\mathfrak{G}$ is finite and m is *odd*, then $\mathfrak{G}$ must be solvable. In this connection, P. Hall and G. Higman [164] have shown that a bound $s(m, r)$ on the orders of finite *solvable* groups on r generators exists if one has the existence of bounds $l(p_i^{e_i}, t)$ for all t, where $m = p_1^{e_1} \ldots p_n^{e_n}$ is the prime factorization of m. Thus *the restricted Burnside problem* is *answered affirmatively if m is a product of distinct odd primes.*

In fact, W. Feit has informed the author that the *restricted Burnside problem is answered affirmatively whenever m is a product of distinct primes,* possibly including 2. Namely, by results of Hall and Higman [164] and that of Kostrikin it suffices to show that there are only finitely many simple non-abelian finite groups of exponent dividing m, and that the group of outer automorphisms of each of these is solvable. Now WALTER [403] has shown that the non-abelian finite simple groups $\mathfrak{G}$ whose 2-Sylow subgroups are elementary abelian (a group of exponent 2 is necessarily abelian) are as follows:

1) $\mathfrak{G} \cong \mathrm{PSL}(2, q)$, $q = p^n$ odd, $q \equiv 3$ or $5 \bmod 8$;

$$|\mathfrak{G}| = \tfrac{1}{2} q(q^2 - 1).$$

2) $\mathfrak{G} \cong \mathrm{PSL}(2, 2^n)$, $n > 1$; $|\mathfrak{G}| = 2^n(2^{2n} - 1)$.

3) $\mathfrak{G} \cong \mathfrak{R}(q)$, $q = 3^{2n+1}$, $n > 0$; $|\mathfrak{G}| = q^3(q^3 + 1)(q - 1)$.

(It is not known at this writing whether such a group is necessarily the group discovered by REE and associated with a Lie algebra of type G_2 [331]).

4) $\mathfrak{G} \cong \mathfrak{J}$, the simple group of JANKO [239]; $|\mathfrak{G}| = 266{\cdot}660$.

For given m as above, it is clear that only a finite number of these groups can have exponent dividing m. It is also known [103, 148, 239] that the groups of outer automorphisms are solvable. Thus the affirmative answer is established.

In various cases, the effect of an Engel condition in Lie algebras has been studied by ZORN [426], GRUENBERG [157], HIGGINS [185], COHN [87] and HIGMAN [186]. ZORN showed that the n-th Engel condition and the maximum condition for Lie subrings imply nilpotency for an arbitrary Lie ring. GRUENBERG showed that nilpotency of a finitely generated Lie ring follows from solvability and the nilpotency of $\mathrm{ad}\, x$ for all x. HIGGINS gave bounds for the nilpotency class of a solvable Lie ring satisfying the n-th Engel condition, all of whose elements have additive order which is infinite or prime to $n!$, and applied these results in the cases $n = 2, 3, 4$. In the case $n = 4$, HIGGINS proved nilpotency of a general Lie ring without 2-, 3-, 5- or 7-torsion; HIGMAN was able to establish the same result in the absence of 2- and 3-torsion. COHN gave an example of a solvable, non-nilpotent Lie algebra over a field of prime characteristic p, satisfying the $(p + 1)$-th Engel condition.

Other (usually more general) connections between central series for groups and associated Lie rings have been studied by MAGNUS (who seems to have initiated the idea) [289, 290], ZASSENHAUS [420], SANOV [340], and LAZARD [278]. In several of these works, the observation of MAGNUS, that the elements $1 + x_\alpha$ of the power series ring in non-commuting indeterminates $\{x_\alpha\}$ (over $\mathbf{Z}$) are free generators for a free group, is used in a fundamental way. This fact and some recent work of GOLOD and SHAFAREVITCH [154] is the basis for an example due to GOLOD [153] of an infinite p-group on finitely many generators, thus providing a fairly simple negative answer to the strong form of the unrestricted Burnside problem. (In this case, the elements of the group have unbounded orders, in contrast to the case studied by NOVIKOV.) For surveys of work on the Burnside problems, see [189, 286].

Another case where Lie ring methods have had some effectiveness in group theory is the study of groups admitting a *regular* automorphism, i.e., one which fixes only the identity. Some results on Lie algebras satisfying the analogous condition have already been cited in Chapter V, § 5. THOMPSON [388, 389] has shown that a *finite* group admitting a regular automorphism of *prime* period is nilpotent. One thus reduces the study of such groups to that of nilpotent Lie rings, at least insofar as orders, nilpotency class, etc. are concerned. HIGMAN [188] has shown that if a nilpotent group G admits a regular automorphism of prime period p, and contains no elements of order p, then the automorphism induced in the obvious way on an associated Lie ring is also regular. He also showed that any Lie ring with a regular automorphism of order p is nilpotent of class at most $k(p)$, where $k(p)$ depends only on p. It seems likely that a finite group with a regular automorphism is solvable [156, 132].

§ 2. Linear algebraic groups and Lie algebras

In this and the next section, we indicate two directions in which an attempt has been made to develop an analogy, for general fields, of the Lie group-Lie algebra correspondence which has proved so useful in the real and complex cases. The first of these directions, in which CHEVALLEY has led much of the way [71, 72, 79; see also 254, 405, 42, 22, 335] consists of replacing the notion of Lie group by that of linear algebraic group. Generally, without attempting to be complete as to definitions, one can describe an algebraic group as an (abstract) algebraic variety $\mathfrak{V}$ whose underlying point-set G admits a group multiplication xy such that $(x, y) \to x y^{-1}$ is a morphism from $\mathfrak{V} \times \mathfrak{V}$ to $\mathfrak{V}$ in the category of algebraic varieties (fields of reference being suppressed here). As in the case of Lie groups, one may define tangent vectors (linear mappings into a specified field of reference $\mathfrak{F}$, *which we assume algebraically closed*, of the local ring of a point, satisfying the differentiation property with respect to products), and of left-invariant (with respect to translations) tangent vector fields. The totality of the latter constitute the *Lie algebra* $\mathfrak{L}$ of the algebraic group, being a linear space over $\mathfrak{F}$ of dimension equal to that of $\mathfrak{V}$ as algebraic variety, closed under the Poisson bracket and under p-th powers if $\mathfrak{F}$ has prime characteristic p. The space $\mathfrak{L} \otimes_{\mathfrak{F}} \mathfrak{F}(\mathfrak{V})$ may also be identified with the Lie algebra of $\mathfrak{F}$-derivations of the field $\mathfrak{F}(\mathfrak{V})$ of rational functions on $\mathfrak{V}$, and the subalgebras of $\mathfrak{L} \otimes_{\mathfrak{F}} \mathfrak{F}(\mathfrak{V})$ stable under conjugation by those automorphisms of $\mathfrak{F}(\mathfrak{V})$ determined by left translations in G are the subspaces of the form $\mathfrak{M} \otimes_{\mathfrak{F}} \mathfrak{F}(\mathfrak{V})$, $\mathfrak{M}$ a subalgebra of $\mathfrak{L}$ (cf. [364]).

In the above degree of generality, the structure of the Lie algebra (as abstract Lie algebra) does not carry enough information about the group G to determine G within any reasonable sense of equivalence. This is seen most strikingly when $\mathfrak{V}$ is *complete*, in which case G is an *abelian variety* [364, 23]. However, an algebraic group G has a unique normal subgroup H which is an *affine* algebraic set, and such that the factor-group G/H carries a natural structure of abelian variety [22, 335, 77]. Moreover every affine algebraic group (irreducibility not assumed) is isomorphic, as algebraic group, to a subgroup of $GL(n, \mathfrak{F})$ for some n, this subgroup being the intersection with $GL(n, \mathfrak{F})$ of an affine algebraic set in $\mathfrak{F}^{n^2}$, regarded as consisting of all n by n matrices. Thus the structure of G reduces to that of the linear group H, the abelian variety G/H, and the extension $1 \to H \to G \to G/H \to 1$. The direct analogy with the Lie theory is most meaningful in the study of the *linear* algebraic groups.

Namely, let G_0 be an irreducible component, containing the identity, of the linear algebraic group G. Then G_0 is a normal subgroup of finite index in G, and the irreducible components of G are the cosets of G_0. Let $\mathfrak{J}$ be the ideal vanishing on G_0 in the polynomial ring in n^2 variables over $\mathfrak{F}$, which variables may be thought of as entries in an "indeterminate" n by n matrix (X_{ij}). Then each n by n $\mathfrak{F}$-matrix $(\alpha) = (\alpha_{ij})$ determines a unique $\mathfrak{F}$-derivation $D_{(\alpha)}$ of $\mathfrak{F}[X_{ij}]$ sending X_{ik} into $\sum_j X_{ij}\alpha_{jk}$. The ideal $\mathfrak{J}$ is a prime ideal and the field $\mathfrak{F}(G_0)$ is the field of quotients of $\mathfrak{F}[X_{ij}]/\mathfrak{J}$. If $\mathfrak{J}D_{(\alpha)} \subseteqq \mathfrak{J}$, then $D_{(\alpha)}$ induces a derivation of $\mathfrak{F}(G_0)$, which commutes with left translations by elements of G_0. In fact, all elements of $\mathfrak{L} = \mathfrak{L}(G_0)$ have this form, and the mapping $(\alpha) \to D_{(\alpha)}$ yields an isomorphism onto $\mathfrak{L}(G_0)$ of the Lie algebra of n by n matrices (α) with $\mathfrak{J}D_{(\alpha)} \subseteqq \mathfrak{J}$. In prime characteristic, this is a restricted isomorphism. Thus one identifies $\mathfrak{L}(G_0)$ with a Lie algebra of n by n matrices, which is also regarded as the Lie algebra $\mathfrak{L}(G)$. In this sense, subgroups of G have as Lie algebras subalgebras of $\mathfrak{L}(G)$, and for irreducible algebraic linear groups $G \underset{\neq}{\supset} H$ implies $\mathfrak{L}(G) \underset{\neq}{\supset} \mathfrak{L}(H)$. If G is solvable, nilpotent, resp. abelian, so is $\mathfrak{L}(G)$. For $U \in G$, the mapping $\operatorname{Ad} U: Y \to U^{-1}YU$ is an automorphism of $\mathfrak{L}(G)$, and Ad is a homomorphism (of algebraic groups) of G into the linear algebraic group $\operatorname{Aut}(\mathfrak{L}(G))$. If H is a normal algebraic subgroup of G, then $\mathfrak{L}(H)$ is stable under $\operatorname{Ad} G$, and is in fact an ideal in $\mathfrak{L}(G)$.

This brief list of elementary properties, which are established in [71], appears to promise a fruitful correspondence between algebraic linear groups and Lie algebras. If we call a Lie algebra $\mathfrak{L}$ of n by n

matrices *algebraic* if $\mathfrak{L} = \mathfrak{L}(G)$ for some algebraic subgroup G of $GL(n, \mathfrak{F})$, it is natural to ask which Lie algebras are algebraic. In the case of characteristic zero, a fairly simple test may be given: For each n by n matrix (α), there is a unique irreducible algebraic group $G_{(\alpha)}$ in $GL(n, \mathfrak{F})$ contained in all algebraic groups G with $(\alpha) \in \mathfrak{L}(G)$; the Lie algebra $\mathfrak{L}(G_{(\alpha)})$ contains (α) and is thus the smallest algebraic Lie algebra containing (α) [71, 199]. Its elements are called *replicas* of (α), and may be described in terms of the (infinitesimal) tensor invariants of (α), or more explicitly in terms of the Jordan canonical form for (α) [68, 71, 199]. If (α) is diagonalizable, $\mathfrak{L}(G_{(\alpha)})$ has dimension equal to the rank of the additive group generated by the characteristic roots; otherwise the dimension exceeds this rank by one. A Lie algebra $\mathfrak{L}$ of n by n matrices is algebraic if and only if $\mathfrak{L}$ contains $\mathfrak{L}(G_{(\alpha)})$ for every $(\alpha) \in \mathfrak{L}$, i.e., if and only if $\mathfrak{L}$ is closed under taking replicas. For prime characteristic p, it is clear that each algebraic Lie algebra $\mathfrak{L}$ containing (α) must contain $(\alpha)^p, (\alpha)^{p^2}, \ldots$ If the notion of replica is defined by tensor invariants as indicated above, the replicas of (α) are just the linear combinations of $(\alpha), (\alpha)^p, \ldots$ [68, 84, 104, 402], and one can show [104] that this commutative Lie algebra $\mathfrak{L}((\alpha))$ is algebraic. Its dimension is the linear dimension over the prime field of the additive group generated by the characteristic roots of (α), plus the number of non-zero matrices $N, N^p, N^{p^2}, \ldots$, where $(\alpha) = S + N$ is the decomposition of (α) into commuting semisimple (S) and nilpotent (N) matrices. It ist *not* in general true that $\mathfrak{L}((\alpha))$ is the Lie algebra of a *unique* irreducible algebraic group $G_{(\alpha)}$ of n by n matrices; hence it is not true that the intersection of all algebraic groups G with $(\alpha) \in \mathfrak{L}(G)$ again has this property. For example, $(\alpha) \doteq \begin{pmatrix} 1 & 0 \\ 0 & 0 \end{pmatrix}$ is a basis for $\mathfrak{L}(G_1) = \mathfrak{L}(G_2)$, where

$$G_1 = \left\{ \begin{pmatrix} \gamma & 0 \\ 0 & 1 \end{pmatrix} \middle| \, \gamma \neq 0 \right\}, \quad G_2 = \left\{ \begin{pmatrix} \gamma & 0 \\ 0 & \gamma^p \end{pmatrix} \middle| \, \gamma \neq 0 \right\}.$$

Chevalley has also displayed ([71]; Chap. II, § 10) two 2-dimensional irreducible algebraic groups of 3 by 3 matrices, one of which is non-commutative and one commutative, having the *same* (commutative) Lie algebra.

The above remarks point out some shortcomings of the group–Lie algebra correspondence in characteristic p. For characteristic zero, the situation is much better; starting with a basis $(\alpha)_1, \ldots, (\alpha)_s$ for $\mathfrak{L}(G)$, G irreducible, one may form a matrix $\prod_i \exp T_i (\alpha)_i$ with entries in the formal power series field $\mathfrak{F} \ll T_1, T_2, \ldots, T_s \gg$, and show that this matrix is a generic point for G [71]. One thus obtains a one-one correspondence between irreducible algebraic subgroups of $GL(n, \mathfrak{F})$ and algebraic Lie algebras in $M_n(\mathfrak{F})$. Abelian, nilpotent, solvable, semi-simple, resp. simple Lie algebras and groups with the analogous properties correspond. In particular, the totality of all simple algebraic groups acting linearly in $\mathfrak{F}^n$ may be read off (to within linear isomorphism) from the list of simple Lie algebras over $\mathfrak{F}$ and the list of their n-dimensional representations. (Although there is no remark to this effect in [71 or 72], it is clear from [71] that such is the case.)

The difficulties in applying the methods of Lie algebras in prime characteristic have been partly responsible for the development of the very rich intrinsic theory of linear algebraic groups, which theory has as one of its most striking features the determination of all simple algebraic linear groups [79]. It turns out, upon comparing the results of this classification with ours in Chapter II, that except for type A_n with $p \mid (n+1)$ (where there are still quite close relations), the Lie algebras of such groups are exactly our classical Lie algebras (characteristic $\neq 2, 3$). In his dissertation [206], HUMPHREYS has made a systematic study of the structure of algebraic (linear) Lie algebras in prime characteristic. In particular, if G is an irreducible semisimple algebraic group, and if $\mathfrak{L} = \mathfrak{L}(G)$, $\mathfrak{Z}$ = center of $\mathfrak{L}$, then $\mathfrak{K}(G) = [\mathfrak{L}\,\mathfrak{L}]/\mathfrak{Z} \cap [\mathfrak{L}\,\mathfrak{L}]$ is a classical Lie algebra, provided that the characteristic is not 2 or 3 (for characteristic 2, $\mathfrak{L}(G)$ can be solvable). The adjoint representation of G in $\mathfrak{L}$ induces an *isogeny* (homomorphism of algebraic groups with finite kernel) of G onto the Chevalley group $G(\mathfrak{K})$ (cf. Chap. III, § 1) of automorphisms of $\mathfrak{K}$, the component of the identity in $\mathrm{Aut}(\mathfrak{K})$. HUMPHREYS is able to establish these facts using only general structural properties of algebraic groups (Exposés 1—17 of [79]) and some recent work of BOREL and SPRINGER [44] on semisimple and nilpotent elements of algebraic Lie algebras.

Maintaining the restriction to characteristics $\neq 2, 3$, and appealing to some general results on isogenies in [79, Exposé 18], HUMPHREYS deduces the principal theorem of CHEVALLEY on existence of isogenies of semisimple algebraic groups [79, Exposé 23, Th. 1] by studying the representations induced in the Lie algebras $\mathfrak{K}(G)$. Since $\mathfrak{K}(G) = \mathfrak{L}(G)$ unless $\mathfrak{K}(G)$ has direct summands of type A_n, $n \equiv -1 \,(\mathrm{mod}\, p)$, it is only in the latter case that one may hope to distinguish non-isomorphic isogenous groups G by the structure of their Lie algebras $\mathfrak{L}(G)$. For the case where $\mathfrak{K}(G)$ is *simple* of this type, HUMPHREYS has shown that $\mathfrak{L}(G)$ must have one of *three* distinct forms, and that each of these reflects divisibility by p of one or both of d and $(n+1)/d$, where d is the "numerical invariant" of G [79, Exposé 20].

From these results and general theory, HUMPHREYS is able to fill a number of gaps in the (irreducible) algebraic group–Lie algebra correspondence in prime characteristic (usually assumed not 2 or 3). For example: G is *reductive* (no non-trivial irreducible normal unipotent subgroups) if and only if $\mathfrak{L}(G)$ has no non-trivial ideals consisting of nilpotent matrices ("*nil ideals*"); G is solvable if and only if $\mathfrak{L}(G)$ is solvable; the *maximal tori* of $\mathfrak{L}(G)$ (i.e., maximal commutative subalgebras consisting of semisimple elements) are the subalgebras $\mathfrak{L}(T)$, where T runs over the maximal tori of G, and the correspondence $T \to \mathfrak{L}(T)$ is one-one if G is reductive. Conjugacy of maximal tori T of G is reflected by conjugacy under the adjoint group of maximal tori $\mathfrak{T}$ in $\mathfrak{L}(G)$, as well as that of their normalizers $\mathfrak{N}(\mathfrak{T})$ in $\mathfrak{L}(G)$. As in Chapter V, § 7, the latter are Cartan subalgebras, and one can show that every Cartan subalgebra $\mathfrak{H}$ of $\mathfrak{L}(G)$ has the form $\mathfrak{N}(\mathfrak{T})$, with $\mathfrak{T}$ the set of semisimple elements of $\mathfrak{H}$. Thus all Cartan subalgebras of $\mathfrak{L}(G)$ are conjugate under the adjoint group (compare with Chap. V, § 7). With G the automorphism group (= Chevalley group) of a classical Lie algebra of type E_8 over $\mathfrak{F}$, one finds that $\mathfrak{L}(G)$ is again a classical algebra of type E_8; the conjugacy of Cartan subalgebras in $\mathfrak{L}(G)$ now enables us to drop the exclusion of characteristic 5 from Th. IV.1.1 and IV.1.2.

In view of the fact that algebraic linear Lie algebras display behavior so much more regular than that of general Lie algebras of prime characteristic, the desirability of a simple useful criterion for a linear Lie algebra $\mathfrak{L}$ to be algebraic is apparent. By the fact that not all restricted Lie algebras are algebraic, whereas restricted Lie algebras with one generator are algebraic, a criterion requiring that $\mathfrak{L}$ contain with each of its members (α) all matrices of some set associated with (α) cannot be sufficient by itself.

§ 3. Formal groups, hyperalgebras and Lie algebras

A second adaptation of Lie theory to general fields has been the study of "formal Lie groups", wherein the notion of local group is abstracted from the n power series in $2n$ variables which, in a Lie group, give the group multiplication in an n-dimensional coordinate neighborhood of the identity. One begins with n formal power series in $2n$ indeterminates $\{x_i, y_j\}$ over the field $\mathfrak{F}$, without constant term, such that the i-th series is congruent to $x_i \bmod (y_1, \ldots, y_n)$ and congruent to $y_i \bmod (x_1, \ldots, x_n)$, and such that the formal associative law holds. The form of these series yields the existence of n formal series in n variables corresponding to the inverse operation. Homomorphisms of formal groups are systems of formal power series, without constant term, satisfying the homomorphism property when combined in the expected way with the families of power series giving the two "group laws". These ideas, which seem to have been initiated by Bochner [40], have been studied in great detail by DIEUDONNÉ [107—111, 113—120, 122].

The theory of formal Lie groups is not unrelated to the theory of algebraic groups; namely, if G is an algebraic group over $\mathfrak{F}$ then all points of G are simple, and in particular the local ring $\mathfrak{L}$ of G at 1 is a regular local ring, as is the local ring $\mathfrak{M}$ of $G \times G$ at $(1, 1)$. Let $u_1, \ldots, u_n$ be a system of "uniformizing parameters" for $\mathfrak{L}$; then the elements $u_i \otimes 1$, $1 \otimes u_i$ form a system of uniformizing parameters for $\mathfrak{M}$. Now the completion $\mathfrak{M}^*$ of $\mathfrak{M}$ is the ring of formal power series in the $2n$ indeterminates $\{u_i \otimes 1, 1 \otimes u_j\}$. The condition that G be an algebraic group yields that if $f \in \mathfrak{L}$, then $f \circ \pi$ (π the product: $G \times G \to G$) is in $\mathfrak{M}$, hence in $\mathfrak{M}^*$; applying this with $f = u_i$, $1 \leqq i \leqq n$, gives the n formal power series $u_i \circ \pi$ in $2n$ indeterminates, defining a formal Lie group as above. The formal Lie groups resulting from different choices of uniformizing parameters are isomorphic by a substitution isomorphism, i.e., by a reversible change of variables in the power series. The associated formal Lie group is thus seen to be an appropriate device for defining and studying the notion of "local isomorphism" for algebraic groups. For a systematic analysis of connections between formal Lie groups and algebraic groups, especially linear algebraic groups, see [117], where DIEUDONNÉ has also given a meaning to such notions as normal subgroup and supremum resp. infimum for a pair of subgroups of a formal Lie group. He has also developed there the notions of isogeny and linear representability for formal Lie groups, and has used the corresponding ideas for algebraic groups [22, 42, 335] to establish the analogues of a number of results on algebraic groups to be found in these papers.

The notion of isogeny, rather than that of isomorphism, appears to provide a context wherein one can best effect a classification of general types of formal Lie groups. For example, DIEUDONNÉ has classified, to within isogeny, all *commutative* formal Lie groups over an algebraically closed field. (An *isogeny* φ from a formal Lie group G_1 to a formal Lie group G_2 may be defined as an epimorphism with finite kernel, where the notion of epimorphism is given in terms of "cancellability". A kernel of φ is a pair consisting of a formal group H, in the more general sense to be given below, and a morphism $\psi: H \to G_1$ such that whenever η is a morphism $G \to G_1$ such that η composes with φ to give the trivial morphism of G into G_2, then there is a unique morphism $\gamma: G \to H$ making the following diagram commutative:

$$\begin{array}{ccc} G & \xrightarrow{\gamma} & H \\ & {\scriptstyle\eta}\searrow \quad \swarrow{\scriptstyle\psi} & \\ & G_1 & \end{array} .$$

The formal group H is *finite* if the algebra of functions on H, also defined below, is finite-dimensional over the field of reference. Two formal groups G, H are *isogenous* if there is a finite sequence $G=G_0, G_1, \ldots, G_n=H$ of formal groups and isogenies φ_i such that, for each i, either $\varphi_i : G_i \to G_{i+1}$ or $\varphi_i : G_{i+1} \to G_i$.) The reader is here referred to [111 and 116]. For some results on an isomorphism-classification of low-dimensional formal groups, see [119, 280, 294, 296]; for connections with algebraic groups, especially in the commutative case, see also [23, 24, 63, 293, 296].

In the case of characteristic zero, BOCHNER [40] showed how to assign a Lie algebra to a formal Lie group, and proved that all the basic theorems of LIE on the (local) Lie group–Lie algebra correspondence have valid counterparts in this setting. One may define the Lie algebra as a family of homogeneous linear first-order differential operators on the ring of formal power series, namely those which are left-invariant relative to the formal Lie group (BOCHNER actually obtains the multiplication table for the Lie algebra by differentiating the group laws). This concept is equally meaningful in the modular case, so that the Lie algebra of a formal Lie group is defined; it is a restricted Lie algebra and, in the case of a formal Lie group obtained as above by "localizing" an algebraic group, it coincides with the Lie algebra of the algebraic group [108, 117]. The inadequacies for algebraic groups of the Lie algebra and the fact that all elements of the Lie algebra vanish on p-th powers of the variables hint rather strongly that one cannot expect the Lie algebra to reflect faithfully the structure and mapping-behavior of the formal Lie group. Indeed, if one denotes by $G^{(s)}$ the formal Lie group obtained from G by replacing by their p^s-th powers the coefficients in the power series giving the group law in G, then the isogeny of G "onto" $G^{(1)}$ which we may denote by $\{x_i\} \to \{x_i^p\}$ has "differential" vanishing on the entire Lie algebra of G.

In an attempt to surmount difficulties of this sort, DIEUDONNÉ has enlarged the class of differential operators to be considered [108]. Roughly speaking, one enlarges the tangent space at the identity to that spanned over the ground field $\mathfrak{F}$ by the formal analogues of the operators $(p^k!)^{-1} \dfrac{\partial^{p^k}}{\partial x_i^{p^k}}$, $k = 0, 1, 2, \ldots$, and extends these operators by left-invariance to differential operators on the ring of formal power series. Monomials over $\mathfrak{F}$ in these operators, with exponents less than p, generate an (infinite-dimensional) space $\mathfrak{H}$, with an increasing filtration $\mathfrak{H}_0 \subset \mathfrak{H}_1 \subset \mathfrak{H}_2 \subset \cdots$: $\mathfrak{H} = \mathfrak{F} \cdot 1 \oplus \bigcup_{i=0}^{\infty} \mathfrak{H}_i$, where $\mathfrak{H}_r$ consists of the left-invariant *semi-derivations of height* r. That is, if $\mathfrak{O}_r$ denotes the subring of formal power series in $\{x_i^{p^r}\}$, elements of $\mathfrak{H}_r$ leave $\mathfrak{O}_r$ stable and satisfy the derivation rule on products of elements from $\mathfrak{O} = \mathfrak{O}_0$, whenever at least one factor is in $\mathfrak{O}_r$. For each $r > 0$, one has an $\mathfrak{F}$-subspace $\mathfrak{S}_r$, $\mathfrak{H}_{r-1} \subset \mathfrak{S}_r \subset \mathfrak{H}_r$, consisting of all elements of $\mathfrak{H}_r$ vanishing on $\mathfrak{O}_r$, and called the space of (left-invariant) *special* semi-derivations of height r. Some compositions admissible in $\mathfrak{H}$ may be summarized as follows: $\mathfrak{H}_r^p \subseteqq \mathfrak{H}_r$; $\mathfrak{S}_r^p \subseteqq \mathfrak{S}_r$; $[\mathfrak{H}_r\, \mathfrak{H}_r] \subseteqq \mathfrak{H}_r$; $[\mathfrak{H}_r, \mathfrak{S}_r] \subseteqq \mathfrak{S}_r$; $\mathfrak{S}_r\, \mathfrak{S}_r \subseteqq \mathfrak{S}_r$. In particular, $\mathfrak{H}$ is an associative algebra over $\mathfrak{F}$. The algebra $\mathfrak{H}$ admits an *augmentation* mapping ε (projection onto $\mathfrak{F} \cdot 1$ in the decomposition $\mathfrak{H} = \mathfrak{F} \cdot 1 \oplus \bigcup_{i=0}^{\infty} \mathfrak{H}_i$), and a diagonal mapping $\Delta: \mathfrak{H} \to \mathfrak{H} \otimes \mathfrak{H}$, which may be regarded as the differential of the "diagonal homomorphism" $x_i \to (x_i, x_i)$ of formal Lie groups: $G \to G \times G$.

In terms of its product, augmentation, diagonal mapping ("coproduct") and filtration, an abstract definition of hyperalgebra has been given by CARTIER [58]. When the filtration is ignored, such a structure is currently referred to as a "bialgebra". We shall adopt this term at what seems to be a convenient point in the sequel. The analogue of the third fundamental theorem of Lie is not valid: it is not true that each abstract hyperalgebra is the hyperalgebra $\mathfrak{H}$ of a formal Lie group, nor is it true that each sub-hyperalgebra of the hyperalgebra of a formal Lie group is the hyperalgebra of a subgroup. Conditions on associated "structure constants" which assure these properties are given in [115 and 116]. However, it is true (from more general results of Cartier indicated below) that a homomorphism of hyperalgebras of formal Lie groups (i.e., a map preserving all the structures indicated above) is the "differential" of a homomorphism of the groups.

The hyperalgebra reflects properties of the formal Lie group more faithfully than does the Lie algebra. For example, the hyperalgebra of a formal Lie group is commutative if and only if the group is, while the Lie algebra of a non-commutative group can be commutative [108]; differentials of distinct homomorphisms of formal Lie groups are distinct, although they can coincide on the Lie algebras [108]. One case in which some information about a formal Lie group G can be obtained from its Lie algebra is that in which the Lie algebra is *simple* (as restricted Lie algebra): then if one has an isogeny ϱ of G onto a formal Lie group G', ϱ defines an isomorphism between G and $G'^{(-h)}$ for some $h \geqq 0$ [108].

A somewhat more general notion of formal group has been given by CARTIER [63], one in which the correspondence with Lie algebras, and especially with bialgebras, is more nearly perfect. Very imprecisely put, one considers substitutions, for the variables in a ring of formal power series (associated with a formal Lie group G), of nilpotent elements from a commutative algebra $\mathfrak{A}$ over $\mathfrak{F}$, and uses the group law of G to define a product in the set $G_{\mathfrak{A}}$ of such substitutions from $\mathfrak{A}$. More generally, a *formal group* G over $\mathfrak{F}$ is a covariant functor $\mathfrak{A} \to G_{\mathfrak{A}}$ from the category of commutative $\mathfrak{F}$-algebras to the category of groups, subject to certain conditions of "representability" [63]. A *function* f on G is a family of mappings $f_{\mathfrak{A}}$, $f_{\mathfrak{A}}: G_{\mathfrak{A}} \to \mathfrak{A}$, such that the diagram

$$\begin{array}{ccc} G_{\mathfrak{A}} & \xrightarrow{G(\sigma)} & G_{\mathfrak{B}} \\ {\scriptstyle f_{\mathfrak{A}}}\downarrow & & \downarrow{\scriptstyle f_{\mathfrak{B}}} \\ \mathfrak{A} & \xrightarrow{\sigma} & \mathfrak{B} \end{array}$$

is commutative whenever $\sigma \in \operatorname{Hom}(\mathfrak{A}, \mathfrak{B})$ (in the category of commutative $\mathfrak{F}$-algebras). An element $u \in G_{\mathfrak{A}}$ defines a mapping $\chi_u: f \to f_{\mathfrak{A}}(u)$ into $\mathfrak{A}$ of the totality $\mathfrak{O}(G)$ of functions on G. The representability conditions guarantee that $\mathfrak{O}(G)$ may be regarded as a set, indeed a commutative $\mathfrak{F}$-algebra, which one topologizes as weakly as possible subject to the conditions that each $\chi_u^{-1}(a)$ $(u \in G_{\mathfrak{A}},\ a \in \mathfrak{A})$ be open. It turns out that $\mathfrak{O}(G)$ is a (Hausdorff) topological ring which is complete.

Now let $\mathfrak{U}(G)$ be the subspace of the ($\mathfrak{F}$-) dual space of $\mathfrak{O}(G)$ consisting of those linear forms with open kernels. Regarding $\mathfrak{O}(G)$ as a space of "test functions" on G, one calls $\mathfrak{U}(G)$ the space of *distributions* on G. As in the analytic theory of distributions, a product (convolution) is defined in $\mathfrak{U}(G)$, and the ordinary product in $\mathfrak{O}(G)$ has as dual mapping a "coproduct" $\Delta: \mathfrak{U}(G) \to \mathfrak{U}(G) \otimes \mathfrak{U}(G)$. The mapping $T \to \langle 1, T\rangle$, where 1 is the unit element of $\mathfrak{O}(G)$, defines an augmentation ε in $\mathfrak{U}(G)$ and completes the introduction in $\mathfrak{U}(G)$ of the structure of a bialgebra, the bialgebra ("hyperalgebra"), *of* G. CARTIER distinguishes two important classes of formal groups: the *separable* formal groups (roughly, those G which are determined by $G_{\mathfrak{K}}$, $\mathfrak{K}$ the algebraic closure of $\mathfrak{F}$), and the *infinitesimal* formal groups (those for which $G_{\mathfrak{K}}$ consists of the identity). Over a perfect field $\mathfrak{F}$, one has a semi-direct factorization of a formal group into unique separable and infinitesimal subgroups, the infinitesimal group being an invariant subgroup [63].

If G is a formal group over $\mathfrak{F}$, the *primitive elements* of $\mathfrak{U}(G)$ are those x for which $\Delta(x) = x \otimes 1 + 1 \otimes x$, where 1 is the unit element of $\mathfrak{U}(G)$ in its convolution product: $\langle f, 1\rangle = f_{\mathfrak{F}}(e_{\mathfrak{F}}) = \chi_e(f)$, e the "identity" of G. The primitive elements form a Lie subalgebra of the associative algebra $\mathfrak{U}(G)$, a restricted subalgebra in the modular case ([58, 305]); this Lie algebra *is the Lie algebra* $\mathfrak{L}(G)$ *of the formal group* G.

Given a bialgebra $\mathfrak{U}$ over $\mathfrak{F}$, one can construct a formal group G from $\mathfrak{U}$ as follows: For each commutative $\mathfrak{F}$-algebra $\mathfrak{A}$, consider the maps Δ', ε_1', ε_2' of $\mathfrak{U} \otimes \mathfrak{A}$ into $\mathfrak{U} \otimes \mathfrak{U} \otimes \mathfrak{A}$, defined respectively as $\Delta' = \Delta \otimes id$, $\varepsilon_i' = \varepsilon_i \otimes id$, $\varepsilon_1(u) = u \otimes 1$, $\varepsilon_2(u) = 1 \otimes u$. Let $\mathfrak{U}^+$ be the augmentation ideal (kernel of ε) in $\mathfrak{U}$; let $G_{\mathfrak{A}}$ be the subset of

$\mathfrak{U} \otimes \mathfrak{A}$ consisting of all Z such that $Z \equiv 1 \,(\mathrm{mod}\, \mathfrak{U}^+ \otimes \mathfrak{A})$, $\Delta'(Z) = \varepsilon_1'(Z)\, \varepsilon_2'(Z)$. The resulting formal group G has $\mathfrak{U}$ as its bialgebra, and is isomorphic with H if $\mathfrak{U} = \mathfrak{U}(H)$. One may express this by saying that *the functor* $G \to \mathfrak{U}(G)$ *is an equivalence of the category of formal groups over* $\mathfrak{F}$ *with that of bialgebras over* $\mathfrak{F}$ [63].

Now let G be an *infinitesimal* formal group over $\mathfrak{F}$; in this case, $\mathfrak{O}(G)$ is a local ring with maximal ideal the kernel $\mathfrak{m}$ of χ_e as above. If $\mathfrak{m}$ is finitely generated, G is said to be of *finite type*, and the powers of $\mathfrak{m}$ form a fundamental system of neighborhoods of zero in $\mathfrak{O}(G)$. When $\mathfrak{F}$ is of characteristic zero, $\mathfrak{O}(G)$ is a ring of formal power series in a minimal set of generators for $\mathfrak{m}$; when $\mathfrak{F}$ is perfect and of prime characteristic p, $\mathfrak{O}(G)$ is obtained from a power-series ring by specializing to zero the p^{n_i}-th power of the i-th variable for some positive integers n_i and some set (possibly empty) of indices i [63]. The infinitesimal formal groups of finite type include the formal Lie groups of DIEUDONNÉ, in the sense that the functor assigning to each G of the latter category the groups $G_{\mathfrak{A}}$ obtained by substitution of elements of $\mathfrak{A}$ for the variables as indicated above sends formal Lie groups to infinitesimal formal groups of finite type, and the associated formal groups are isomorphic if and only if the original formal Lie groups are isomorphic.

If one begins with a Lie algebra $\mathfrak{L}$ over $\mathfrak{F}$, the universal associative algebra $\mathfrak{U}(\mathfrak{L})$ carries the structure of bialgebra in a natural way, as does the "u-algebra", written here $\mathfrak{U}_p(\mathfrak{L})$, of a restricted Lie algebra over a field of prime characteristic p [58, 305]. Together with the constructions above, this yields a functor from the category of $\mathfrak{F}$-Lie algebras to that of formal groups over $\mathfrak{F}$, and one from the category of restricted Lie algebras to that of formal groups. Since a formal group is not usually determined by its Lie algebra, these functors cannot be expected to be equivalences with the entire category of formal groups. In particular, $\mathfrak{L}(G)$ is always *restricted* in prime characteristic, so that one cannot expect to recover any but the restricted Lie algebras from the formal groups. In the case of characteristic zero, $\mathfrak{L}$ is indeed the set of primitive elements of $\mathfrak{U}(\mathfrak{L})$; in the restricted case, $\mathfrak{L}$ is the set of primitive elements of $\mathfrak{U}_p(\mathfrak{L})$ [58, 305]. The equivalence thus established between the category of Lie algebras (of restricted Lie algebras, in the modular case) and its "image" under the functor $\mathfrak{L} \to \mathfrak{U}(\mathfrak{L})$ (or $\mathfrak{L} \to \mathfrak{U}_p(\mathfrak{L})$) yields, by composition with the equivalence $\mathfrak{U} \to G$ of bialgebras and formal groups described above, an equivalence between the category of Lie algebras (resp. restricted Lie algebras) and a subcategory of the category of formal groups over $\mathfrak{F}$. It remains to distinguish these subcategories more explicitly. In the case of characteristic zero, the corresponding formal groups are simply the infinitesimal groups [63]; that those of finite type correspond to the finite-dimensional

Lie algebras is essentially a reformulation of the results of BOCHNER cited above.

Now consider the case of a modular field $\mathfrak{F}$. Each commutative $\mathfrak{F}$-algebra $\mathfrak{A}$ carries another structure of commutative $\mathfrak{F}$-algebra $\mathfrak{A}_p$ in which the ring-operations are unchanged, but with the structure of $\mathfrak{A}$ as $\mathfrak{F}$-module replaced by a new scalar multiplication $\lambda \cdot a$ defined as $\lambda^p a$ $(\lambda \in \mathfrak{F}, a \in \mathfrak{A})$ in the original module-structure. Then the mapping $a \to a^p$ is in $\mathrm{Hom}(\mathfrak{A}, \mathfrak{A}_p)$, and yields for each formal group G over $\mathfrak{F}$ a "Frobenius homomorphism" $\varphi_{\mathfrak{A}} = G(a \to a^p)$ in $\mathrm{Hom}(G_{\mathfrak{A}}, G_{\mathfrak{A}_p})$. When G is obtained from $\mathfrak{U}_p(\mathfrak{L})$ as above, one has $\varphi_{\mathfrak{A}}(x) = e_{\mathfrak{A}_p}$ for all $\mathfrak{A}$ and all $x \in G_{\mathfrak{A}}$, i.e., φ is trivial. Conversely if G is a formal group over $\mathfrak{F}$ for which the Frobenius homomorphism is trivial, the algebra $\mathfrak{U}(G)$ is the u-algebra of $\mathfrak{L}(G)$ [63]. By this means, one obtains an equivalence between the category of restricted Lie algebras over $\mathfrak{F}$ and that of formal groups with trivial Frobenius homomorphism. In particular, finite-dimensional restricted Lie algebras correspond to finite infinitesimal groups. These results indicate a class of formal groups which may be of special interest, namely those corresponding to the simple restricted Lie algebras of Chapter V, § 4. It appears likely that the constructions of [265] are relevant in this connection.

This brief survey has neglected to take into account direct approaches to formal groups, not utilizing bialgebras or Lie algebras. Notable progress along such lines has been made, especially in the commutative case. The reader is referred to the works of DIEUDONNÉ [111, 116], of LAZARD [279, 280], and of MANIN [296]. A generalized and somewhat more detailed exposition of the results of [63] has been given by GABRIEL [142a].

§ 4. Lie derivation algebras and purely inseparable extensions

Let $\mathfrak{K}$ be a field extension of $\mathfrak{F}$, of prime characteristic p, of degree p^n, with $\mathfrak{K} = \mathfrak{F}(\xi_1, \ldots, \xi_n)$ where $\xi_i^p \in \mathfrak{F}$. That is, $\mathfrak{K}/\mathfrak{F}$ is a finite purely inseparable extension *of exponent one* ($\mathfrak{K}^p \subseteqq \mathfrak{F}$), and $\{\xi_i\}$ is a *p-basis* for $\mathfrak{K}/\mathfrak{F}$ (no ξ_i is in $\mathfrak{F}(\{\xi_j \mid j \neq i\})$). Let $\mathfrak{D} = \mathfrak{D}(\mathfrak{K}/\mathfrak{F})$ denote the set of derivations of $\mathfrak{K}$ as $\mathfrak{F}$-algebra. Then $\mathfrak{D}$ is a restricted Lie algebra over $\mathfrak{F}$, and is in addition a right $\mathfrak{K}$-module under the action $D \to D\lambda$ defined by $\xi(D\lambda) = (\xi D)\lambda$ $(\xi, \lambda \in \mathfrak{K},\ D \in \mathfrak{D})$. $\mathfrak{D}$ is an $\mathfrak{F}$-form of the split Jacobson–Witt algebra $\mathfrak{W}_n$ over $\mathfrak{K}$, has dimension n over $\mathfrak{K}$ and has dimension $n p^n$ over $\mathfrak{F}$. For $\lambda, \mu \in \mathfrak{K}$, $D, E \in \mathfrak{D}$, one has $[D\lambda, E\mu] = [DE]\lambda\mu + D((\lambda E)\mu) - E((\mu D)\lambda)$, $(D\lambda)^p = D^p\lambda^p + D(\lambda(D\lambda)^{p-1})$. It follows that the right $\mathfrak{K}$-module generated by $D, D^p, D^{p^2}, \ldots \in \mathfrak{D}$ is a restricted Lie subalgebra of $\mathfrak{D}$. If $\xi \in \mathfrak{K}$, $\xi \notin \mathfrak{F}$, one may take ξ as a member of a p-basis for $\mathfrak{K}$ over $\mathfrak{F}$ and obtain $D \in \mathfrak{D}(\mathfrak{K}/\mathfrak{F})$ with $\xi D \neq 0$.

Thus $\mathfrak{F}$ is the set of $\mathfrak{D}$-*constants* of $\mathfrak{K}$, i.e., $\mathfrak{F} = \{\xi \in \mathfrak{K} \mid \xi D = 0\}$. (For these and most of the other results of this section, cf. [223, Chap. IV].) BAER (cf. [223, p. 185]) has shown that there exists $D \in \mathfrak{D}$ such that the set of D-constants of $\mathfrak{K}$ is $\mathfrak{F}$. Thus $\mathfrak{F}$ is the field of constants for the $\mathfrak{K}$-module generated by $D, D^p, D^{p^2}, \ldots$ as above, and this is a restricted Lie subalgebra $\mathfrak{L}$ (over $\mathfrak{F}$) of $\mathfrak{D}(\mathfrak{K}/\mathfrak{F})$.

The ring $\mathfrak{A}$ of (additive) endomorphisms of $\mathfrak{K}$ generated by D and right multiplications by elements of $\mathfrak{K}$ (i.e., by $\mathfrak{L}$) evidently acts irreducibly in $\mathfrak{K}$, and has as its centralizer the left multiplications by elements of $\mathfrak{F}$; thus $\mathfrak{A}$ is the full ring of $\mathfrak{F}$-linear transformations of $\mathfrak{K}$, has dimension $[\mathfrak{K}:\mathfrak{F}]^2 = p^{2n}$ over $\mathfrak{F}$, and is of dimension $[\mathfrak{K}:\mathfrak{F}] = p^n$ over $\mathfrak{K}$. Corresponding conclusions hold with $\mathfrak{L}$ replaced by $\mathfrak{D}(\mathfrak{K}/\mathfrak{F})$. Therefore $\mathfrak{A}$ is the enveloping $\mathfrak{K}$-algebra of $\mathfrak{D}(\mathfrak{K}/\mathfrak{F})$; but $[\mathfrak{A}:\mathfrak{K}]$ is the degree of the minimum polynomial over $\mathfrak{K}$ of D, and this degree is at most p^n by the fact that all $D^{p^i} \in \mathfrak{D}(\mathfrak{K}/\mathfrak{F})$, which has dimension n over K. From $[\mathfrak{A}:\mathfrak{K}] = p^n$, it follows that $D, D^p, \ldots, D^{p^{n-1}}$ generate $\mathfrak{D}(\mathfrak{K}/\mathfrak{F})$ as $\mathfrak{K}$-module, or that $\mathfrak{L} = \mathfrak{D}(\mathfrak{K}/\mathfrak{F})$. That is, *$\mathfrak{D}(\mathfrak{K}/\mathfrak{F})$ is the $\mathfrak{K}$-module generated by $D, D^p, \ldots$, where $D \in \mathfrak{D}(\mathfrak{K}/\mathfrak{F})$ has $\mathfrak{F}$ as its constants.*

This last fact has been used by GERSTENHABER [146] as the basis for an extension of previous results of JACOBSON [211, 219]. Gerstenhaber's theorem reads as follows: *Let $\mathfrak{K}$ be a field of characteristic p, and let $\mathfrak{L}$ be a right $\mathfrak{K}$-submodule of the derivations $\mathfrak{D}(\mathfrak{K})$ of $\mathfrak{K}$, closed under p-th powers; let $\mathfrak{F}$ be the field of constants of $\mathfrak{L}$. Then $[\mathfrak{L}:\mathfrak{K}]$ is finite if and only if $[\mathfrak{K}:\mathfrak{F}]$ is; when this is the case, $\mathfrak{L}$ is a Lie ring, and $\mathfrak{L} = \mathfrak{D}(\mathfrak{K}/\mathfrak{F})$. The correspondence $\mathfrak{L} \leftrightarrow \mathfrak{F}$ is a duality between finitely generated $\mathfrak{K}$-submodules, closed under p-th powers, of $\mathfrak{D}(\mathfrak{K})$ and subfields $\mathfrak{F}$ of $\mathfrak{K}$ such that $[\mathfrak{K}:\mathfrak{F}] < \infty$ and $\mathfrak{K}^p \subseteq \mathfrak{F}$.*

By the fact that all elements of $\mathfrak{K}^p$ are constants for all $D \in \mathfrak{D}(\mathfrak{K})$, it is clear that $\mathfrak{K}^p \subseteq \mathfrak{F}$, where $\mathfrak{F}$ is the field of $\mathfrak{L}$-constants of $\mathfrak{K}$, and that $[\mathfrak{K}:\mathfrak{F}] < \infty$ implies $[\mathfrak{L}:\mathfrak{K}] \leqq [\mathfrak{D}(\mathfrak{K}/\mathfrak{F}):\mathfrak{K}] < \infty$. Gerstenhaber's generalization of the generation of $\mathfrak{D}(\mathfrak{K}/\mathfrak{F})$ by a single D as above lies in showing that if D', D'' are in $\mathfrak{D}(\mathfrak{K})$; if $\xi_1, \ldots, \xi_m \in \mathfrak{K}$ are p-independent over the constants of D'; if η is a D'-constant, but not a D''-constant, then there is $D \in \mathfrak{L}(D', D'')$ (the smallest $\mathfrak{K}$-module of derivations of $\mathfrak{K}$ closed under p-th powers and containing D' and D'') such that $\xi_1, \ldots, \xi_m, \eta$ are p-independent over the constants of D. From this it follows that $[\mathfrak{L}:\mathfrak{K}] < \infty$ implies $[\mathfrak{K}:\mathfrak{F}] < \infty$. In this case, $\mathfrak{L} = \mathfrak{D}(\mathfrak{K}/\mathfrak{F})$ by earlier remarks and the fact that $\mathfrak{L}$ must contain D having exactly $\mathfrak{F}$ as constants (using the result above on $\mathfrak{L}(D', D'')$). The remainder of the theorem now follows. The original formulation of JACOBSON set up a duality of intermediate fields between $\mathfrak{K}$ and $\mathfrak{F}$ ($[\mathfrak{K}:\mathfrak{F}] < \infty$, $\mathfrak{K}^p \subseteq \mathfrak{F}$) and restricted Lie $\mathfrak{F}$-algebras $\mathfrak{L}$ in $\mathfrak{D}(\mathfrak{K}/\mathfrak{F})$ which are $\mathfrak{K}$-submodules; the above shows that $\mathfrak{K}$-submodules closed under p-th powers are $\mathfrak{F}$-Lie subalgebras.

By analogy with the infinite Galois theory of KRULL (cf. [223, Chap. IV]), one may extend the above theory to include infinite-dimensional cases. One must restrict attention to the class of $\mathfrak{K}$-submodules of $\mathfrak{D}(\mathfrak{K})$ closed under p-th powers and *closed in the finite topology* on the set of mappings $\mathfrak{K}^{\mathfrak{K}}$ (the topological product of the discrete topologies on the $|\mathfrak{K}|$ copies of $\mathfrak{K}$ involved in forming $\mathfrak{K}^{\mathfrak{K}}$). One then has the same kind of duality between the set of these *closed* restricted $\mathfrak{K}$-submodules of $\mathfrak{D}(\mathfrak{K})$, and *all* subfields $\mathfrak{F}$ of $\mathfrak{K}$ with $\mathfrak{K}^p \subseteqq \mathfrak{F}$ [147, 322].

Attempts to generalize the Galois duality above to purely inseparable extensions of arbitrary exponent lead to consideration of *higher derivations*, which are sequences of mappings serving as substitutes for the operators $k!^{-1} D^k$, $k = 0, 1, 2, \ldots$, where D is a derivation (cf. [169, 223, Chap. IV], and especially [410], where WEISFELD gives a characterization of purely inseparable extensions $\mathfrak{K}/\mathfrak{F}$ such that $\mathfrak{F}$ is the field of constants of a single higher derivation of $\mathfrak{K}$). These are rather clumsy in their algebraic structure, and it seems more convenient to analyze such extensions as sequences of extensions of exponent one or else by means of the general linear Galois theory (cf. [223, Chap.I]). The *semi-derivations* of DIEUDONNÉ discussed in § 3 seem first to have been introduced by him to study purely inseparable extensions of arbitrary exponent [101]. WEISFELD [409] has established a similar Galois duality for division rings, analogous to the Galois theory in terms of automorphisms as presented in Chapter VII of [226].

The existence of a derivation D of $\mathfrak{K}$ (as above) with $\mathfrak{F}$ as constant field suggests an analogy with cyclic Galois extensions. This analogy is quite complete in the theory of normal simple associative algebras $\mathfrak{A}$ over $\mathfrak{F}$ having $\mathfrak{K}$ as a maximal subfield, or of the Brauer group $B_{\mathfrak{F}}^{\mathfrak{K}}$ of similarity classes of normal simple algebras over $\mathfrak{F}$ split by $\mathfrak{K}$. Each such class has a representative which contains $\mathfrak{K}$ as a maximal subfield, and the normal simple algebras over $\mathfrak{F}$ with $\mathfrak{K}$ as maximal subfield have been described by JACOBSON [212] as follows: Let $D^{p^n} + D^{p^{n-1}}\tau_1 + \cdots + D\tau_n = 0$ be the first $\mathfrak{F}$-linear relation among $D, D^p, \ldots$ The derivation D is the restriction to $\mathfrak{K}$ of an inner derivation of $\mathfrak{A}$ [211], say ad$d$, $d \in \mathfrak{A}$. It follows that $d^{p^n} + d^{p^{n-1}}\tau_1 + \cdots + d\tau_n = \gamma \in \mathfrak{K}$, and, from the fact that $\gamma\,\mathrm{ad}d = 0 = \gamma D$, that $\gamma \in \mathfrak{F}$. Then $\mathfrak{A}$ consists of all right $\mathfrak{K}$-combinations of $1, d, d^2, \ldots, d^{p^n-1}$, with $\lambda d = [\lambda d] + d\lambda = d\lambda + \lambda D (\lambda \in \mathfrak{K})$, and with d^{p^n} expressed as a combination of lower powers of d as indicated above. This construction may be interpreted as taking the factor-ring by a certain ideal of the ring of *differential polynomials* in one variable over $\mathfrak{K}$ with respect to the derivation D [314, 14]. For each choice of $\gamma \in \mathfrak{F}$, the construction gives a normal simple algebra $(\mathfrak{K}, D, \gamma)$ over $\mathfrak{F}$ split by $\mathfrak{K}$ and having $\mathfrak{K}$ as a maximal subfield; thus, by remarks above, the mapping $\gamma \to (\mathfrak{K}, D, \gamma)$ yields a mapping of $\mathfrak{F}$ *onto* the Brauer group $B_{\mathfrak{F}}^{\mathfrak{K}}$. Now this mapping is in fact a *homomorphism* of the additive group of $\mathfrak{F}$ onto $B_{\mathfrak{F}}^{\mathfrak{K}}$ [202], and its kernel is the set of those $\gamma \in \mathfrak{F}$ of the form $V(\lambda)$, $\lambda \in \mathfrak{K}$, where V is the additive mapping

$$\lambda \to V_{p^n}(\lambda) + V_{p^{n-1}}(\lambda)\, \tau_1 + \cdots + V_1(\lambda)\, \tau_n,$$

with the τ_i as above, and

$$V_{p^i}(\lambda) = \lambda^{p^i} + (\lambda D^{p-1})^{p^{i-1}} + \cdots + (\lambda D^{p^i-1}).$$

Thus an isomorphism $B_{\mathfrak{F}}^{\mathfrak{K}} \cong \mathfrak{F}^+/V(\mathfrak{K}^+)$ results, wherein the right-hand side may be regarded as an additive analogue of $\mathfrak{F}^*/N(\mathfrak{K}^*)$, where $\mathfrak{K}$ is a cyclic galois extension of $\mathfrak{F}$ (cf. [100, Chap. V; 5, Chap. VII; 226, Chap. VI]) and N is the norm mapping. Thus V can be thought of as an analogue of the norm mapping N. Whereas the mapping N is independent of the choice of a generator for the Galois group, the mapping V seems to depend on the choice of generating derivation D (cf. [202]).

For an "invariant" description of $B_{\mathfrak{F}}^{\mathfrak{K}}$, it is more useful to appeal for analogy to the general case where $\mathfrak{K}/\mathfrak{F}$ is (finite) galois, with group $\mathfrak{G}$; then $B_{\mathfrak{F}}^{\mathfrak{K}}$ is isomorphic with the group of extensions of $\mathfrak{K}^*$ by $\mathfrak{G}$, or with the second (multiplicative) cohomology group $H^2(\mathfrak{G}, \mathfrak{K})$ ([226], Chap. VI). (When $\mathfrak{G}$ is cyclic, the isomorphism with $\mathfrak{F}^*/N(\mathfrak{K}^*)$ follows from this by the periodicity of the cohomology of $\mathfrak{G}$ and the identification of $\hat{H}_0(\mathfrak{G}, \mathfrak{K})$ with $\mathfrak{F}^*/N(\mathfrak{K}^*)$—cf. [56, Chap. XII].) An approach along these lines has been made by HOCHSCHILD [195], who established, for $\mathfrak{K}/\mathfrak{F}$ finite and purely inseparable of exponent one, an isomorphism between $B_{\mathfrak{F}}^{\mathfrak{K}}$ and the group of equivalence classes of "*regular*" extensions (as restricted Lie algebras over $\mathfrak{F}$) of $\mathfrak{K}$ by $\mathfrak{D} = \mathfrak{D}(\mathfrak{K}/\mathfrak{F})$ where $\mathfrak{K}$ is regarded as an abelian restricted Lie algebra over $\mathfrak{F}$ with its natural p-th power. A *regular extension* of $\mathfrak{K}$ by $\mathfrak{D}$ is a restricted Lie algebra $\mathfrak{S}$ over $\mathfrak{F}$, containing $\mathfrak{K}$, together with a (restricted) homomorphism φ of $\mathfrak{S}$ onto $\mathfrak{D}$ having kernel $\mathfrak{K}$, such that $\mathfrak{S}$ can be given the structure of a (right) $\mathfrak{K}$-module satisfying

$$[x\,\lambda, y\,\mu] = [x\,y]\,\lambda\,\mu + x([\lambda\,y]\,\mu) - y([\mu\,x]\,\lambda),$$

$(x\,\lambda)^p = x^p\,\lambda^p + x(\lambda(\mathrm{ad}\,x\,\lambda)^{p-1})$ for all $x, y \in \mathfrak{S}$, $\lambda, \mu \in \mathfrak{K}$ (compare the identities for $[D\,\lambda, E\,\mu]$ and $(D\,\lambda)^p$ at the beginning of this section, where $D, E \in \mathfrak{D}$). It should be noted that the above does *not* establish an isomorphism between $B_{\mathfrak{F}}^{\mathfrak{K}}$ and a subgroup of $H_*^2(\mathfrak{D}, \mathfrak{K})$, as defined in Chapter V, § 3, since $H_*^2(\mathfrak{D}, \mathfrak{K})$ is identified with extensions of $\mathfrak{K}$ by $\mathfrak{D}$, where $\mathfrak{K}$ is regarded as an abelian Lie algebra with *trivial* p-th power. (A unifying cohomology theory for fields has been given by AMITSUR [17]; for relations with the above results in the case of purely inseparable extensions of exponent one, cf. [334].) In another paper [196], HOCHSCHILD has refined this theory to study the structure of normal simple algebras having maximal commutative subrings which are purely inseparable extensions (of *arbitrary* exponent) of the center. His methods yield a proof that if $\mathfrak{K}$ is a finite purely inseparable extension of $\mathfrak{F}$, the operation of extending the ground field to $\mathfrak{K}$ induces a homomorphism of the Brauer group $B_{\mathfrak{F}}$ *onto* $B_{\mathfrak{K}}$ (cf. also [202]). His work involves an analysis of "regular extensions" of $\mathfrak{A}$ by $\mathfrak{D}$, where $\mathfrak{A}$ is a normal simple algebra over $\mathfrak{K}$, a finite purely inseparable extension of $\mathfrak{F}$ of exponent one, and where $\mathfrak{D} = \mathfrak{D}(\mathfrak{K}/\mathfrak{F})$ as before.

The Galois correspondence for purely inseparable extensions of exponent one has been used by several authors [23, 61, 291, 364] in studying isogenies of algebraic groups, especially of abelian varieties. A *purely inseparable isogeny of height one* is a homomorphism φ (in the sense of algebraic groups over $\mathfrak{E}$) of the irreducible algebraic group G (defined over the algebraically closed field $\mathfrak{E}$) onto the algebraic group G', such that if $\mathfrak{F}$ is the field of G', regarded as embedded by means of φ in the field $\mathfrak{K}$ of G, $\mathfrak{K}/\mathfrak{F}$ is (finite and) purely inseparable of exponent one. The field $\mathfrak{F}$ essentially determines G' and the isogeny φ [364]. Those fields $\mathfrak{F}$ between $\mathfrak{K}$ and $\mathfrak{K}^p$ ($\mathfrak{K}/\mathfrak{K}^p$ is finite since $\mathfrak{K}$ is finitely generated over $\mathfrak{E}$) associated with isogenies correspond to certain restricted $\mathfrak{K}$-submodules (actually $\mathfrak{F}$-Lie subalgebras) of $\mathfrak{D}(\mathfrak{K}/\mathfrak{E}) = \mathfrak{D}(\mathfrak{K}/\mathfrak{K}^p)$, and these correspond to restricted subalgebras of the Lie algebra $\mathfrak{L}(G)$ which are stable under the adjoint representation of G on $\mathfrak{L}(G)$. This correspondence of purely inseparable isogenies of height one with certain ideals in $\mathfrak{L}(G)$ has been used especially effectively in [364] when G is *commutative*. For other applications and related questions, see [60, 62].

§ 5. Infinite-dimensional analogues of the classical Lie algebras

Let $\mathfrak{A}$ be a simple associative ring, not necessarily having a unit element. As $\mathfrak{A}$ is a generalization of a normal simple finite-dimensional associative algebra, so the derived Lie ring of $\mathfrak{A}$ and the quotient of the derived ring by its center are generalizations of the normal simple Lie algebras of type A_{I} of Chapter IV, § 3. Likewise, if $\mathfrak{A}$ has an involution η, one obtains generalizations of the other simple Lie algebras of Chapter IV, § 3 by considering the Lie ring $\mathfrak{S}$ of skew elements, its derived Lie ring, and the quotient of the latter by its center. Both these cases can be subsumed under the second if we assume that $(\mathfrak{A}, \eta)$ is a *simple involutorial ring* in the sense analogous to that of Chapter IV, § 3. If $\mathfrak{A}$ is not simple, one has $\mathfrak{A} = \mathfrak{B} \oplus \mathfrak{B}^\eta$, where $\mathfrak{B}$ is a simple ideal, and the Lie ring $\mathfrak{S}$ of skew elements is isomorphic with the Lie ring $\mathfrak{B}$ as in Chapter IV, § 3. The structure of $\mathfrak{S}$ and the structure of $\mathfrak{A}$ as

Lie $\mathfrak{S}$-module have been studied by HERSTEIN [175—180] and BAXTER [25—28]. In the case of characteristic 2, results on the structure of $\mathfrak{S}(\cong \mathfrak{B})$ have been obtained when $\mathfrak{A} = \mathfrak{B} \oplus \mathfrak{B}^\eta$; in particular, the quotient of $[\mathfrak{B}\,\mathfrak{B}]$ by its center is a simple Lie ring except when the center $\mathfrak{Z}$ of $\mathfrak{B}$ is a field and $[\mathfrak{B}:\mathfrak{Z}] = 4$ [177, 25]. Other results of this kind have involved the assumption of a *characteristic different from* 2, and will be summarized below under this hypothesis.

The general form of Herstein's results on Lie ideals in $\mathfrak{S}$ is that every Lie ideal $\mathfrak{U}$ of $\mathfrak{S}$ is either contained in the center of $\mathfrak{A}$ or contains $[\mathfrak{S}\,\mathfrak{S}]$, and that $[\mathfrak{S}\,\mathfrak{S}]$ is not abelian, hence is equal to its derived ring. These assertions are valid when one of the following holds: (1) the center $\mathfrak{Z}$ of $\mathfrak{A}$ is zero; (2) $\mathfrak{A} = \mathfrak{B} \oplus \mathfrak{B}^\eta$, as above; (3) $\mathfrak{A}$ is a simple ring, $\mathfrak{Z}$ a field fixed under η, $[\mathfrak{A}:\mathfrak{Z}] > 16$; (4) $\mathfrak{A}$ is a simple ring, $\mathfrak{Z}$ is not fixed under η, and $[\mathfrak{A}:\mathfrak{Z}] > 4$. BAXTER showed that $[\mathfrak{S}\,\mathfrak{S}]/(\mathfrak{Z} \cap [\mathfrak{S}\,\mathfrak{S}])$ is a simple Lie ring under any of these conditions (in cases (1) and (3), this means that $[\mathfrak{S}\,\mathfrak{S}]$ is simple). These authors and others have also considered the (associative) subrings of $\mathfrak{A}$ generated by $\mathfrak{S}$ and by $[\mathfrak{S}\,\mathfrak{S}]$ [178, 179, 26, 15, 351, 253]. Some of these papers also contain results on the Jordan ring $\mathfrak{H}$ of symmetric elements of $\mathfrak{A}$ and its actions on $\mathfrak{A}$.

In the case corresponding to $\mathfrak{A} = \mathfrak{B} \oplus \mathfrak{B}^\eta$, most of these questions amount to questions about the Lie structure of a simple ring $\mathfrak{B}$. Even if $\mathfrak{B}$ does not have a unit element, one may define *inner automorphisms* of $\mathfrak{B}$ as mappings $b \to b - a'\,b - b\,a + a'\,b\,a$, where $a + a' - a\,a' = 0 = a + a' - a'\,a$, and one may study additive subgroups, subrings, etc., which are stable under all inner automorphisms. AMITSUR [15] has shown that if $\mathfrak{B}$ contains an idempotent $e \neq 0, 1$, then such subgroups are exactly the Lie ideals of $\mathfrak{B}$, so that the results of HERSTEIN and BAXTER may be applied; less general results on subgroups and subrings stable under the full group of automorphisms were obtained by HATTORI [170], KASCH [248], and BAXTER [25]. A slight extension of the method of AMITSUR may be used to show that if $\mathfrak{B}$ contains two non-zero orthogonal idempotents, whose sum is not 1, and if the center $\mathfrak{Z}$ of $\mathfrak{B}$ is sufficiently large, then the $\mathfrak{Z}$-submodules of $\mathfrak{B}$ which are stable under the commutator subgroup of the inner automorphisms are those which are Lie ($\mathfrak{Z}$-) submodules under the adjoint action of $[\mathfrak{B}\,\mathfrak{B}]$. A tempting conjecture, to which the author knows no counter-example, is that the commutator subgroup above is simple under such conditions. Since this group is a projective special linear group when $\mathfrak{B}$ is a simple Artinian ring, the conjecture is correct for these rings. It is not evident how the structure of $\mathfrak{B}$ as $[\mathfrak{B}\,\mathfrak{B}]$-module can be used to advantage in studying the internal structure of the group of inner automorphisms.

As in Chapter IV, one may ask whether all automorphisms (or, more generally, homomorphisms) of the Lie structure discussed above arise from mappings closely connected with their original involutorial rings; e.g., are all automorphisms of $[\mathfrak{S}\,\mathfrak{S}]/\mathfrak{Z} \cap [\mathfrak{S}\,\mathfrak{S}]$ induced by automorphisms of $(\mathfrak{A}, \eta)$? Results in several special cases have been obtained by HUA [204], MARTINDALE [297], and KLOTZ [253]. Hua determined the Lie automorphisms of a simple Artinian ring $\mathfrak{B}$ whose unit element is the sum of at least three non-trivial orthogonal idempotents. These automorphisms arise from automorphisms and anti-automorphisms of $\mathfrak{B}$, to which one may add homomorphisms, vanishing on $[\mathfrak{B}\,\mathfrak{B}]$, of the additive group of $\mathfrak{B}$ into that of the center of $\mathfrak{B}$. Martindale extended these results to *primitive* rings with unit, subject to the same assumption on idempotents and the exclusion of characteristics 2 and 3. His arguments deal with the Lie isomorphisms of two such rings, and yield that the rings must be isomorphic or anti-isomorphic. KLOTZ considered Lie $\mathfrak{F}$-isomorphisms of $[\mathfrak{S}_1\,\mathfrak{S}_1]$ onto $[\mathfrak{S}_2\,\mathfrak{S}_2]$, where $(\mathfrak{A}_i, \eta_i)$ is a normal simple involutorial $\mathfrak{F}$-algebra and where $\mathfrak{S}_i$ is the set of η_i-skew elements of $\mathfrak{A}_i$, under the following restrictions: (1) The characteristic of $\mathfrak{F}$ is not 2; (2) $(\mathfrak{A}_1, \eta_1)$ contains a finite-dimensional normal simple involutorial subalgebra $(\mathfrak{A}_0, \eta_1)$ of the type A_r, B_r, C_r or D_r (with $r > 4$ for D_r, $r + 1$ prime to the characteristic for A_r), the unit element of $\mathfrak{A}_1$ being that of $\mathfrak{A}_0$; (3) $[\mathfrak{S}_1\,\mathfrak{S}_1]$ is a simple Lie ring. He proved that each such Lie isomorphism is the restriction of an isomorphism of $(\mathfrak{A}_1, \eta_1)$ onto $(\mathfrak{A}_2, \eta_2)$. The effect of the assumption (2) is to introduce into $\mathfrak{A}_1$ not only an adequate number of idempotents, but also some associated matrix units on which the behavior of η_1 can be rather well described.

For related work, see [151, 171, 182, 298, 404, 427].

Bibliography

The following abbreviations of names of journals are those which, in the judgment of the author, might require explanation for some readers:

A.M.S. Transl.	Translations of the American Mathematical Society
Alg. i Log. Sem.	Akademija Nauk SSSR, Sibirskoe Otdelenie, Institut Matematiki, Algebra i Logika, Seminar (Novosibirsk)
Am. J. Math.	American Journal of Mathematics
Am. Math. Monthly	American Mathematical Monthly
Ann. di Mat.	Annali di Matematica Pura ed Applicata (Bologna)
Ann. Sci. E.N.S.	Annales Scientifiques de l'École Normale Supérieure (Paris)
Arch. Math.	Archiv der Mathematik (Basel-Stuttgart)
Bull. A.M.S.	Bulletin of the American Mathematical Society
Bull. S.M. France	Bulletin de la Société Mathématique de France
Camb. Ph. Trans.	Cambridge Philosophical Transactions
Can. J. Math.	Canadian Journal of Mathematics
Comm. Math. Helv.	Commentarii Mathematici Helvetici (Zuerich)
C. R. Paris	Comptes Rendus Hebdomadaires des Séances de l'Académie des Sciences (Paris)
Dokl. Ak. Nauk SSSR	Doklady Akademii Nauk SSSR
Duke M. J.	Duke Mathematical Journal (Durham, N. C.)
Hamb. Abh.	Abhandlungen aus dem Mathematischen Seminar der Universität Hamburg
I.H.E.S. Publ. Math.	Institut des Hautes Études Scientifiques, Publications Mathématiques (Paris)
Ill. J. Math.	Illinois Journal of Mathematics
Indag. Math.	Koninklijke Nederlandse Akademie van Wetenschappen, Indagationes Mathematicae ex Actis Quibus Titulus (= Nederl. Akad. Wetensch. Proceedings, Series A)
Izv. Vys. Uchebn. Zav. Mat.	Izvestija Vysshih Uchebnyh Zavedenii, Matematika (Kazan)
J. Chinese M. S.	Journal of the Chinese Mathematical Society
J. für Math.	Journal für die reine und angewandte Mathematik (Crelles Journal)
J. Ind. M. S.	The Journal of the Indian Mathematical Society
J. London M. S.	The Journal of the London Mathematical Society
J.M.S. Japan	Journal of the Mathematical Society of Japan
J. Math. Mech.	Journal of Mathematics and Mechanics (Bloomington, Ind.)

Jap. J. Math.	Japanese Journal of Mathematics
Jour. de Math.	Journal de Mathématiques Pures et Appliquées (Paris)
Mat. Sb. N.S.	Matematicheskii Sbornik, Novaja Serija (Moscow)
Math. Ann.	Mathematische Annalen
Math. Z.	Mathematische Zeitschrift
Mem. A.M.S.	Memoirs of the American Mathematical Society
Mich. M.J.	The Michigan Mathematical Journal (Ann Arbor, Mich.)
Nagoya M.J.	Nagoya Mathematical Journal
Pac. J. Math.	Pacific Journal of Mathematics (Berkeley, Calif.)
Proc. A.M.S.	Proceedings of the American Mathematical Society
Proc. Camb. Ph.S.	Proceedings of the Cambridge Philosophical Society
Proc. London M.S.	Proceedings of the London Mathematical Society
Proc. N.A.S.	Proceedings of the National Academy of Sciences of the United States of America
Publ. M.S. Japan	Publications of the Mathematical Society of Japan
Rend. Palermo	Rendiconti del Circolo Matematico di Palermo
Rend. Torino	Università e Politecnico di Torino. Rendiconti del Seminario Matematico
Ric. Mat.	Ricerche di Matematica (Naples)
Sibirsk Mat. Z.	Sibirskii Matematicheskii Zhurnal
Soviet Math.	Soviet Mathematics, Doklady (English translation of Dokl. Ak. Nauk SSSR.) (Providence, R. I.)

Titles of papers in languages other than English, French, German and Italian have been translated.

[1] Abe, E.: On the groups of C. Chevalley. J.M.S. Japan **11,** 15—41 (1959).

[2] —: Groupes simples de Chevalley. Tôhoku Math. J. (2) **13,** 253—267 (1961).

[3] —, and T. Kanno: Some remarks on algebraic groups. Tôhoku Math. J. (2) **11,** 376—384 (1959).

[4] Ado, I. D.: The representation of Lie algebras by matrices. Uspehi Mat. Nauk (N.S.) **2,** 159—173 (1947); A.M.S. Transl. Ser. 1, No. 2 (1949).

[5] Albert, A. A.: Structure of Algebras. Amer. Math. Soc., Providence, 1939 and 1961.

[6] —: A structure theory for Jordan algebras. Ann. of Math. **48,** 546—567 (1947).

[7] —: On simple alternative rings. Can. J. Math. **4,** 129—135 (1952).

[8] —: A construction of exceptional Jordan division algebras. Ann. of Math. **67,** 1—28 (1958).

[9] —, and M. S. Frank: Simple Lie algebras of characteristic p. Rend. Torino **14,** 117—139 (1954/55).

[10] —, and N. Jacobson: On reduced exceptional simple Jordan algebras. Ann. of Math. **66,** 400—417 (1957).

[11] Allen, H. P.: Jordan Algebras and Lie Algebras of Type D_4. Yale dissertation, 1965 (to appear in J. Alg.).

[12] —: Jordan algebras and Lie algebras of type D_4. Bull. A.M.S. **72,** 65—67 (1966).

[13] Alperin, J. L.: Automorphisms of solvable groups. Proc. A.M.S. **13,** 175—180 (1962).

[14] Amitsur, S. A.: Differential polynomials and division algebras. Ann. of Math. **59,** 126—136 (1954).

[15] —: Invariant submodules of simple rings. Proc. A.M.S. **7,** 987—989 (1956).

[16] —: Derivations in simple rings, Proc. Lond. M. S. (3) **7,** 87—112 (1957).

[17] —: Simple algebras and cohomology groups of arbitrary fields. Trans. A.M.S. **90,** 73—112 (1959).

[18] Artin, E.: Geometric Algebra. New York: Interscience 1957.

[19] Barnes, D. W.: Nilpotency of Lie algebras. Math. Z. **79,** 237—238 (1962).

[20] —: Conditions for nilpotency of Lie rings, ibid., 289—296.

[21] Barnes, R. T.: On derivation algebras and Lie algebras of prime characteristic. Yale dissertation, 1963.

[21a] —: On splitting fields for certain Lie algebras of prime characteristic. Proc. A.M.S. **17,** 930—935 (1966).

[22] Barsotti, I.: Structure theorems for group varieties. Ann. di Mat. (4) **38,** 77—119 (1955).

[23] —: Abelian varieties over fields of positive characteristic. Rend. Palermo (2) **5,** 145—169 (1956).

[24] —: Analytical methods for abelian varieties in positive characteristic. Colloq. Th. des Gpes. Algs., Brussels 1962, pp. 77—85.

[25] Baxter, W. E.: Lie simplicity of a class of associative rings. Proc. A.M.S. **7,** 855—863 (1956).

[26] —: II. Trans. A.M.S. **87,** 63—75 (1958).

[27] —: Concerning strong Lie ideals. Proc. A.M.S. **11,** 393—395 (1960).

[28] —: Concerning the commutator subgroup of a ring. Proc. A.M.S. **16,** 803—805 (1965).

[29] Berkson, A. J.: The *u*-algebra of a restricted Lie algebra is Frobenius. Proc. A.M.S. **15,** 14—15 (1964).

[30] Bernat, P.: Sur le corps des quotients de l'algèbre enveloppante d'une algèbre de Lie. C. R. Paris **254,** 1712—1714 (1962).

[31] Birkhoff, G.: Representability of Lie algebras and Lie groups by matrices. Ann. of Math. **38,** 526—532 (1937).

[32] Block, R.: On torsion-free abelian groups and Lie algebras. Proc. A.M.S. **9,** 613—620 (1958).

[33] —: New simple Lie algebras of prime characteristic. Trans. A.M.S. **89,** 421—449 (1958).

[34] —: On Lie algebras of classical type. Proc. A.M.S. **11,** 377—379 (1960).

[35] —: Trace forms on Lie algebras. Can. J. Math. **14,** 553—564 (1962).

[36] —: On Lie algebras of rank one. Trans. A.M.S. **112,** 19—31 (1964).

[37] —: The Lie algebras with a quotient trace form. Ill. J. Math. **9,** 277—285 (1965).

[38] —: On the Mills-Seligman axioms for Lie algebras of cla sical type. Trans. A.M.S. **121,** 378—392 (1966).

[39] —, and H. Zassenhaus: The Lie algebras with a nondegenerate trace form. Ill. J. Math. **8,** 543—549 (1964).

[40] Bochner, S.: Formal Lie groups. Ann. of Math. **47,** 192—201 (1946).

[41] Bokut, L. A.: Embedding Lie algebras into algebraically closed Lie algebras. Alg. i Log. Sem. **1,** No. 2, 47—53 (1962).

[42] Borel, A.: Groupes linéaires algébriques. Ann. of Math. **64,** 20—82 (1956).

[43] —, and G. D. Mostow: On semi-simple automorphisms of Lie algebras. Ann. of Math. **61,** 389—405 (1955).

[44] —, and T. A. SPRINGER: Rationality properties of linear algebraic groups, to appear in Proc. Symp. Pure Math. **9** (1967).
[45] —, and J. TITS: Groupes réductifs. I.H.E.S. Publ. Math. **27,** 55—150 (1965).
[46] BOURBAKI, N.: Algèbre, Chap. III, Algèbre multilinéaire. Paris: Hermann 1948.
[47] —: Groupes et Algèbres de Lie, Chap. I, Algèbres de Lie. Paris: Hermann 1960.
[48] BRAUN, H., and M. KOECHER: Jordan-Algebren. Berlin/Heidelberg/New York: Springer 1965.
[49] BROWN, G.: On commutators in a simple Lie algebra. Proc. A.M.S. **14,** 763—767 (1963).
[50] BRUHAT, F.: Sur une classe de sous-groupes compacts maximaux des groupes de Chevalley sur un corps $\mathfrak{p}$-adique. I.H.E.S. Publ. Math. **23,** 45—74 (1964).
[51] CAMPBELL, H. E.: On the Casimir operator. Pac. J. Math. **7,** 1325—1331 (1957).
[52] CAROLL, C. L.: Normal Simple Lie Algebras of Type D and Order 28 over a Field of Characteristic 0. Univ. N. Carolina dissertation, 1943.
[53] CARTAN, É.: Sur la structure des groupes de transformations finis et continus. Paris 1894 and 1933.
[54] —: Les groupes projectifs qui ne laissent invariante aucune multiplicité plane. Bull. S.M. France **41,** 53—96 (1913).
[55] —: Les représentations linéaires des groupes de Lie. Jour. de Math. **17,** 1—12 (1938).
[56] CARTAN, H., and S. EILENBERG: Homological algebra. Princeton 1956.
[57] CARTER, R. W.: Simple groups and simple Lie algebras. J. London M. S. **40,** 193—240 (1965).
[58] CARTIER, P.: Hyperalgèbres et groupes de Lie formels. Séminaire "Sophus LIE" II. Paris 1957.
[59] —: Remarques sur le théorème de Birkhoff-Witt. Ann. Pisa (3) **12,** 1—4 (1958).
[60] —: Questions de rationalité des diviseurs en géométrie algébrique. Bull. S.M. France **86,** 177—251 (1958).
[61] —: Isogénies des variétés de groupes. Bull. S.M. France **87,** 191—220 (1959).
[62] —: Isogenies and duality of abelian varieties. Ann. of Math. **71,** 315—351 (1960).
[63] —: Groupes algébriques et groupes formels. Coll. Th. des Gpes. algs. Brussels 1962, pp. 87—111.
[64] —, et al: Théorie des algèbres de Lie. Topologie des groupes de Lie. Séminaire "Sophus LIE" I, Paris 1955.
[65] CHANG, B.: On Engel rings of exponent $p-1$ over GF(p). Proc. London M.S. **11,** 203—212 (1961).
[66] CHANG, H. J.: Über Wittsche Lie-Ringe. Hamb. Abh. **14,** 151—184 (1941).
[67] CHEVALLEY, C.: An algebraic proof of a property of Lie groups. Am. J. Math. **63,** 785—793 (1941).
[68] —: A new kind of relationship between matrices. Am. J. Math. **65,** 521—531 (1943).
[69] —: Sur le groupe exceptionnel (E_6). C. R. Paris **232,** 1991—1993 (1951).
[70] —: Sur une variété algébrique liée à l'étude du groupe (E_6), ibid., 2168—2170.
[71] —: Théorie des groupes de Lie, Tome II, Groupes algébriques. Paris: Hermann 1951.
[72] —: Tome III. Théorèmes généraux sur les algèbres de Lie. Paris: Hermann 1955.

[73] —: The algebraic theory of Spinors. New York: Columbia 1954.

[74] —: The construction and study of certain important algebras. Publ. M.S. Japan I, 1955.

[75] —: Fundamental concepts of algebra. New York: Academic Press 1956.

[76] —: Sur certains groupes simples. Tôhoku Math. J. (2) **7,** 14—66 (1956).

[77] —: La théorie des groupes algébriques. Proc. Int. Cong. Math. 1958, Cambridge 1960, pp. 53—68.

[78] —: Certains schémas de groupes semi-simples. Sém. Bourbaki 1961, exp. 219.

[79] —, et al.: Classification des groupes de Lie algébriques, 2 vols. Paris: Sém. C. Chevalley 1958.

[80] —, and S. Eilenberg: Cohomology theory of Lie groups and Lie algebras. Trans. A.M.S. **63,** 85—124 (1948).

[81] —, and R. D. Schafer: The exceptional Lie algebras F_4 and E_6. Proc. N.A.S. **36,** 137—141 (1950).

[82] Chwe, B. S.: Relative homological algebra and homological dimension of Lie algebras. Trans. A.M.S. **117,** 477—493 (1965).

[83] —: On the commutativity of restricted Lie algebra. Proc. A.M.S. **16,** 547 (1965).

[84] Cohen, I. S.: Note on a note of H. F. Tuan. Bull. A.M.S. **52,** 175—177 (1946).

[85] Cohn, P. M.: Generalization of a theorem of Magnus. Proc. London M.S. (3) **2,** 297—310 (1952).

[86] —: Sur le critère de Friedrichs pour les commutateurs dans une algèbre associative libre. C. R. Paris **239,** 743—745 (1954).

[87] —: A non-nilpotent Lie ring satisfying the Engel condition and a non-nilpotent Engel group. Proc. Camb. Ph. S. **51,** 401—405 (1955).

[88] —: Simple rings without zero-divisors, and Lie division rings. Mathematika **6,** 14—18 (1959).

[89] —: On the embedding of rings in skew fields. Proc. London M.S. **11,** 511—530 (1961).

[90] —: A remark on the Birkhoff-Witt theorem. J. London M. S. **38,** 197—203 (1963).

[91] —: The embedding of Lie algebras in restricted Lie algebras. J. London M. S. **39,** 277—287 (1964).

[92] Curtis, C. W.: A note on the representations of nilpotent Lie algebras. Proc. A.M.S. **5,** 813—824 (1954).

[93] —: Modular Lie algebras. I. Trans. A.M.S. **82,** 160—179 (1956).

[94] —: II, ibid. **86,** 91—108 (1957).

[95] —: Representations of Lie algebras of classical type with applications to linear groups. J. Math. Mech. **9,** 307—326 (1960).

[96] —: On the dimensions of the irreducible modules of Lie algebras of classical type. Trans. A.M.S. **96,** 135—142 (1960).

[97] —: On projective representations of certain finite groups. Proc. A.M.S. **11,** 852—860 (1960).

[98] Curzio, M.: Sui trasportatori nei gruppi e nelle algebre di Lie. Ric. Mat. **11,** 3—23 (1962).

[99] Davis, R. L.: Torsion in Engel modules. Proc. A.M.S. **10,** 679—685 (1959).

[100] Deuring, M.: Algebren. Berlin: Springer 1935.

[101] Dieudonné, J.: Les semi-dérivations dans les extensions radicielles. C. R. Paris **227,** 1319—1320 (1948).

[102] —: Sur les Groupes Classiques. Paris: Hermann 1948.

[103] —: On the automorphisms of the classical groups. Mem. A.M.S. **2,** 1—95 (1951).

[104] —: Sur les groupes de Lie algébriques sur un corps de caractéristique $p > 0$. Rend. Palermo (2) **1,** 380—402 (1952).
[105] —: On semi-simple Lie algebras. Proc. A.M.S. **4,** 931—932 (1953).
[106] —: Sur les multiplicateurs des similitudes. Rend. Palermo (2) **3,** 398—408 (1954).
[107] —: Sur quelques groupes de Lie abéliens sur un corps de caractéristique $p > 0$. Arch. Math. **5,** 274—281 (1954).
[108] —: Groupes de Lie et hyperalgèbres de Lie sur un corps de caractéristique $p > 0$. Comm. Math. Helv. **28,** 87—118 (1954).
[109] —: III. Math. Z. **63,** 53—75 (1955).
[110] —: V. Bull. S.M. France **84,** 207—239 (1956).
[111] —: VII. Math. Ann. **134,** 114—133 (1957).
[112] —: La Géométrie des Groupes Classiques. Berlin/Göttingen/Heidelberg: Springer 1955 and 1963.
[113] —: Witt groups and hyperexponential groups. Mathematika **2,** 21—31 (1955).
[114] —: Sur la notion de variables canoniques. An. Ac. Bras. Ciencias **27,** 251—258 (1955).
[115] —: Lie groups and Lie hyperalgebras over a field of characteristic $p > 0$. II. Am. J. Math. **77,** 218—244 (1955).
[116] —: IV., ibid. 429—452.
[117] —: VI. Am. J. Math. **79,** 331—388 (1957).
[118] —: VIII. Am. J. Math. **80,** 740—772 (1958).
[119] —: On a theorem of Lazard. Am. J. Math. **78,** 675—676 (1956).
[120] —: Sur les groupes formels abéliens unipotents. Rend. Palermo (2) **5,** 170—180 (1956).
[121] —: Les algèbres de Lie simples associées aux groupes simples algébriques sur un corps de caractéristique $p > 0$. Rend. Palermo (2) **6,** 198—204 (1957).
[122] —: Hyperalgèbres et groupes formels. Inst. Naz. di alta Mat. Roma. Sem. 1962/63, pp. 512—524.
[123] Dixmier, J.: Sur les algèbres dérivées d'une algèbre de Lie. Proc. Camb. Ph. S. **51,** 541—544 (1955).
[124] —: Cohomologie des algèbres de Lie nilpotentes. Acta Szeged **16,** 246—250 (1955).
[125] —: Homologie des anneaux de Lie. Ann. Sci. E.N.S. (3) **74,** 25—83 (1957).
[126] —, and W. G. Lister: Derivations of nilpotent Lie algebras. Proc. A.M.S. **8,** 155—158 (1957).
[127] Drazin, M. P., and K. W. Gruenberg: Commutators in associative rings. Proc. Camb. Ph. S. **49,** 590—594 (1953).
[128] Dynkin, E. B.: The structure of semi-simple Lie algebras. Uspehi Mat. Nauk (N.S.) **2,** 59—127 (1947); A.M.S. Transl. Ser. **1,** No. 17 (1950).
[129] Eichler, M.: Quadratische Formen und Orthogonale Gruppen. Berlin/Göttingen/Heidelberg: Springer 1952.
[130] Feit, W., and J. G. Thompson: Solvability of groups of odd order. Pac. J. Math. **13,** 775—1029 (1963).
[131] Ferrar, J. C.: On Lie algebras of type E_6. Yale dissertation, 1966.
[132] Fischer, B.: Finite groups admitting a fixed-point-free automorphism of period $2p$. J. Alg. **3,** 99—114 (1966).
[133] Frank, M. S.: A new class of simple Lie algebras. Proc. N.A.S. **40,** 713—719 (1954).
[134] —: On a theory relating matric Lie algebras of characteristic p and subalgebras of the Jacobson–Witt algebra . . . Univ. of Minnesota 1960.

[135] —: Two new classes of simple Lie algebras. Trans. A.M.S. **112,** 456—482 (1964).

[136] —: A unifying principle for simple Lie algebras of characteristic p (to appear).

[137] FREUDENTHAL, H.: Oktaven, Ausnahmegruppen und Oktavengeometrie. Utrecht 1951.

[138] —: Sur le groupe exceptionnel E_7. Indag. Math. **15,** 81—89 (1953).

[139] —: Sur le groupe exceptionnel E_8., ibid. 95—98.

[140] —: Zur ebenen Oktavengeometrie, ibid. 195—200.

[141] —: Beziehungen der E_7 und E_8 zur Oktavenebene. I. Indag. Math. **16,** 218—230 (1954); II. ibid. 363—368; III. ibid. **17,** 151—157 (1955); IV. ibid. 277—285; V—VII. ibid. **21,** 165—201 (1959); VIII., IX. ibid. 447—474.

[142] —: Zur Berechnung der Charaktere der halbeinfachen Lieschen Gruppen. I. Indag. Math. **16,** 369—376 (1954); II. ibid. 487—491; III. ibid. **18,** 511—514 (1956).

[142a] GABRIEL, P.: Étude infinitesimale des schémas en groupe et groupes formels (Exp. VIIA); Groupes formels (Exp. VIIB). In Schémas en Groupes, report of I.H.E.S. Séminaire de Géométrie Algébrique, 1963/64, Fascicule 2B.

[143] GAINOV, A. T.: Identical relations for binary Lie rings. Uspehi Mat. Nauk (N.S.) **12,** No. 3 (75), 141—146 (1957).

[144] —: Derivations of reduced free algebras. Alg. i. Log. Sem. **1,** No. 6, 20—25 (1962/63).

[145] GANTMACHER, F.: Canonical representation of automorphisms of a complex semi-simple Lie group. Mat. Sbornik **5,** 101—146 (1939).

[146] GERSTENHABER, M.: On the Galois theory of inseparable extensions. Bull. A.M.S. **70,** 561—566 (1964).

[147] —: On infinite inseparable extensions of exponent one, ibid. **71,** 878—881 (1965).

[148] GLAUBERMAN, G.: On the automorphism group of a core-free group (to appear).

[149] GOLDBERG, S. I.: Extensions of Lie algebras and the third cohomology group. Can. J. Math. **5,** 470—476 (1953).

[150] —: On the Euler characteristic of a Lie algebra. Am. Math. Monthly **62,** 239—240 (1955).

[151] GOLDHABER, J. K.: A note on Lie k system automorphisms. Am. J. Math. **75,** 859—863 (1953).

[152] GOLDIE, A. W.: The structure of prime rings. Proc. London M.S. (3) **8,** 589—608 (1958).

[153] GOLOD, E. S.: On nil algebras and finitely approximable p-groups. Izv. Ak. Nauk SSSR. Ser. Mat. **28,** 273—276 (1964); A.M.S. Transl. Ser. 2, v. **48,** 103—106 (1965).

[154] —, and I. R. SHAFAREVITCH: On class field towers. Izv. Ak. Nauk SSSR. Ser. Mat. **28,** 261—272 (1964); A.M.S. Transl. Ser. 2, v. **48,** 91—102 (1965).

[155] GOPALA KRISHNAN, N. S.: A note on global dimension. Proc. Indian Ac. Sci. Sec. A. **57,** 250—253 (1963).

[156] GORENSTEIN, D., and I. N. HERSTEIN: Finite groups admitting a fixed-point-free automorphism of order 4. Am. J. Math. **83,** 71—78 (1961).

[157] GRUENBERG, K. W.: Two theorems on Engel groups. Proc. Camb. Ph. S. **49,** 377—380 (1953).

[158] GUREVITCH, G. B.: Certain properties of metabelian Lie algebras. Dokl. Ak. Nauk SSSR **138,** 998—1001 (1961).

[159] HALL, M.: A basis for free Lie rings and higher commutators in free groups. Proc. A.M.S. **1,** 575—581 (1950).

[160] —: Solution of the Burnside problem for exponent six. III. J. Math. **2,** 764—786 (1958).

[161] —: The theory of groups. New York: Macmillan 1959.

[162] —: Generators and relations in groups—the Burnside Problem. Lectures in Modern Math. (T. L. Saaty, ed.) vol. II, pp. 42—92. New York: J. Wiley and Sons 1964.

[163] HALL, P.: Some word-problems. J. London M. S. **33,** 482—496 (1958).

[164] —, and G. HIGMAN: On the p-length of p-soluble groups and reduction theorems for Burnside's problem. Proc. London M.S. (3) **6,** 1—42 (1956).

[165] HARISH-CHANDRA: Faithful representations of Lie algebras. Ann. of Math. **50,** 68—76 (1949).

[166] —: Some applications of the universal enveloping algebra of a semi-simple Lie algebra. Trans. A.M.S. **70,** 28—96 (1951).

[167] HARRIS, B.: Centralizers in Jordan algebras. Pac. J. Math. **8,** 757—790 (1958).

[168] —: Cohomology of Lie triple systems and Lie algebras with involution. Trans. A.M.S. **98,** 148—162 (1961).

[169] HASSE, H., and F. K. SCHMIDT: Noch eine Begründung der Theorie der höheren Differentialquotienten in einem algebraischen Funktionenkörper einer Unbestimmten. J. für Math. **177,** 215—237 (1937).

[170] HATTORI, A.: On the multiplicative group of simple algebras and orthogonal groups of three dimensions. J.M.S. Japan **4,** 205—217 (1952).

[171] —, and G. TOYODA: On the multiplicative group of simple algebras. J.M.S. Japan **6,** 262—265 (1954).

[172] HEINEKEN, H.: Eine Bemerkung über Engelsche Elemente. Arch. Math. **11,** 321 (1960).

[173] —: Endomorphismenringe und Engelsche Elemente. Arch. Math. **13,** 29—37 (1962).

[174] —: Liesche Ringe mit Engel-Bedingung. Math. Ann. **149,** 232—236 (1962/63).

[175] HERSTEIN, I. N.: On the Lie ring of a division ring. Ann. of Math. **60,** 571—575 (1954).

[176] —: The Lie ring of a simple associative ring. Duke M. J. **22,** 471—476 (1955).

[177] —: On the Lie and Jordan rings of a simple associative ring. Am. J. Math. **77,** 279—285 (1955).

[178] —: Lie and Jordan systems in simple rings with involution. Am. J. Math. **78,** 629—649 (1956).

[179] —: Certain submodules of simple rings with involution. Duke M. J. **24,** 357—364 (1957).

[180] —: Lie and Jordan structures in simple associative rings. Bull. A.M.S. **67,** 517—531 (1961).

[181] —: Sugli annelli soddisfacenti ad una condizione di Engel. Atti Ac. Naz. Lincei Rend. Cl. Sci. Fis. Mat. Nat. **32,** 177—180 (1962).

[182] —, and E. KLEINFELD: Lie mappings in characteristic 2. Pac. J. Math. **10,** 843—852 (1960).

[183] HERTZIG, D.: Forms of algebraic groups. Proc. A.M.S. **12,** 657—660 (1961).

[184] —: The structure of Frobenius algebraic groups. Am. J. Math. **83,** 421—431 (1961).

[185] HIGGINS, P. J.: Lie rings satisfying the Engel condition. Proc. Camb. Ph. S. **50,** 8—15 (1954).

[186] HIGMAN, G.: On finite groups of exponent five. Proc. Camb. Ph. S. **52,** 381—390 (1956).

[187] —: Le problème de Burnside, Colloque d'algèbre supérieure. Brussels 1956, pp. 123—128.

[188] —: Groups and rings having automorphisms without non-trivial fixed elements. J. London M. S. **32,** 321—334 (1957).

[189] —: Lie ring methods in the theory of finite nilpotent groups. Proc. Int. Cong. Math. 1958, pp. 307—312.

[190] Hijikata, H.: A remark on the groups of type G_2 and F_4. J.M.S. Japan **15,** 159—164 (1963).

[191] Hochschild, G.: Representations of restricted Lie algebras of characteristic p. Proc. A.M.S. **5,** 603—605 (1954).

[192] —: Cohomology of restricted Lie algebras. Am. J. Math. **76,** 555—580 (1954).

[193] —: Lie algebra kernels and cohomology. Am. J. Math. **76,** 698—716 (1954).

[194] —: Cohomology classes of finite type and finite dimensional kernels for Lie algebras. Am. J. Math. **76,** 763—778 (1954).

[195] —: Simple algebras with purely inseparable splitting fields of exponent 1. Trans. A.M.S. **79,** 477—489 (1955).

[196] —: Restricted Lie algebras and simple associative algebras of characteristic p. Trans. A.M.S. **80,** 135—147 (1955).

[197] —: Relative homological algebra. Trans. A.M.S. **82,** 246—269 (1956).

[198] —: Note on Lie algebra kernels in characteristic p. Proc. A.M.S. **7,** 551—557 (1956).

[199] —: On the algebraic hull of a Lie algebra. Proc. A.M.S. **11,** 195—199 (1960).

[200] —: An addition to Ado's theorem. Proc. A.M.S. **17,** 531—533 (1966).

[201] —, and J. P. Serre: Cohomology of Lie algebras. Ann. of Math. **57,** 591—603 (1953).

[202] Hoechsmann, K.: Simple algebras and derivations. Trans. A.M.S. **108,** 1—12 (1963).

[203] —: Algebras split by a given purely inseparable field. Proc. A.M.S. **14,** 768—776 (1963).

[204] Hua, L. K.: A theorem on matrices over a field and its applications. J. Chinese M. S. (N.S.) **1,** 110—163 (1951).

[205] —: A note on the total matrix ring over a non-commutative field. Ann. Soc. Polon. Math. **25,** 188—193 (1952).

[206] Humphreys, J. E.: Algebraic Lie algebras over fields of prime characteristic. Yale dissertation, 1966 (Mem. A.M.S. **71** (1967)).

[207] Iwahori, N.: On some matrix operators. J.M.S. Japan **6,** 76—105 (1954).

[208] Iwasawa, K.: On the representations of Lie algebras. Jap. J. Math. **19,** 405—426 (1948).

[209] Jacobson, F. D., and N. Jacobson: Classification and representations of semi-simple Jordan algebras. Trans. A.M.S. **65,** 141—169 (1949).

[210] Jacobson, N.: Rational methods in the theory of Lie algebras. Ann. of Math. **36,** 875—881 (1935).

[211] —: Abstract derivation and Lie algebras. Trans. A.M.S. **42,** 206—224 (1937).

[212] —: p-algebras of exponent p. Bull. A.M.S. **43,** 667—670 (1937).

[213] —: Simple Lie algebras over a field of characteristic zero. Duke M. J. **4,** 534—551 (1938).

[214] —: Cayley numbers and simple Lie algebras of type G. Duke M. J. **5,** 775—783 (1939).

[215] —: Restricted Lie algebras of characteristic p. Trans. A.M.S. **50,** 15—25 (1941).

[216] —: Classes of restricted Lie algebras of characteristic p. I. Am. J. Math. **63,** 481—515 (1941).

[217] —: II. Duke M. J. **10,** 107—121 (1943).

[218] —: The Theory of Rings. Amer. Math. Soc., New York 1943.

[219] —: Galois theory of purely inseparable fields of exponent one. Am. J. Math. **66,** 645—648 (1944).

[220] —: Derivation and multiplication systems of semi-simple Jordan algebras. Ann. of Math. **50,** 866—874 (1949).

[221] —: A note on Lie algebras of characteristic p. Am. J. Math. **74,** 357—359 (1952).

[222] —: Lectures in Abstract Algebra, v. II: Linear Algebra. New York: van Nostrand 1953.

[223] —: v. III: Theory of fields and Galois theory. Princeton: van Nostrand 1964.

[224] —: A note on automorphisms and derivations of Lie algebras. Proc. A.M.S. **6,** 281—283 (1955).

[225] —: Commutative restricted Lie algebras, ibid. 476—481.

[226] —: Structure of Rings. Amer. Math. Soc., Providence 1956.

[227] —: Exceptional Lie algebras. Yale mimeographed notes, 1957.

[228] —: A note on three-dimensional simple Lie algebras. J. Math. Mech. **7,** 823—831 (1958).

[229] —: Composition algebras and their automorphisms. Rend. Palermo (2) **7,** 55—80 (1958).

[230] —: Nilpotent elements in semi-simple Jordan algebras. Math. Ann. **136,** 375—386 (1958).

[231] —: Some groups of transformations defined by Jordan algebras. I. J. für Math. **201,** 178—195 (1959).

[232] —: II. J. für Math. **204,** 74—98 (1960).

[233] —: III. J. für Math. **207,** 61—85 (1961).

[234] —: Lie Algebras. New York: Interscience 1962.

[235] —: A note on automorphisms of Lie algebras. Pac. J. Math. **12,** 303—315 (1962).

[236] —: Triality and Lie algebras of type D_4. Rend. Palermo (2) **13,** 1—25 (1964).

[237] —: Forms of algebras. Yeshiva Sci. Confs. **1,** 41—71 (1966).

[238] —: Jordan Algebras, to appear.

[239] Janko, Z.: A new finite simple group with abelian Sylow 2-subgroups and its characterization. J. Alg. **3,** 147—186 (1966).

[240] Jenner, W. E.: On non-associative algebras associated with bilinear forms. Pac. J. Math. **10,** 573—575 (1960).

[241] Jennings, S. A.: On rings whose associated Lie rings are nilpotent. Bull. A.M.S. **53,** 593—597 (1947).

[242] —: Radical rings with nilpotent associated groups. Trans. Roy. Soc. Canada Sec. III (3) **49,** 31—38 (1955).

[243] —, and R. Ree: On a family of Lie algebras of characteristic p. Trans. A.M.S. **84,** 192—207 (1957).

[244] Kanno, T.: On the representations of Lie algebras. Tôhoku Math. J. (2) **8,** 46—53 (1956).

[245] —: On replicas. Tôhoku Math. J. (2) **11,** 287—298 (1959).

[246] Kaplansky, I.: Seminar on simple Lie algebras. Bull. A.M.S. **60,** 470—471 (1954).

[247] —: Lie algebras of characteristic p. Trans. A.M.S. **89,** 149—183 (1958).

[248] Kasch, F.: Invariante Untermoduln des Endomorphismenrings eines Vektorraums. Arch. Math. **4,** 182—190 (1953).

[249] Kawada, Y.: On the derivations in simple algebras. Sci. Papers Coll. Gen. Ed. Univ. Tokyo **2,** 1—8 (1952).

[250] Killing, W.: Die Zusammensetzung der stetigen endlichen Transformationsgruppen, I—IV; I, Math. Ann. **31,** 252—290 (1888); II, ibid. **33,** 1—48 (1889); III, ibid. **34,** 57—122 (1889); IV, ibid. **36,** 161—189 (1890).

[251] Kleinfeld, E.: Simple alternative rings. Ann. of Math. **58,** 544—547 (1953).

[252] —: A note on Moufang-Lie rings. Proc. A.M.S. **9,** 72—74 (1958).

[253] Klotz, E. A.: Isomorphisms of simple Lie rings. Yale dissertation, 1965.

[254] Kolchin, E. R.: Algebraic matrix groups and Picard-Vessiot theory of homogeneous linear ordinary differential equations. Ann. of Math. **49,** 1—42 (1948).

[255] —, and S. Lang: Existence of invariant bases. Proc. A.M.S. **11,** 140—148 (1960).

[256] Kostrikin, A. I.: Solution of the restricted problem of Burnside for exponent 5. Izv. Ak. Nauk SSSR Ser. Mat. **19,** 233—244 (1955).

[257] —: On the connection between periodic groups and Lie rings. Izv. Ak. Nauk SSSR Ser. Mat. **21,** 289—310 (1957); A.M.S. Transl. Ser. 2, No. **45,** 165—190 (1965).

[258] —: Lie rings satisfying the Engel condition. Izv. Ak. Nauk SSSR Ser. Mat. **21,** 515—540 (1957); A.M.S. Transl. Ser. 2, No. **45,** 191—220 (1965).

[259] —: On the Burnside problem. Izv. Ak. Nauk SSSR Ser. Mat. **23,** 3—34 (1959); A.M.S. Transl. Ser. 2, No. **36,** 63—100 (1964).

[260] —: On Engel properties of groups with the identical relation $x^{p^\alpha} = 1$. Dokl. Ak. Nauk SSSR **135,** 524—526 (1960); (transl.) Soviet Math. **1,** 1282—1284 (1961).

[261] —: Simple Lie p-algebras. Trudy Mat. Inst. Steklov **64,** 79—89 (1961).

[262] —: Strong degeneracy of simple Lie p-algebras. Dokl. Ak. Nauk SSSR **150,** 248—250 (1963); (transl.) Soviet Math. **4,** 637—640 (1963).

[263] —: Lie algebras and finite groups. Proc. Int. Cong. Math. 1962, 264—269; A.M.S. Transl. **31,** 40—46 (1963).

[264] —, and V. A. Kreknin: Lie algebras with regular automorphisms. Dokl. Ak. Nauk SSSR **149,** 249—251 (1963); (transl.) Soviet Math. **4,** 355—358 (1963).

[265] —, and I. R. Shafarevitch: Cartan pseudo-groups and Lie p-algebras. Dokl. Ak. Nauk SSSR **168,** 740—742 (1966); (transl.) Soviet Math. **7,** 715—718 (1966).

[266] Koszul, J. L.: Homologie et cohomologie des algèbres de Lie. Bull. S.M. France **78,** 65—127 (1950).

[267] Kreknin, V. A.: Solvability of Lie algebras with a regular automorphism of finite period. Dokl. Ak. Nauk SSSR **150,** 467—469 (1963); (transl.) Soviet Math. **4,** 683—685 (1963).

[268] Kurosh, A. G.: The present status of the theory of rings and algebras. Uspehi Mat. Nauk (N.S.) **6,** 3—15 (1951).

[269] —: Lectures on general algebra. Moscow 1962; (transl.) New York: Chelsea 1963.

[270] Lang, S.: Algebraic groups over finite fields. Am. J. Math. **78,** 555—563 (1956).

[271] —: Introduction to algebraic geometry. New York: Interscience 1958.

[272] —: Abelian varieties. New York: Interscience 1959.

[273] LATYSHEV, V. N.: On zero-divisors in finite-dimensional anticommutative algebras. Izv. Vys. Uchebn. Zav. Mat. No. **2** (21), 100—108 (1961).

[274] —: Algebras with identical relations. Dokl. Ak. Nauk SSSR **146**, 1003—1006 (1962); (transl.) Soviet Math. **3**, 1423—1427 (1962).

[275] —: Lie algebras with identical defining relations. Sibirsk Mat. Z. **4**, 821—829 (1963).

[276] —: Divisors of zero and nilelements in a Lie algebra. Sibirsk Mat. Z. **4**, 830—836 (1963).

[277] LAZARD, M.: Sur les algèbres enveloppantes universelles de certaines algèbres de Lie. Publ. Sci. Univ. Algér. Sér. A **1**, 281—294 (1954).

[278] —: Sur les groupes nilpotents et les anneaux de Lie. Ann. Sci. E.N.S. (3) **71**, 101—190 (1954).

[279] —: Lois de groupes et analyseurs. Ann. Sci. E.N.S. (3) **72**, 299—400 (1955).

[280] —: Sur les groupes de Lie formels à un paramètre. Bull. S.M. France **83**, 251—274 (1955).

[281] LEGER, G. F.: On cohomology of Lie algebras. Proc. A.M.S. **8**, 1010—1020 (1957).

[282] —: A particular class of Lie algebras. Proc. A.M.S. **16**, 293—296 (1965).

[283] —, and S. TÔGÔ: Characteristically nilpotent Lie algebras. Duke M. J. **26**, 623—628 (1959).

[284] LEVI, F. W., and B. L. VAN DER WAERDEN: Über eine besondere Klasse von Gruppen. Hamb. Abh. **9**, 154—158 (1933).

[285] LISSNER, D.: Automorphisms of generalized Witt algebras have fixed points. Proc. A.M.S. **14**, 205—210 (1963).

[286] LYNDON, R. C.: Burnside groups and Engel rings. Proc. Symp. Pure Math. **1**, 4—14 (1959).

[287] MACLANE, S.: Homology. Berlin/Göttingen/Heidelberg: Springer 1963.

[288] MAGNUS, W.: Beziehungen zwischen Gruppen und Idealen in einem speziellen Ring. Math. Ann. **111**, 259—280 (1935).

[289] —: Über Gruppen und zugeordnete Liesche Ringe. J. für Math. **182**, 142—149 (1940).

[290] —: A connection between the Baker–Hausdorff formula and a problem of Burnside. Ann. of Math. **52**, 111—126 (1950).

[291] MANIN, JU. I.: The Hasse–Witt matrix of an algebraic curve. Izv. Ak. Nauk SSSR Ser. Mat. **25**, 153—172 (1961); A.M.S. Transl. Ser. 2, No. **45**, 245—264 (1965).

[292] —: A remark on Lie p-algebras. Sibirsk Mat. Z. **3**, 479—480 (1962).

[293] —: On the theory of abelian varieties over fields of finite characteristic. Izv. Ak. Nauk SSSR Ser. Mat. **26**, 281—292 (1962); A.M.S. Transl. Ser. 2, **50**, 127—140 (1966).

[294] —: Two dimensional formal abelian groups. Dokl. Ak. Nauk SSSR **143**, 35—37 (1962); (transl.) Soviet Math. **3**, 335—337 (1962).

[295] —: Commutative formal groups and abelian varieties. Dokl. Ak. Nauk SSSR **145**, 280—283 (1962); (transl.) Soviet Math. **3**, 992—994 (1962).

[296] —: The theory of commutative formal groups over fields of finite characteristic. Uspehi Mat. Nauk **18**, 3—90 (1963); (transl.) Russian Math. Surveys **18**, 1—80 (1963).

[297] MARTINDALE, W. S.: Lie isomorphisms of primitive rings. Proc. A.M.S. **14**, 909—916 (1963).

[298] —: Lie derivations of primitive rings. Mich. M. J. **11**, 183—187 (1964).

[299] Mauler, H.: Eine Darstellung für Identitäten zwischen den Kommutatoren eines Ringes. Math. Ann. **126,** 410—417 (1953).

[300] May, J. P.: The cohomology of restricted Lie algebras and of Hopf algebras. Bull. A.M.S. **71,** 372—377 (1965).

[301] —: The cohomology of the Steenrod algebra; stable homotopy groups of spheres. Bull. A.M.S. **71,** 377—380 (1965).

[302] —: The cohomology of restricted Lie algebras and of Hopf algebras. J. of Alg. **3,** 123—146 (1966).

[303] Mills, W. H.: Classical type Lie algebras of characteristic 5 and 7. J. Math. Mech. **6,** 559—566 (1957).

[304] —, and G. B. Seligman: Lie algebras of classical type. J. Math. Mech. **6,** 519—548 (1957).

[305] Milnor, J. W., and J. C. Moore: On the structure of Hopf algebras. Ann. of Math. **81,** 211—264 (1965).

[306] Moran, S.: Unrestricted nilpotent products. Acta Math. **108,** 61—88 (1962).

[307] Mori, M.: On the three-dimensional cohomology group of Lie algebras. J.M.S. Japan **5,** 171—183 (1953).

[308] Nielsen, G. M.: A Determination of the minimal right ideals in the enveloping algebra of a Lie algebra of classical type. U. of Wisconsin dissertation, 1963.

[309] Novikov, P. S.: On periodic groups. Dokl. Ak. Nauk SSSR **127,** 749—752 (1959); A.M.S. Transl. Ser. 2, v. **45,** 19—22 (1965).

[310] Oehmke, R. H.: On a class of Lie algebras. Proc. A.M.S. **16,** 1107—1113 (1965).

[311] Ono, T.: Sur les groupes de Chevalley. J.M.S. Japan **10,** 307—313 (1958).

[312] —: An identification of Suzuki groups with groups of generalized Lie type. Ann. of Math. **75,** 251—259 (1962).

[313] Ore, O.: Linear equations in non-commutative fields. Ann. of Math. **32,** 463—477 (1931).

[314] —: Theory of non-commutative polynomials. Ann. of Math. **34,** 480—508 (1933).

[315] Paige, L. J.: A note on noncommutative Jordan algebras. Portugal. Math. **16,** 15—18 (1957).

[316] Palais, R. S.: The cohomology of Lie rings. Proc. Symp. Pure Math. III, 130—137 (1961).

[317] Patterson, E. M.: Note on nilpotent and solvable algebras. Proc. Camb. Ph. S. **51,** 37—40 (1955).

[318] —: On regular automorphisms of certain classes of rings. Quart. J. Math. (2) **12,** 127—133 (1961).

[319] Pavaman Murthy, M., and R. Sridharan: On the global dimension of some algebras. Math. Z. **81,** 108—111 (1963).

[320] Plotkin, B. I.: Algebraic sets of elements in groups and Lie algebras. Uspehi Mat. Nauk **13**, No. 6 (84), 133—138 (1958).

[321] Poincaré, H.: Sur les groupes continus. Camb. Ph. Trans. **18,** 220—255 (1899).

[322] Ponomarenko, P.: The Galois theory of infinite purely inseparable extensions. Bull. A.M.S. **71,** 876—877 (1965).

[323] Raffin, R.: Remarques sur certains algèbres de Lie. Symp. de top. alg. Mexico City 1958, 83—86.

[324] Ramanan, S.: A note on the global dimension of some algebras. Math. Z. **81,** 392—394 (1963).

[325] Ree, R.: On generalized Witt algebras. Trans. A.M.S. **83,** 510—546 (1956).

[326] —: On some simple groups defined by C. Chevalley. Trans. A.M.S. **84,** 392—400 (1957).

[327] —: Lie elements and an algebra associated with shuffles. Ann. of Math. **68,** 210—220 (1958).

[328] —: The simplicity of certain nonassociative algebras. Proc. A.M.S. **9,** 886—892 (1958).

[329] —: Note on generalized Witt algebras. Can. J. Math. **11,** 345—352 (1959).

[330] —: A family of simple groups associated with the simple Lie algebra of type (F_4). Am. J. Math. **83,** 401—420 (1961).

[331] —: A family of simple groups associated with the simple Lie algebra of type (G_2). ibid., 432—462.

[332] —: Construction of certain semi-simple groups. Can. J. Math. **16,** 490—508 (1964).

[333] Rinehart, G. S.: Differential forms on general commutative algebras. Trans. A.M.S. **108,** 195—222 (1963).

[334] Rosenberg, A., and D. Zelinsky: On Amitsur's complex. Trans. A.M.S. **97,** 327—356 (1960).

[335] Rosenlicht, M.: Some basic theorems on algebraic groups. Am. J. Math. **78,** 401—443 (1956).

[336] Ross, L. E.: Representations of graded Lie algebras. Trans. A.M.S. **120,** 17—23 (1965).

[337] Sagle, A. A.: Malcev algebras. Trans. A.M.S. **101,** 426—458 (1961).

[338] Sanov, I. N.: Solution of Burnside's problem for exponent 4. Ann. Leningrad St. Univ. Ser. Mat. **10,** 166—170 (1940).

[339] —: On a certain system of relations in periodic groups with period a power of a prime number. Izv. Ak. Nauk SSSR Ser. Mat. **15,** 477—502 (1951).

[340] —: Establisment of a connection between periodic groups with period a prime number and Lie rings, ibid. **16,** 23—58 (1952).

[341] Schafer, R. D.: Alternative algebras over an arbitrary field. Bull. A.M.S. **49,** 549—555 (1943).

[342] —: The exceptional simple Jordan algebras. Am. J. Math. **70,** 82—94 (1948).

[343] —: Nodal noncommutative Jordan algebras and simple Lie algebras of characteristic p. Trans. A.M.S. **94,** 310—326 (1960).

[344] —: An introduction to nonassociative algebras. New York: Academic Press 1966.

[345] —: On the simplicity of the Lie algebras E_7 and E_8. Indag. Math. **28,** 64—69 (1966).

[346] —, and M. L. Tomber: On a simple Lie algebra of characteristic 2. Mem. A.M.S. **14,** 11—14 (1955).

[347] Schellekens, G. J.: On a hexagonic structure. I and II. Indag. Math. **24,** 201—234 (1962).

[348] Schenkman, E.: A theory of subinvariant Lie algebras. Am. J. Math. **73,** 453—474 (1951).

[349] —: Infinite Lie algebras. Duke M. J. **19,** 529—535 (1952).

[350] —: On the derivation algebra and the holomorph of a nilpotent Lie algebra. Mem. A.M.S. **14,** 15—22 (1955).

[351] —: On a theorem of Herstein. Proc. A.M.S. **10,** 236—238 (1959).

[352] Schue, J. R.: Symmetry for the enveloping algebra of a restricted Lie algebra. Proc. A.M.S. **16,** 1123—1124 (1965).

[353] Seligman, G. B.: On Lie algebras of prime characteristic. Mem. A.M.S. **19** (1956).

[354] —: Characteristic ideals and the structure of Lie algebras. Proc. A.M.S. **8,** 159—164 (1957).

[355] —: Some remarks on classical Lie algebras. J. Math. Mech. **6,** 549—558 (1957).

[356] —: A survey of Lie algebras of characteristic p. Report Conf. Linear Algebras, NAS-NRC Pub. No. 502, Washington 1957, pp. 24—32.

[357] —: On automorphisms of Lie algebras of classical type. Trans. A.M.S. **92,** 430—448 (1959).

[358] —: II. ibid. **94,** 452—482 (1960).

[359] —: III. ibid. **97,** 286—316 (1960).

[360] —: The complete reducibility of certain modules. Mimeographed, Yale Univ. 1961.

[361] —: On nilpotent skew transformations. Mimeographed, Yale Univ. 1962.

[362] —: Normal simple Lie algebras and the classical groups. Mimeographed, Yale Univ. 1963.

[363] On the split exceptional Lie algebra E_7. Mimeographed, Yale Univ. 1963.

[364] Serre, J. P.: Quelques propriétés des variétés abéliennes en caractéristique p. Am. J. Math. **80,** 715—739 (1958).

[365] —: Lie Algebras and Lie Groups. New York: Benjamin 1965.

[366] Shirshov, A. I.: On the representation of Lie rings in associative rings. Uspehi Mat. Nauk (N.S.) **8,** No. 5, 173—175 (1953).

[367] —: Subalgebras of free Lie algebras. Mat. Sb. N.S. **33** (75), 441—452 (1953).

[368] —: On free Lie rings, ibid. **45** (87), 113—122 (1958).

[369] —: On the bases of a free Lie algebra. Alg. i Log. Sem. **1,** No. 1, 14—19 (1962).

[370] Smith, D. A.: On fixed points of automorphisms of classical Lie algebras. Pac. J. Math. **14,** 1079—1089 (1964).

[371] —: Chevalley bases for Lie modules. Trans. A.M.S. **115,** 283—299 (1965).

[372] Soda, D. C.: Groups of Type D_4 Defined by Jordan Algebras. Yale dissertation 1964.

[373] —: Some groups of type D_4 defined by Jordan algebras. J. für Math. **223,** 150—163 (1966).

[374] Springer, T. A.: The projective octave plane. I and II. Indag. Math. **22,** 74—101 (1960).

[375] —: The classification of reduced exceptional simple Jordan algebras, ibid. 414—422.

[376] —: On the geometric algebra of the octave planes, ibid. **24,** 451—468 (1962).

[377] Sridharan, R.: Filtered algebras and representations of Lie algebras. Trans. A.M.S. **100,** 530—550 (1961).

[378] Steinberg, R.: Variations on a theme of Chevalley. Pac. J. Math. **9,** 875—891 (1959).

[379] —: The simplicity of certain groups, ibid. **10,** 1039—1041 (1960).

[380] —: Automorphisms of finite linear groups. Can. J. Math. **12,** 606—615 (1960).

[381] —: Automorphisms of classical Lie algebras. Pac. J. Math. **11,** 1119—1129 (1961).

[382] —: Générateurs, relations et revêtements de groupes algébriques. Coll. Th. des Gpes. algs., Brussels 1962, pp. 113—127.

[383] —: Representations of algebraic groups. Nagoya M. J. **22,** 33—56 (1963).

[384] Sterling, D. J.: Coverings of algebraic groups and Lie algebras of classical type. Pac. J. Math. **14,** 1449—1462 (1964).

[385] SUL'DIN, A. V.: On irreducible representations of Lie algebras over fields of characteristic p. Kazan Gos. Univ. Uch. Zap. 114, No. **2,** 167—168 (1954).

[386] SUZUKI, M.: A new type of simple groups of finite order. Proc. N.A.S. **46,** 868—870 (1960).

[387] TAMARI, D.: On the embedding of Birkhoff–Witt rings in quotient fields. Proc. A.M.S. **4,** 197—202 (1953).

[388] THOMPSON, J. G.: Finite groups with fixed-point-free automorphisms of prime order. Proc. N.A.S. **45,** 578—581 (1959).

[389] —: Normal p-complements for finite groups. Math. Z. **72,** 332—354 (1960).

[390] TITS, J.: Le plan projectif et les groupes exceptionnels E_6 et E_7. Ac. Roy. Belg. Bull. Cl. Sci. (5) **40,** 29—40 (1954).

[391] —: Sur les analogues algébriques des groupes semi-simples complexes. Coll. d'Alg. Sup., Brussels 1956, pp. 261—289.

[392] —: Les "formes réelles" des groupes de type E_6. Sém. Bourbaki. Exp. **162** (1958).

[393] —: Groupes algébriques semi-simples et géometries associées. Proc. Coll. Alg. and Top. Foundations of Geometry. Utrecht 1959, pp. 175—192.

[394] —: Sur la trialité et certains groupes qui s'en deduisent. Publ. Math. I.H.E.S. **2,** 14—60 (1959).

[395] —: Une classe d'algèbres de Lie en relation avec les algèbres de Jordan. Indag. Math. **24,** 530—535 (1962).

[396] —: Groupes semi-simples isotropes. Coll. Th. des Gpes. algs., Brussels 1962, pp. 137—147.

[397] —: Algèbres alternatives, algèbres de Jordan et algèbres de Lie exceptionnels. Chicago: Mimeographed 1962.

[398] —: Algebraic and abstract simple groups. Ann. of Math. **80,** 313—329 (1964).

[399] TÔGÔ, S.: Derivations of Lie algebras. Bull. A.M.S. **72,** 690—692 (1966).

[400] TOMBER, M. L.: Lie algebras of type F. Proc. A.M.S. **4,** 759—768 (1953).

[401] —: Lie algebras of types A, B, C, D and F. Trans. A.M.S. **88,** 99—106 (1958).

[402] TUAN, H. F.: A note on the replicas of nilpotent matrices. Bull. A.M.S. **51,** 305—312 (1945).

[403] WALTER, J. H.: Paper to appear on determination of simple finite groups with abelian 2-Sylow subgroups.

[404] WAN, C-H.: On the matrix Lie ring defined by a Hamiltonian or skew-Hamiltonian matrix. Acta Math. Sinica **7,** 451—470 (1957).

[405] WEIL, A.: Variétés abéliennes et courbes algébriques. Paris: Hermann 1948.

[406] —: Algebras with involutions and the classical groups. J. Ind. M. S. (N.S.) **24,** 589—623 (1960).

[407] —: Foundations of algebraic geometry, 2nd ed. Amer. Math. Soc., Providence 1962.

[408] WEINER, L. M.: Lie admissible algebras. Univ. Nac. Tucuman Rev. Ser. A **11,** 10—24 (1957).

[409] WEISFELD, M.: On derivations in division rings. Pac. J. Math. **10,** 335—343 (1960).

[410] —: Purely inseparable extensions and higher derivations. Trans. A.M.S. **116,** 435—449 (1965).

[411] WEYL, H.: Theorie der Darstellung kontinuierlicher halb-einfacher Gruppen durch lineare Transformationen. I, II, III: I, Math. Z. **23,** 271—304 (1925); II, ibid. **24,** 328—376 (1926); III, ibid. 377—395.

[412] WHITEHEAD, J. H. C.: On the decomposition of an infinitesimal group. Proc. Camb. Ph. S. **32,** 229—237 (1936).

[413] —: Certain equations in the algebra of a semi-simple infinitesimal group. Quart. J. Math. **8**, 220—237 (1937).

[414] Winter, D. J.: On groups of automorphisms of Lie algebras (to appear).

[415] Witt, E.: Treue Darstellung Liescher Ringe. J. für Math. **177**, 152—160 (1937).

[416] —: Spiegelungsgruppen und Aufzählung halbeinfacher Liescher Ringe. Hamb. Abh. **14**, 289—337 (1941).

[417] —: Treue Darstellung beliebiger Liescher Ringe. Collectanea Math. **6**, 107—114 (1953).

[418] —: Die Unterringe der freien Lieschen Ringe. Math. Z. **64**, 195—216 (1956).

[419] Zassenhaus, H.: Über Liesche Ringe mit Primzahlcharakteristik. Hamb. Abh. **13**, 1—100 (1939).

[420] —: Ein Verfahren, jeder endlichen p-Gruppe einen Lie-Ring mit der Charakteristik p zuzuordnen, ibid. 200—207.

[421] —: Darstellungstheorie nilpotenter Lie-Ringe bei Charakteristik $p > 0$. J. für Math. **182**, 150—155 (1940).

[422] —: The representations of Lie algebras of prime characteristic. Proc. Glasgow Math. Assoc. **2**, 1—36 (1954).

[423] —: On an application of the theory of Lie algebras to group theory. Proc. Symp. Pure Math. **1**, 105—108 (1959).

[424] —: On trace bilinear forms on Lie algebras. Proc. Glasgow Math. Assoc. **4**, 62—72 (1959).

[425] Zorn, M.: Theorie der alternativen Ringe. Hamb. Abh. **8**, 123—147 (1930).

[426] —: On a theorem of Engel. Bull. A.M.S. **43**, 401—404 (1947).

[427] Zuev, I. I.: Lie ideals of associative rings. Uspehi Mat. Nauk **18**, No. 1 (109), 155—158 (1963).

Index

721/42/67 — III/18/203

Ergebnisse der Mathematik und ihrer Grenzgebiete